T4-AHE-037

THE EVOLUTIONARY ECOLOGY OF ANIMALS

STUDIES IN SOVIET SCIENCE

LIFE SCIENCES

1973

MOTILE MUSCLE AND CELL MODELS
N. I. Arronet
PATHOLOGICAL EFFECTS OF RADIO WAVES
M. S. Tolgskaya and Z. V. Gordon
CENTRAL REGULATION OF THE PITUITARY-ADRENAL COMPLEX
E. V. Naumenko

1974

SULFHYDRYL AND DISULFIDE GROUPS OF PROTEINS
Yu. M. Torchinskii
MECHANISMS OF GENETIC RECOMBINATION
V. V. Kushev

1975

THYROID HORMONES: Biosynthesis, Physiological Effects, and Mechanisms of Action
Ya. Kh. Turakulov, A. I. Gagel'gans, N. S. Salakhova, A. K. Mirakhmedov, L. M. Gol'ber, V. I. Kandror, and G. A. Gaidina

1977

THE EVOLUTIONARY ECOLOGY OF ANIMALS
S. S. Shvarts
HEMATOPOIETIC AND LYMPHOID TISSUE IN CULTURE
E. A. Luriya
STRUCTURE AND BIOSYNTHESIS OF ANTIBODIES
R. S. Nezlin

PROTEIN METABOLISM OF THE BRAIN
A. V. Palladin, Ya. V. Belik, and N. M. Polyakova

STUDIES IN SOVIET SCIENCE

THE EVOLUTIONARY ECOLOGY OF ANIMALS

S. S. Shvarts
Academy of Sciences of the USSR, Urals Branch
Sverdlovsk, USSR

Translated from Russian and edited by

Ayesha E. Gill
University of California, Los Angeles

With new material by the author and editor

CONSULTANTS BUREAU • NEW YORK AND LONDON

Library of Congress Cataloging in Publication Data

Shvarts͡, Stanislav Semenovich.
The evolutionary ecology of animals.

(Studies in Soviet science)
Translation of Évoli͡ut͡sionnai͡a ėkologii͡a zhivotnykh.
Includes bibliographies and index.
1. Evolution. 2. Zoology—Ecology. 3. Population genetics. I. Title. II. Series.
QH371.S4813 575 76-50647
ISBN 0-306-10920-4

The original Russian text of *The Evolutionary Ecology of Animals* was published by the Urals Branch of the Academy of Sciences of the USSR in Sverdlovsk as Volume 65 of the Proceedings (Trudy) of the Institute of Plant and Animal Ecology in 1969. It was corrected by the author for the present edition. This translation is published under an agreement with the Copyright Agency of the USSR (VAAP).

ЭВОЛЮЦИОННАЯ ЭКОЛОГИЯ ЖИВОТНЫХ
С. С. **ШВАРЦ**

EVOLYUTSIONNAYA EKOLOGIYA ZHIVOTNYKH
S. S. Shvarts

A Division of Plenum Publishing Corporation
227 West 17th Street, New York, N.Y. 10011

Printed in the United States of America

Editor's Foreword

While translating this book, I was in close communication with the author, S. S. Shvarts (Schwarz), who read and commented on the entire translated manuscript. In particular, any ambiguities as to the identity of organisms described only by common names in the original text were removed, because the author kindly supplied the Latin names in all such cases. Common names are retained in the translation, but the Latin names are also added where needed.

Some of the terminology used in the Russian is a transliteration from English words employed now more by European workers than Americans. I have defined these terms or noted their more common equivalents used in current American literature where it seemed useful in the text.

A final chapter, "Recent Work on the Evolutionary Ecology of Animals," is presented as Appendix II to the translation of the original text. I have written this chapter in order to update the material presented in the original edition published in 1969. The chapter discusses important recent contributions relevant to the subject matter presented by Shvarts. I would like to thank W. Z. Lidicker, Jr., and Y. B. Linhart for reading this final chapter and providing very helpful suggestions and comments.

I am particularly grateful to the author, S. S. Shvarts, for his careful reading of the translated manuscript.

Ayesha E. Gill

Los Angeles, California

[Prof. Shvarts died May 12, 1976]

Contents

Preface		3
Introduction		5
Chapter I.	The Genetic Basis of the Reorganization of Populations	21
Chapter II.	An Ecological Estimate of Interpopulation Differences	33
Chapter III.	The Reorganization of Populations—Homeostatic Alterations of the Genetic Structure of Populations and Microevolution	49
Chapter IV.	Ecological Mechanisms for the Maintenance of the Genetic Heterogeneity of a Population	75
Chapter V.	Ecological Mechanisms for the Reorganization of the Genetic Structure of Populations	91
	1. Role of the Dynamics of Population Age Structure in the Reorganization of Its Genetic Makeup	91
	2. On the Significance of Nonselective Elimination	110
	3. The Spatial Structure of Populations—A Factor in Microevolution	120
Chapter VI.	Speciation	145
Chapter VII.	The Ecological Essence of Macroevolution	193
Conclusions		215
References and Bibliography		223
Appendix I.	The Species Problem and New Methods of Systematics	257
Appendix II.	Recent Work on the Evolutionary Ecology of Animals (by A. E. Gill)	275

The Evolutionary Ecology of Animals

Ecological Mechanisms of the Evolutionary Process

S. S. Shvarts

Preface

Over a period of 20 years, the specific rules determining the ecological and morphophysiological features of populations of different species under different environmental conditions were studied in the Zoological Laboratories of the Institute of Biology, Urals Branch, of the Academy of Sciences of the USSR.* All classes of terrestrial vertebrates were studied. Research was conducted both under natural conditions in diverse biomes (from the steppes to the tundra) and experimentally on model populations. The material accumulated during this period led us to conclusions concerning the mechanisms of the evolutionary modifications of animals. Data were obtained that indicated the possibility of utilizing the animals' reactions in the alteration of environmental conditions during an elaboration of the most difficult problems of speciation. We presented the synthesis of these data in a monograph (Shvarts, 1959). Another series of investigations was devoted to the study of the dynamics of population structure. These investigations led us to the conclusion that alteration of the intrapopulation structure of a species leads to an alteration in the genetic makeup of intraspecific groupings—to microevolutionary modifications. The results of this work were published in several journals (Shvarts, 1963*a–c*, 1965, and others). A general analysis of the data obtained in our laboratories along the two paths indicated made evident the deep inner connection between the morphophysiological reactions of different forms and the alteration of population structure. On this basis, the representation of the ecological mechanisms of the evolutionary process—the study of which constitutes, in our opinion, the main task of evolutionary ecology—matured. Synthesis of

* Since 1965, the Laboratory of the Population Ecology of Vertebrates of the Institute of Plant and Animal Ecology.

the results of morphophysiological and population–ecological investigations, however, required the study (both experimental and theoretical) of certain questions that until recently had not attracted our attention. The most important of these questions are the evolutionary significance of the dynamics of population spatial structure, the correlation of micro- and macroevolution, and the reality of taxa above the species level. The results of these investigations are published for the first time in this book. The basic task of the book is to present in a well-grounded way the key significance of ecological mechanisms in the evolutionary process. We view these mechanisms, together with natural selection, as the basic motive force of evolution. Both the structure of the book as a whole and the structure of individual chapters are subordinate to this task. In particular, we use data from the literature only insofar as they are necessary for the objective analysis of the basic problems of evolutionary ecology. We did not strive to give a comprehensive summary of the literature on all the subjects broached. There is no necessity to do this, since a critical review of the literature on the evolutionary genetics of populations has been published by Dubinin (1966*a*).

This book is, to a considerable degree, the theoretical result of the work of a large collective of scientific workers and laboratory assistants. Particularly great contributions to the general work, which I aspired to synthesize in this book, were made by V. N. Pavlinin, V. S. Smirnov, V. E. Bergovyi, V. N. Bol'shakov, L. N. Dobrinskii, Z. D. Epifantseva, V. G. Ishchenko, N. A. Ovchinnikova, V. G. Olenev, A. V. Pokrovskii, O. A. Pyastolova, L. M. Syuzyumova, and L. K. Yashkova.

I deeply thank all my colleagues, collaborators from the Laboratory of the Population Ecology of Vertebrates, whose creative work made it possible to accumulate material for the statement of certain principal questions of evolutionary theory.

Introduction

The majority of contemporary biologists regard evolution as a process of progressive mastery of the arena of life by animals and plants, as progressive adaptation to diverse conditions of existence. In this respect, even the adherents of concepts of spontaneous generation (in any of their variants) represent no exception, since the result of spontaneous generation is, in the final analysis, evaluated and corrected in the process of the organism's interaction with its environment. Hence, it is clear that any evolutionary conception broaches, to some degree or other, the subject of the interaction of an organism with its environment. It is natural, also, that any evolutionary theory is obliged to rest on ecological laws, for, irrespective of the well-known transformation of biologists' views on the purpose and method of ecology, its basic task has remained unchanged now for the course of a hundred years. This task is to investigate the lives of animals and plants in their natural habitats, in nature.

If one speaks not of evolutionary theory in general, but of the only evolutionary theory that has withstood the test of time and satisfied the demands of modern science—of Darwinism—then its ecological basis is clear. Petrusewicz (1959) is correct in entitling one of his theoretical works "Darwin's Theory of Evolution Is an Ecological Theory" [in Polish]. The author correctly emphasizes that evolution is in essence an ecological process, accomplished on the basis of laws governed by group, not individual, phenomena. Orians (1962) is even more categorical, asserting that the general theory of ecology is a theory of natural selection.

The attention paid by evolutionists to ecological rules has intensified especially in recent times in connection with the establishment of certain general laws of ecology, reflecting in its most general form the interrelationship of organisms with their environment. On the other hand, the role of ecology in the development of evolutionary studies has grown continuously in proportion to the development of population genetics, following the laws of modification of a population's genetic makeup under changing environmental conditions (changes in the direction or intensity of selection), during changes in the numbers of animals, or as a result of the action of stochastic processes. The book by Ford (1964) and the synoptic article by Lerner (1965), both of which bear the same title, "Ecological Genetics," are significant in this respect. It is symptomatic that both Ford's book and Lerner's article could have been called, with equal right, "Evolutionary Ecology" or "Genetic Ecology," since they analyze such questions as these: the dependence of the dynamics of size on the genetic makeup of natural and model populations; the correlation among the size of populations, the dynamics of population size, and the effectiveness of selection; the mutual adaptations of animals at different trophic levels, and so forth. It is especially interesting that in both publications, the evolutionary significance not only of the rules of single populations, but also of multiple populations, is considered. The study of the "host–parasite" and "predator–prey" relationships at the level of multiple-species systems is significant in this respect. Thus, Lerner (1965) showed that the outcome of competition between *Tribolium confusum* and *T. castaneum* is determined to a large extent by the specific genetic features of the populations studied and by the environmental conditions. Usually, cultures of *T. confusum* develop normally on corn meal; *T. castaneum* cannot develop normally on this type of food. When both species are maintained together, however, the latter species destroys the former, and with the reduction in the number of *T. confusum,* both species die out. Such research could be found quite naturally in any modern ecology course. It is not by chance that in his theoretical works, an author is guided equally by the research of both geneticists and ecologists (Schlager, 1963; F. W. Robertson, 1960; Sheppard, 1958; Haldane, 1956; Lewontin, 1958; Birch, 1960; Odum, 1963). Lerner rightly observes that the tasks of ecological genetics have not been defined exactly even yet, but he considers it an independent synthetic science, not the application of genetics to ecology (or vice versa), and not a simple study of the genetic bases of the phenotypic expression of characteristics of different populations in different environments. Since the essence of ecological genetics consists in the study of genetic rules in the modification of populations under changing environmental conditions that elicit changes in size or changes in the direction or intensity of selection, this new biological discipline could with equal right be called evolutionary ecology, which properly uses the techniques of genetic analysis on the results obtained.

Another line of investigation, for which the theoretical task is especially clearly formulated in the work of Orians (1962) and Lack (1965), is usually understood as evolutionary ecology. Evolutionary ecology (in the understanding of these authors) includes in the sphere of its investigations study of the evolutionary (phylogenetic) conditionality of specific ecological characters of separate species and forms (the dynamics of population size, the biology of reproduction, territoriality, distribution according to biotope*). It studies the origin and development of ecological adaptation. According to Lack, it is necessary to make a strict distinction between changes in animals' ecological characteristics that appear to be direct answers to specific conditions of their existence, and ecological adaptations, in the true sense of the word, that are consolidated in the process of evolution. Ecological adaptations can show up in combination within a single population of ecologically different animals, e.g., migratory and resident birds. If such a combination is advantageous, selection maintains and stabilizes it. Lack thinks that ecological adaptations arise under the influence of individual natural selection. He denies the significance of group selection. He sees the distinction between evolutionary ecology and population genetics to be that the former studies traits that are polygenically determined. This observation has, of course, a particular character, and does not reflect a principal difference between the two related biological disciplines. Rather, the shortcomings of a definite stage in the development of population genetics are noted by Lack. But they are noted correctly, since it is risky to extend to the least degree the outcome of the results obtained during an analysis of the rules by which traits controlled by a single gene spread in a population to complex polygenic traits (and it is precisely these traits that determine the specific character of the majority of populations). We turn our attention to this particular question, since Lack clearly shows that evolutionary ecology (even in the narrow sense) makes it possible to overcome the well-known one-sidedness in the development of population genetics. In the opinion of the authors cited, however, the content of evolutionary ecology is in essence the study of the origin and development of ecological adaptations. This assertion is debatable. Since *ecology* is usually understood to mean the study of the manner of life of animals, the study of changes in the manner of life occurring in the process of evolution may be called *evolutionary ecology*. One would think this term the more legitimate, since the names of many disciplines are, to a significant degree, relative. Since modern ecology is not simply the study of the mode of animals' lives, however, the definition cited suffers one-sidedness at least. With great truth, we can call research that applies itself to the task of studying changes in the interrelationships of organisms with the environment in the process of the phylogenetic development of separate groups evolutionary ecological research. As will be shown in the concluding chapters

* A region with particular environmental conditions and particular populations of animals and plants, for which it is the habitat—Ed.

of this book, such research is of exceptionally great interest, and undoubtedly enters into the sphere of evolutionary ecology's tasks, but does not exhaust its problems. Moreover, it seems to us that the main channel for the development of evolutionary ecology goes in a different direction.

As is known, the modern stage of development of evolutionary studies is quite closely linked with the progress of genetics. This question of the history of science is so important that we devote an entire chapter to it (Chapter I). Here, it should be noted that the synthesis of modern genetics with the basic principles of Darwinism led to the creation of the so-called synthetic theory of evolution (Neo-Darwinism), which at present is enjoying the widest recognition and diffusion. The basic postulates of this theory can be summarized as follows: the elementary unit of the evolutionary process is the population; a change in the genetic structure of a population is the initial stage of microevolution; the rate of the genetic modification of populations (i.e., of microevolutionary modifications) is determined not only by the strength of selection, but also by the position of the population in the species system (dimensions of the population, degree of isolation from neighboring populations, and so forth); the direction of microevolutionary modifications is determined not only by the direction of selection, but also by random changes in population structure in conjunction with changes in size or with the chance settling of new areas in the arena of life by genetically distinctive individuals.

Neo-Darwinism successfully surmounted a series of difficulties in evolutionary studies. Its basic position—the population is the elementary unit of the evolutionary process; the elementary evolutionary act consists in the modification of the population—is unconditionally true. Indeed, it is corroborated by a vast quantity of facts (see Chapter I), and itself signifies the progress of evolutionary studies as a whole. The synthetic theory of evolution also successfully coped with a series of particular difficulties. It made clear the role of the chorological* structure of a species and the dynamics of size in microevolution. Finally, the synthetic theory of evolution makes it possible to explain the evolutionary process on the basis of firmly established genetic mechanisms, without resorting to auxiliary hypotheses that themselves still demand substantiation. This theory gave concrete substance to the most important concept of Darwinism—"indeterminate variability."

Giving a great deal of attention to different manifestations of genetic automatic processes† (see Chapter I), the synthetic theory of evolution explains well the results of experiments on model populations, and gives a

* Zoogeographical and phytogeographical—Ed.

† Briefly, these are processes that occur in the development of an isolated population started by a small, randomly determined number of individuals (founders). The distinctive development of such a population depends on the action of selection on its relatively impoverished gene pool, as well as on stochastic processes—Ed.

natural explanation of many phenomena observed in fully or partially isolated natural populations. It can be said that Neo-Darwinism successfully explains the details of the evolutionary process, but it far from always copes with those difficulties that arise in the necessity to explain the process of mastering large subdivisions of the arena of life such as biomes, the process of adaptation to cardinal changes in the conditions of existence, or adaptive radiation of broad biological significance. None of these processes can proceed in the "hothouse" conditions of isolation, in impoverished biocenoses,* or under the conditions of a weakened struggle for existence. Genetic automatic processes, of course, can also play a substantial role, but by no means a leading role, in these conditions. This creates serious, sometimes insuperable, difficulties when it is necessary to coordinate the real rates of the evolutionary process with the possible effectiveness of natural selection. These difficulties of the synthetic theory compel many specialists to seek other paths for elucidation of the evolutionary process.

On the other hand, the synthetic theory of evolution, in giving maximum attention to investigations of the initial stages of evolution, leaves the most important stage of evolution—speciation—in the background, tacitly acknowledging the formation of new species as a simple continuation of intraspecific differentiation. The connection between these processes (intraspecific differentiation and speciation) is undoubted, and its establishment can be boldly classified as being among the most important accomplishments of biology. Nevertheless, species are not simply sharply differentiated intraspecific forms. They are stages of development of the organic world, and therefore the rules of speciation cannot be reduced to the rules of intraspecific differentiation. The process of speciation is prepared by the previous history of one population or several populations of an ancestral species. The opposite point of view inevitably leads to active antihistoricism in biology and saltationism in all its manifestations. Denial of qualitative differences between species and intraspecific forms, however, distorts the real picture of development in animate nature. Darwin's greatness consists precisely in that he saw the key to the solution of evolutionary problems in the origin of species—the elementary units of life. Species are a product of the development of living matter, but besides that, species are a prerequisite of progressive evolution.

Life is discrete. The nature of this discreteness is dual. It is manifested not only in that all life on our planet is embodied in separate individuals, but also in that these individuals are grouped into species, within which there is free interbreeding. The discreteness of life is one of the most fundamental laws of nature.

* Biocenosis: an assemblage of diverse organisms inhabiting a common biotope; a biotic community—Ed.

The smallest and most primitive living creature is a full-valued bearer of life, possessing variability and heredity. Hence, it follows that the discreteness of life in its first manifestation provides material for selection and, consequently, for progressive adaptation to the changing conditions of the external world. The sexual process (its existence is now demonstrated even for such primitive creatures as bacteria) leads to the continuous and, in principle, limitless enrichment of that genetic material from which natural selection creates a startling diversity of living organisms. Completely unlimited interbreeding, however, could very quickly be transformed from a factor of progress into its impediment. It would lead to the gradual combination of genotypes having differences that would make the harmonious development of their offspring impossible. The sexual process was to have been limited to a group of individuals sufficiently close that the joining of their sex cells would give a new viable organism, but sufficiently diverse to provide continuous enrichment of the common gene pool. Hence, it follows that *a species is not only the basic category of taxonomy, but is also the basic form of the existence of animate matter, and the isolation of a new species is a stage in the development of life.*

The process of speciation is the most important stage of evolution, and it cannot be directly inferred from the rules of intraspecific divergence. This chapter of the evolutionary teachings of Neo-Darwinism has not been worked out well. Finally, the problems of macroevolution remained, in essence, outside the field of vision of the synthetic theory of evolution. It is no coincidence, therefore, that the synthetic theory of evolution, despite its major contribution to the development of theoretical biology, was and is subject to serious criticism even though it is the most widespread, but by no means universally recognized, theory of evolution. This is not a book on the history of science; for us, therefore, it is important only to point out the basic line of criticism of Neo-Darwinism, which contains in itself a constructive, positive beginning. The criticism of Neo-Darwinism based on openly antiscientific notions holds no interest for us, although from the viewpoint of the history of science, it also deserves examination.

One of the newest of the works that are sharply critical of the synthetic theory of evolution is the work of Russel (1962). The author maintains that Neo-Darwinism cannot explain adaptive specialization and diversification of phyla. There are no connecting links among phyla. Diversification of phyla does not have, in the opinion of the author, anything in common with the genesis of adaptation, and therefore Neo-Darwinism cannot give a natural explanation for the main event of evolution (the origin of phyla). Russel maintains that a sober examination of the facts leads to the acknowledgment of orthogenesis. Such assertions are by no means a rarity in the contemporary literature. They are symptomatic. However, those doubts that arise in authors who on the whole adhere to Neo-Darwininst conceptions are even more interesting.

In this respect, the most interesting collection, *Genetics, Paleontology and Evolution,* published in 1963 under the editorship of three of the most eminent representatives of Neo-Darwinism—Jepsen, Mayr, and Simpson—is highly instructive. Nevertheless, in almost all the articles of the collection, dissatisfaction with the modern theory of evolution makes itself known. Watson (1963), following the paleontological bases of evolutionary teaching, concludes his article with a rhetorical question: cannot separate species arise as a result of chance mutations? Stern (1963) thoroughly analyzes the genetic nature of intrapopulational and intraspecific differences and surmises that without the help of Goldschmidt's principle (systemic mutations), cognition of the mechanisms of the evolutionary process seems impossible. Ford (1963)—an investigator who has done an especially great deal for the experimental study of the beginning stages of microevolution under natural conditions—shows that a change in the genetic makeup of a population, which is often accompanied by a change in numerical size, is explained not by genetic automatic processes, but by selection. The author concludes: "The subdivision of a species into isolated groups is favorable to rapid evolutionary change. This is due not to random survival in populations of small size but to selection adjusting them to the varied environments to which such colonies are exposed" (p. 314). This assertion by no means coincides with the basic postulates of the synthetic theory of evolution, and certainly is based on almost unique experimental data. Finally, the conclusion of one of the collection's editors, Simpson (1963), is of the greatest interest. He arrives at the conclusion that evolution will be truly apprehended only in a "utopian age," but that for the present, we ought to study attentively every possible factor of the evolutionary process. This assertion from one of the creators of the synthetic theory of evolution is worth emphasizing. It is characteristic of truly outstanding scholars. Apologists for the synthetic theory of evolution often pass it off as truth in the final analysis. It is wrong to do so. It is not difficult to see that criticism of the synthetic theory of evolution ensues, to a significant degree, from those of its defects to which we are turning our attention in a very general form and that we will yet have the opportunity to examine in detail. Nevertheless, it is necessary to note here that this criticism often leads to still more serious mistakes, and to the discarding of undoubtedly valuable contributions that Neo-Darwinism has introduced into science. This has to do, first of all, with the attempts to revive Lamarckism in our day. The positive role that Lamarck's teachings played at the dawn of the history of evolutionism is too well known for us to dwell on it. In our time, however, Lamarckism is a step backward in comparison with Darwin, not only because it does not find corroboration in special experiments and contradicts many well-demonstrated experimental data from genetics, and not only because Lamarckian principles, *a priori,* are inapplicable to a large sphere of natural phenomena (the evolution of social insects and other well-known examples), but chiefly because it is found in opposition to the basic achievements of modern biology—to historicism, to the idea of the succession of development. Therefore, the new variants of Lamarckism, allegedly based on the achievements of science and attempting to explain the origin of heritable variability on the basis of the principles of immunogenesis (Winterbert, 1962), are just as archaic in essence as even the naïve notions of Lamarck on the transformation of animal organs under the action of "internal impulses." The positive aspect of criticism of Neo-Darwinism from the position of Lamarckism, therefore, can scarcely be of scientific value. Its negative aspect is not bereft of interest, however, since it involuntarily compels us to think about the role of phenotypic reactions in animal evolution. It now seems clear that this role is not at all expressed in the form in which Lamarck and Lamarckians conceived it, but the indirect role of phenotypic mechanisms in evolution is probably more significant than it is represented as being by Neo-Darwinism. Besides some general considerations about this role, there is also a fairly large number of facts concerned with the problem of "selection in pure lines" (Barnett, 1965; Lozina-Lozinskii, 1966) and "the direction of selection and phenotypic variability" (Schmalhausen, 1946; Danielli, 1953; C. H. Waddington, 1958; Milkman, 1965; and others). The inattention of the synthetic theory of evolution to questions of the physiology of development, to the phenotypic realization of the genotype, and to epigenetics should probably also be numbered among the shortcomings of the theory, but these shortcomings cannot be overcome on the basis of Lamarckian conceptions. On the contrary, the application of certain principles of ecology to the resolution of this shortcoming may turn out to be very fruitful.

The opposition to the synthetic theory of evolution by openly antievolutionist conceptions has still less significance (in comparison with Lamarckism). For example, the curious symposium on (one ought to say against) evolution that was held in Spain in 1955 (see Simpson, 1956) testifies that these conceptions have in no way been transformed into the property of the history of science as yet. The basic conclusion of the "evolutionists" assembled at this symposium was: macroevolution is accomplished by way of saltation; evolution, on the whole, is finalistic; its motive force is a deity. Whereas, in this conception, evolution still exists, even though as a manifestation of a higher force, some contemporary authors banish evolution from nature altogether. Thus, Zdansky (1962), arbitrarily interpreting the data of modern biochemistry, postulates the constancy of species (this in 1962!), motivating this archaic conclusion by the reason that the variability of DNA allegedly cannot exceed the limits of the species norm.

It seems to us that the synthetic theory of evolution, despite the shortcomings mentioned, corresponds most fully to the modern level of biology. Its shortcomings can be overcome not on the basis of discarding valuable principles of Neo-Darwinism, but on the basis of developing this theory as a whole. It ought to be examined as the development of Darwinism, since the synthetic theory itself is the development of Darwinism, based on utilizing the data of modern genetics. It seems to us that its further development should proceed on the basis of the synthesis of Darwinism with ecology.

It should be noted that the synthetic theory of evolution, in its most widespread modern variant, already successfully uses certain ecological rules, in particular the rules of the dynamics of numbers and of the exchange of genetic information among populations. Since the elementary evolutionary act consists in the modification of populations, however, and the elementary unit to which natural selection applies its force is also the population, all the rules governing the dynamics of populations (and not only the dynamics of size) should have a direct relationship to the microevolutionary process. *We regard the rules reflecting the interdependence between the dynamics of the structure of populations and the modification of their genetic makeup as the most important factor of the evolutionary process.* The study of these rules is one of the most important tasks of evolutionary ecology. Naturally, this important position can be sufficiently fully substantiated in the conclusions, but not in the introduction. It is therefore sufficient for us to show here that such an understanding of the tasks of evolutionary ecology ensues from contemporary ideas about the tasks of ecology in general, which an ever-greater number of biologists view as the science that studies the interrelationships of animals with the environment at the population level.

Since the latter conclusion has a direct bearing on the theme of this book, it is advisable to dwell on it in more detail. This will save us later on from unnecessary digressions connected with the necessity of elucidating the author's position concerning the principal questions of ecology. On the other hand, the general propositions of the theory of populations are the basis on which an analysis of the basic problems broached in our work is constructed.

Earlier ecologists saw in their science a complex of the knowledge of the interrelationships of organisms with their environment, the study of adaptations, and the process of cognition of natural relationships. This point of view prevailed almost completely until the end of the 1940s. It was most fully reflected in the three editions of *Ecology* by D. N. Kashkarov, the textbooks on which the modern generation of ecologists grew up. Kashkarov (1945) saw the task of ecology "as the study of adaptations—morphological, physiological, and behavioral; and indeed also of the contradictions between an organism and the environment; the study of the life history of a species (or of a community)." N. P. Naumov (1955) justly notes that "this definition, which is broadly disseminated at the present time, is incorrect, since the study of the interrelationships of the environment and organism in their historical development is not the privilege of ecology, but constitutes the basic substance of all of Soviet biology." He sees the main task of ecology as being the study of "survival and its dependence on the conditions of existence," and he regards the population as the unit of ecology. At present, many Soviet ecologists adhere to this or a very similar viewpoint of ecology as the science of populations. It was reflected in *Resolution III of the Ecological Conference,* Kiev (1954), and is dominant in a great series of works devoted to the solution of important practical tasks. The necessity to develop ecology in a modern sense dictates the logic of development of all biology as the science of life in all its manifestations.

A full understanding of the phenomena of life demands its study at different levels. The modern complex of biological sciences answers this demand. Cytology and histology study the phenomena of life at the level of cells and tissues; anatomy and physiology, at the level of individual organisms; biogeography and biocenology, at the level of the biosphere. In any case, however, life is impossible in its highest manifestations not only in the form of cells and tissues, but also in the form of separate individuals. Life is also immediately impossible in the form of biocenoses, since biocenotic ties are the ties among populations of the same species, not the ties among separate individuals.*

Ecology as the science of populations fills an existing gap in the full understanding of life on earth. The population is basic, and for higher animals, it is the only form of existence of a species. Just as the existence of a cell of a multicellular organism outside the organism is inconceivable, so also is the existence of an individual outside a population inconceivable. This does not mean, of course, that the population is an organism of higher order. It means that the population is a definite organization (a structural whole) of individuals, outside of which they cannot exist. From the ecological point of view, a population is a group of individuals of the same

* Beklemishev (1951) wrote that "the study of the interrelationships (which are multiform in their contradictoriness) of populations of different species also constitutes the essence of biocenological investigation."

species that are living together and are united in a unity of life activity. Not every group of animals living together, however, can be called a population—far from it. To be a population, it should possess a complex of properties ensuring its independent existence and development over a long (theoretically—unlimited) period of time.

Independent existence and development is the sole objective criterion by which it is possible, in principle, to differentiate populations (the form of existence of species!) from temporary groupings of animals. On the other hand, this criterion promotes the synthesis of ideas from population genetics and ecology. Population genetics teaches us that the most important feature of a population is a balanced (but not invariable) gene pool, reflecting the conditions of existence of the species in a specific environment. A certain amount of time, measured in generations, is necessary for the formation of this gene pool. Therefore, only permanent populations satisfy the demands that the geneticist makes of them. Ignoring the differences between populations and intrapopulational groupings of animals can lead to mistakes in theory and practice. Indeed, it would be a mistake to ignore intrapopulational groupings of animals, which constitute the most important element of the ecological structure of populations.

That a population is said to have a certain ecological structure is understood to mean that it has a certain ratio of age classes, a certain sex ratio, a combination of resident and migrant animals, and the presence of family, herd, and other groupings. The more complex the structure of a population, the greater its adaptive possibilities (for details, see Shvarts, 1960*b*, 1965). The unity of the population's adaptive reactions is accomplished with the help of a complex system of signals and relationships that inform the individuals about the state of the population as a whole. This system of information is based on the animals' ecological and physiological reactions to external stimuli of a very different nature (chemical, change in the external environment, change in the frequency and degree of intrapopulational contacts, auditory and visual signals, and so forth). The sum of these reactions welds the individuals of the population together into *a single functioning system that ensures the maintenance of the numbers of the species in diverse habitats*.

A population functions as a unity, but this does not mean that it is the smallest biochorological unit. The majority of (but not all!) populations naturally subdivide into *micropopulations,* which differ from populations in that they are not independent forms of existence for the species, are not capable of prolonged independent existence, and exist only as part of the whole. As an example, let us look at the population structure of forest mice (*Apodemus sylvaticus*) in the forest steppe. Mice are encountered here in woodpiles and in fields, as well as along the shores of reservoirs. Among these groupings of mice are found both ecological (dynamics of size, age structure) and morphological differences, which are often very significant. Despite these differences, the settlements of mice in the sown fields and along the shores of reservoirs are not populations, since they are not capable of maintaining their numbers over a prolonged period (in the fields, agrotechnical measures destroy the mice; along the shores of reservoirs, high floods, ice crusts, and so forth). They exist in these habitats only so long as there are woodpiles in proximity, where the conditions of existence are more stable and where there are always survival habitats (usually the undergrowth of shrubs and berry bushes, with an abundance of forage and good protective conditions, and high and well-aerated snowy cover). In our example, however, even the settlements of mice in the woodpiles cannot be called independent populations, since their numbers depend to a significant degree on the propagation of mice in the sown fields and along the shores of reservoirs. Micropopulations of the three types mentioned constitute a single population, the characteristic feature of which is its subdivision into settlements, which develop in distinctive conditions. The tie among micropopulations is accomplished mainly during the autumn and spring displacements; it can be more or less close, depending on conditions in the different years.

In the example analyzed, the micropopulations constituting the population are timed to different biotopes and therefore can be called *biotopic populations*. In other cases, however, micropopulations can occupy several very different biotopes or, the other way around, parts of a biotope. We therefore prefer to operate with a single term, *"micropopulation," which is understood to mean the temporary settlements of animals that are the structural elements of the population*. This term proves to be suitable in equal measure for any species: from the protozoa to the amphibians, for which the settlements of animals in temporary reservoirs can be regarded as micropopulations (down to the separate pools, which can in no sense be called biotopes), to the ungulates, the micropopulations of which are often commensurate, according to occupied territory, with large subdivisions of the biomes. These rules should be expressed in the descriptions of the population structures of specific species, but as a single term for the description of intrapopulation chorological groupings, "micropopulation" seems to us more convenient than biotopic or elementary populations.

It is very important that the population structures of diverse species can be different under the same conditions, while the structure of populations of one and the same species changes abruptly during a change in the conditions of existence. Under optimal conditions, when population size is high, the species occupies all the sections of the territory suitable for settlement. Under deteriorated conditions, the animals die out in large parts of the territory and remain only on small sections (reservation stations) on which, even in suboptimal years, as far as the conditions of existence are concerned, they can maintain their numbers. The dynamics of numbers in reservation stations is correlated to a significant degree with the dynamics of the populations' spatial structure.* The concrete expression of this dependence can be different in different species, but it is manifested in some form or other in all animals. It is important to emphasize, therefore, that the totality of micropopulations, which are tied together by community of origin (the reservation stations) and general rules of the dynamics of population size, should be called a population in the strict sense of the word. Any micropopulation (including those that exist for a short period of time) exerts an influence on the genetic and ecological structure of the population. Study of these micropopulations is therefore of tremendous interest.

It is clear from what has been said that the population is the elementary form of existence of the species—it is in no way a simple biological phenomenon. It possesses a complex structure, ensuring in the final analysis that the species flourishes. The specific limits of the populations in each individual case are determined in accordance with the biological characteristics of the species and the local conditions. Principal considerations regarding the general structure of the species and the structure of its populations can serve only as general reference points during the execution of this most complex work.† Natural populations can often be naturally divided into structural units of higher rank. Their mutual resemblance is manifested in ecological characteristics (similar position in a system of similar biogeocenoses,‡ type of dynamics of numbers, phenology of vital activities), as well as in morphophysiological characteristics. As a rule, like

* This correlation is shown especially clearly by Polyakov (1964 and others) in small mouselike rodents.

† In practice, the limits of the populations can be determined in many cases by studying regularities in the dynamics of the species' numbers (a single type of dynamics of numbers is characteristic for a population). In other cases, the limits of the populations coincide with the limits of the biogeocenoses or the plant associations (Gilyarov, 1954*a,b*; "The population is the aggregate of individuals of a given species that enters into a defined biocenosis."). This, however, is not the general rule. The territory occupied by populations of the reindeer includes both its summer pasture (tundra) and its winter pasture (forest-tundra, northern taiga). Such a state of affairs is natural, since biocenoses are not secluded systems. They are welded together in a system of higher rank and, in the final analysis, in the biosphere. Similar populations are the mechanism that welds the living layer of the earth into a single whole.

‡ A biogeocenosis (or biogeocenose) is the elementary primary structural unit of the biosphere. It is now used as a synonym for ecosystem. "The biogeogeocenoses (ecosystems) are parts of land or water surface, homogeneous in respect to topographic, microclimatic, botanical, zoological, pedological, hydrological and geochemical conditions" (*Use and Conservation of the Biosphere,* 1970, p. 16)—Ed.

groups of similar populations inhabit a common environment in physicogeographical respects—hence the very widespread expression *geographical populations* (N. P. Naumov, 1963; S. P. Naumov, 1966). They are united not by a functional unity, but by a similarity that arose as a result of development in similar habitats. Hence, it follows that populations and "populations" of higher rank (it is possible to construct a hierarchical system of any degree of complexity) are different concepts in principle: "geographical populations" are distinguished on the basis of similarity; populations, on the basis of functional unity.

A group of neighboring populations is not a single functional system. Their similarity is determined by development in a similar (or identical) environment and therefore is manifested even when there are no immediate ties among populations. The best examples are certain mountain species. Populations inhabiting separate mountain ridges often possess no less similarity than neighboring populations of plains forms, despite practically full isolation. There are also other well-known examples showing that a significant degree of similarity is observed even when the exchange of genetic information among them is reduced to a minimum. In other cases, the similarity among populations is determined to a significant degree by the exchange of genetic information.

In different species and in one and the same species in different environments, the populations can be substantially different. A population can occupy territory commensurate in area with the mainland (populations of the polar fox or wild ducks), or it can be limited to several square meters (certain amphibians or mollusks); populations can unite millions of individuals (mosquitos) or, in all, several dozens of animals (large predators); a population can be represented in a great number of micropopulations, synchronized to different biotopes, but in spatial respects united; the numbers of populations can be relatively stable, but can change tens of thousands of times; and so forth.

The general aspect of a population is determined not only by its genetic characteristics, but also by the conditions of existence. It is well known that the phenotypic distinctions among animals are not in strict correspondence with their different genotypes. The phenotype characteristic of an animal is formed under the influence of the environment.

In recent times, a series of works by different investigators has shown that this regularity is distinctly manifested in natural populations, and finds its expression in significant morphological and physiological differences among groups of individuals not living under the same conditions. These differences affect such animal characteristics as rate of growth and development, general size, fertility, and so forth.

Therefore, even in those cases in which the genotypic community of two neighboring settlements does not arouse doubts, their phenotypic characteristics deserve very great attention. They are indicators of differences in the conditions of existence of the groups of individuals compared. They determine the morphophysiological characteristics of animals and their variability, and ensure a unity of seasonal recurrence of their vital activity. This, in turn, determines the type of expression of the basic life phenomena (reproductive cycles, molting, change in habitats, and so forth) that is characteristic for the population as a whole and, in the end, determines the characteristic features of the dynamics of its population size.

The age structure of a population is regarded as one of its basic features. It involves more than a correlation of the various age groups and of animals of different sex. It includes at least the following range of phenomena: a change in the age makeup of the population in different seasons of the year, depending on the specific combination of yearly conditions; the biological specificity of animals of different ages and features of their reactions to a change in the external conditions; the rate of renewal of the population (separately for males and females) under different environmental conditions; the regularity of mate selection under natural conditions (the age balance of reproductives); the biological specificity of animals born in different times of the year and their role in the maintenance of the numbers of the population; and so forth. The change of a population's age structure often has a preadaptive character, being a phenomenon of population homeostasis.

Periodic (seasonal) and aperiodic changes in external conditions elicit a series of complex changes in animal organisms. On the approach of cold weather, the nature of the interchange is altered, both qualitatively and quantitatively. The accumulation of energy reserves of nutrients and vitamins begins. A change in the activity of the gonads and endocrine glands occurs, molting begins, and among many species, there is a shift in the preferred habitat, often evidenced in the form of significant migrations.

In many cases, the study of an aggregate of similar phenomena that occur in response to a change in the external environment entails significant difficulties. Nevertheless, there has already been accumulated a significant body of facts showing that the complex of reactions to a change in external factors proceeds in different ways in different populations of a species. Often, it is precisely in the reactions to a change in external conditions that the specificity of populations is manifested. But even within one population, separate groups of animals conduct themselves differently. From the totality of reactions of separate groups, the general reaction of a population as a whole is established.

The most important characteristic of a population is the type of dynamics of numbers. Therefore, the study of the rules of the dynamics of a species' numbers is the final theoretical and practical task of ecology, which is broken down into a series of subordinate particular questions: the presence or absence of periodic fluctuations in numerical size; the causes of periodicity—external (periodic changes in the external environment or changes of the environment by the animals themselves during their attainment of a certain size) and internal (a change in the population's viability with an increase in the density of the settlement); dependence of the species' numbers on changes in the external environment and the specificity of population structure; a change in the intensity of the birth and death rates, depending on different factors.

The periodicity of the maximum and minimum numbers of a species, which has been ascertained statistically by many years of observations (frequently by the inventory of fur purchases), is purely an external expression of the complex processes proceeding in the population. A change in numbers is inevitably accompanied by a change in the quality of the population.

Our all too brief sketch of the basic problems and basic ideas of modern ecology clearly shows that in the process of coping with different habitats and in the process of adaptation to a change in living conditions, there occur not only the well-known changes in the morphophysiological traits of the species, but also a change in its characteristics at the population level—of its ecological characteristics in the strict sense of the word. This creates the preconditions for the development of population homeostasis, changes the operating conditions and the effectiveness of the action of natural selection in its classic form (individual selection) and directs the possible course of group selection, determines the final result of nondirected (nonselective) elimination, promotes the exchange of genetic information among neighboring populations, creates the conditions for the relative stabilization of the population's optimal gene pool, and aids its enrichment during a change in external conditions. *The population is a biological unity, the genetic and ecological manifestation of which is interconditional. The interconnection of the ecological and genetic in the population is the background against which the elementary evolutionary phenomena are displayed.*

Chapter I

The Genetic Basis of the Reorganization of Populations

The first step in the evolutionary process is the origin within a species of biologically specific populations with characteristics that are genetically based. This being the case, it follows that any evolutionary theory should be founded on modern ideas about the laws of the genetic reorganization of populations. It is therefore necessary to preface an analysis of the ecological mechanisms of the evolutionary process with a short sketch of the basic conclusions of population genetics.

Population genetics is in essence the study of laws of the dynamics of genetic variants (genes and genotypes) in natural and experimental populations of animals, plants, and microorganisms. From a general biological point of view, this task is no less important than the central problem of general genetics—the study of the structure of heritable units (genes) and the mechanisms of their action. In order to understand the ever-increasing role of this young science in the development of evolutionary thought, it is necessary to turn to its history.

The birth of modern genetics is tied to the rediscovery of Mendel's laws, which was by no means met with enthusiasm by Darwinists. This lack of enthusiasm is understandable, since the essence of Darwinist theory consists of the creative role of natural selection. The geneticists, in contrast, asserted that the only known form of heritable variability is mutations. But selection does not require the emergence of mutations. They show up spontaneously in formed species, and, as earlier geneticists thought, the characteristics of mutations are expressed fairly abruptly. The most that selection can do is cut out the obviously unfavorable mutations. The role of sieve is not a very honorable role for the leading mechanism of the evolutionary process. The reaction of the Darwinists was also determined by this consideration. They sensed no mighty ally in genetics, and denied the role of mutations as the source of variability of living organisms, on the basis of which natural selection created all the variety of living organisms. The spread of this point of view was also fostered by the circumstance, of no small importance, that almost all the mutations known at that time turned out to be deleterious, and in the theoretical works of the geneticists of that period, mechanistic tendencies were clearly displayed. On the other hand, the early Darwinists did not yet understand that the law of segregation, independent of the distribution of traits and the relative constancy of the units of heredity (genes), removes a series of difficulties in Darwin's theory.

It is self-evident that in the works of various scholars, tendencies in opposition to Darwinism and Mendelism appeared in different degrees and in different forms, but on the whole, nearly the entire first three decades of the twentieth century can be regarded as a period of independent development of evolutionary study and genetics. In 1926, S. S. Chetverikov took the decisive step toward the mutual enrichment of these leading divisions of natural science in the article "On Some Features of the Evolutionary Process from the Viewpoint of Modern Genetics" (Chetverikov, 1965). It is not our task to enter into an analysis of this remarkable article, which initiated a new direction in both genetics and evolutionary study. We will note only the main points. Chetverikov disclosed the role of discrete hereditary units in the variability of a population and showed that on the basis of mutations in natural populations of animals, an enormous potential of variability—"hidden variability"—accumulated. The mechanism of formation of this hereditary potential conforms fully to the laws of genetics, and it is the material on which natural selection operates. Chetverikov's basic conclusion was confirmed and developed in the works of a galaxy of remarkable geneticists, among whom Russian investigators (N. P. Dubinin, N. V. Timofeev-Resovskii, and others) occupy a conspicuous position. The history of science still makes this very interesting period in the development of biology a subject of special analysis. After some decades, genetics, which prior to Chetverikov's work had been regarded as a dubious fellow traveler of Darwinism, became its basis, whereas evolutionary works that had not taken into consideration the achievements of genetics suddenly became mere anachronisms. It became clear that mutations are precisely the material on which evolution is based; consequently, all the natural laws concerned with the cytological bases of heredity and the phenotypic manifestations of mutations have a direct relation to Darwinism, to the theory of natural selection. Symptomatically, the very title of one of the most important works of this period is *The Genetical Theory of Natural*

Selection (Fisher, 1930). A new system of ideas arose, the essence of which can be summarized as follows: the source of variability is mutations of the genes and their recombinations. The exceptional role of recombinations is especially emphasized. If the inheritance of an animal turns out to be 1000 genes, each of which can appear in 10 allelic forms, this means that on the basis of recombinations, 10^{1000} genotypes can arise. This number exceeds the number of electrons in the visible part of the universe.

What, exactly, determines the spread of individual genes in a population, what determines their combination in specific genotypes, adapted in the best way to the conditions of the external environment? What role does mutation play in this process? It proved to be impossible to answer these questions by the methods of classic genetics. A new direction in genetics arose—population genetics.

From the viewpoint of genetics, a population is the elementary unit of individuals of a species among whom there is free interbreeding, which accomplishes the free exchange of genes. This exchange determines the genetic distinctiveness of the population, its difference from other populations of the species. However, the population's genetic unity is always combined with its genetic diversity.* Even brothers and sisters are not genetically identical. The descendants of one pair of animals represent a population in miniature, the genetic diversity of which remains significant. This being the case, it follows that any interbreeding leads to the emergence of new genetic variants.

This position differs substantially from notions of geneticists at the beginning of the century. Large, striking hereditary changes (macromutations, e.g., shortleggedness in sheep, sharp distinctions in the coloration of birds or mammals, changes in the structure of hair or feathers) do not determine the distinctions among individuals of a population. Macromutations, as a rule, are deleterious, and appear relatively rarely in natural populations. The genetic individuality of animals is determined by small, neutral mutations, insignificant changes in the hereditary apparatus of the cell, that determine such insignificant characteristics of animals as a just barely lighter or darker coloration, a scarcely larger or smaller body size, insignificant differences in the level of metabolism, and so forth. Studying the external manifestations of similar small mutations, geneticists arrived at conclusions that have primary significance for perception of the mechanisms of the evolutionary process. It turned out that the vast majority of traits that differentiated animals within a population were determined not

* It has been shown experimentally that even strong selection pressure does not lead to genetic homogeneity of populations. Genetic diversity was maintained in an experimental population of *Drosophila* even after 70 generations of selection (Stebbins, 1965).

by one gene, but by a complex of genes (polygenic determination of traits), and in addition by a large number of gene modifiers. This ensures the organism against random deleterious changes of the genotype. If a given characteristic of the organism is determined by the molecular structure of one of the thousands of loci on the chromosome, then a change at this locus (as a result of mutation or recombination during interbreeding) inevitably produces an observable change in the morphophysiological characteristics of the animal. If the trait is determined by the structure of many loci, however, especially if they are situated on different chromosomes, the possibility of a random disturbance of normal development is reduced to a minimum. It is clear, therefore, that the most important characteristics of the organism (in particular, its physiological characteristics) are especially reliably ensured in the manner shown.* The properties of the part (the gene) are determined by the property of the whole (the genome) (and very probably also by the property of the sex cell as a whole). There are data indicating that additional hereditary information can be coded at a much higher level (in comparison with the triplets of DNA) (Sand, 1965). In this way, modern genetics overcomes the well-known mechanism of its early representatives. These discoveries have important consequences in population genetics. Since small mutations do not, in the majority of cases, manifest a visible influence on the viability of animals, the conditions are created for their accumulation in the population's common gene pool. A vast reserve of genetic potential arises in the population.

Another, no less important, discovery of population genetics is the heightened viability of the heterozygote. The combination in a single genotype of different alleles (different genetic potentials inherited from the father and the mother) increases the organism's viability. Even in the case in which a mutation proves to be harmful, therefore, it does not disappear from the population, since the increased loss of homozygotes is balanced by a reduction in the deaths of heterozygotes. This retention of mutations leads, for example, to the maintenance in human populations of genes that give rise to serious diseases in the homozygous condition (on the whole, their presence in the population proves useful). The value of the individual is defined by his contribution to the population (Stebbins, 1965).

A special aspect of this problem, of particular interest to the ecologist, is noted in the research of Lewontin (1955) with pure and mixed cultures of *Drosophila*. It was shown that the viability of individual genotypes depends on the genetic structure of the population as a whole, and cannot be determined on the basis of a study of viability in pure cultures of individual genotypes. In this regard, the investigations on the genetics of mimicry are

* That this is the case was shown with particular persuasiveness in the posthumously published book by Schmalhausen (1964).

of exceptional interest. A study of different species of butterflies showed that mimicking coloration is usually determined by single dominant genes, but the perfection of mimicry depends on the genome as a whole (Sheppard, 1965). Studying the polymorphic species *Papilio dardanus,* Ford (1963) established that only an approximate resemblance of the mimic to the model arises as a result of mutations, and that its perfection is the result of the reorganization of the genotype. If the major gene is introduced into a genotype of another race, the perfection of mimicry is reduced. The degree of dominance of the major gene is also reduced. That it is indicates that dominance itself was worked out by selection of different intraspecific forms along the path of reorganization of the genotype (the author suggests that this is connected to the action of supergenes).

A third conclusion of the direction in population genetics that is being examined has perhaps even greater significance for the development of evolutionary studies. Since the most important biological characteristics of an organism are determined not by single genes, but by gene complexes, and since harmful (including lethal) mutations can prove beneficial in the heterozygous condition, it follows that the value of any mutation is determined not by its individual properties, but by the properties of the genotype as a whole. But since interbreeding leads to constant mixing of genes and to constant change in genotypes, the value of separate mutations is determined in the long run by the properties of the common gene pool of the population as a whole. It is very important that direct experimental evidence had already been obtained a fairly long time ago that heterozygotes possess precise homeostatic reactions that heighten the adaptability of animals to change in environmental conditions (Dobzhansky and Levene, 1955).

This conclusion has exceptional significance. It shows that *a population is a single integral system*: the alteration of individual genotypes has an effect on the population's common gene pool, but the alteration of the gene pool also modifies the role of the individual genotypes and even of the individual genes in the development of the system. Hence the uniquely important conclusions: *the elementary units of the evolutionary process are not separate individuals, but populations*. This position became the basis of modern evolutionary studies; its correctness is confirmed each year by more and more new experimental investigations.

On the other hand, it has already been pointed out why variability of organisms has a continuous character, even though individual hereditary changes—mutations—are by their nature discrete. This continuity is a consequence of the cumulative action of a great many genes and their modifiers.

The spread of genetic ideas from the level of individuals to the level of populations compelled biologists to give more attention to stochastic processes and, from historical necessity, elicited the appearance of a cycle of work that can be called the mathematical theory of the reorganization of populations (Fisher, 1930; Haldane, 1954, 1957; S. Wright, 1948, 1955). This theory, for the first time, utilized mathematical models as a means to analyze evolutionary processes, and it revealed the most general rules determining the relative role of selection pressure, population size, degree of isolation, and the type of dynamic by which numbers alter the genetic structure of the population. The essence of the mathematical trend in population genetics is well formulated by Wright. He shows that the mathematical theory ties together the genetics of individuals and the genetics of the population in a single system. It is based on setting up models for the structure of the population. The theory is corrected by comparing the models with field observations. The most important result of this line of endeavor was prediction of the phenomenon of so-called "genetic drift," the analysis of which led to a study of genetic automatic processes.* A series of special works (Li, 1955; Dobzhansky, 1954, 1955, and others; Lerner, 1965; Falconer, 1965) and symposia in Italy in 1953, the United States in 1960, and other countries was devoted to investigation of the degree of correspondence between the mathematical models and phenomena occurring in nature. Nevertheless, as we will attempt to show in the following chapters, the gap between the models and reality still remains significant. It engendered several mistaken notions that have great significance for the development of evolutionary theory.

It is necessary to bring the mathematical models of evolutionary reorganizations closer to the real natural situation observed long ago. Williamson (1957) successfully formulated this approach. He showed that the mathematical theory of intraspecific relations issued from simplified notions about the uniformity of individuals of one species and about the uniformity and seclusion of their habitats. Sheppard (1965) writes about the well-known fact that modern investigations show that selection operates with greater effectiveness than is indicated by the mathematical models.

The direction of population genetics that has been examined deals with investigation of the laws for the formation of the population's gene pool. Another direction involves the study of the genetic reorganization of a population during a change in external environmental conditions (a change in the direction of selection). In correspondence with a change in the direction of selection, there occurs a change in the frequencies of different genotypes and a change in the mean standard variability of the population as a whole. This process can be most easily understood from a concrete

* The bases of this study were in essence laid down by Chetverikov in his views on the role of "waves of life" in the microevolutionary process.

example. Streptomycin in concentrations of 25 mg/kg stops the growth of the intestinal bacterium *Escherichia coli.* If several million bacteria are grown on nutritional medium containing streptomycin, however, an observer soon discerns that in several generations, the *E. coli* begin to grow again, and the growth is not stopped even by high concentrations of the antibiotic. Special analysis showed that among the millions of bacteria (remember that populations of bacteria, just like populations of any other organism, are genetically heterogeneous), there were some that possessed a hereditary resistance to streptomycin. Naturally, they were not destroyed, and produced new generations immune to streptomycin.

In exactly the same way, poison-resistant populations of various insects were created under experimental conditions. Also in exactly the same way, they arise in nature when standard chemical poisons are indiscriminately used. In regions regularly treated with DDT, the resistance of some insects to the poison has increased 100-fold in 10 years. The resistance of the frog to DDT has also increased abruptly (Ferguson, 1963). It is important to note that the resistance of insects to DDT is determined by genes spread throughout many chromosomes (Crow, 1960).

The experimental investigations in this direction at the present time number many hundreds (the total number of works on population genetics long ago passed 2000). Their major results reduce to a few positions that are so firmly demonstrated that they would deserve elevation to the ranks of laws of population biology:

1. The greater the genetic heterogeneity of a population and the larger its gene pool, the greater are its viability and its ecological plasticity, and the more rapid and complete is its reorganization under the influence of a changed environment and a corresponding change in the direction of selection. It has been experimentally shown that in the course of 15 generations, selection changes the resistance of some experimental populations of *Drosophila melanogaster* to DDT 600-fold (Benett, 1960).

2. It has been shown that in individual cases, the reorganization of a population elicits a change in the frequencies of polygenically determined traits.* In a significantly greater proportion of cases, a change in selection

* The most striking example of change in the genetic structure of a population of this type is the origin of the so-called "industrial melanism" of *Biston betularia* (Kettlewell, 1956, and others) in England. At the beginning of the nineteenth century, the light form of this moth (*ss*), the coloring of which harmonized with the color of the light lichens on the birch bark, was still widespread. As the birch bark began to darken due to industrial smoke, however, the dark mutant, *carbonaria* (*SS*), began to gain the advantage. It was ascertained (Kettlewell, 1956) that the light form is actually eaten more often by birds than are the protectively colored ones, *SS* and *Ss*.

In individual (probably rare) cases, biochemical traits also prove to be polygenically determined. This was shown, specifically, during an analysis of differences among lines of mice in the biosynthesis of different corticosteroids (Badr and Spickett, 1965).

leads to a change in the combination of breeding individuals and to a change in the selective value of a different *combination* of genes acting in common. Thus, on the basis of the initial gene pool, there arise new genotypes that were absent in the initial populations.

3. Selection in the course of many generations (the vast majority of natural populations exist hundreds and thousands of years) created balanced genotypes in the best way, a balanced population gene pool. In this balanced system, the viability of individual genotypes is determined by the complex of others. This means that in a population changed as a result of selection, new mutations will have new significance, since they serve as the basis for the formation of new genotypes, even if neither the mutation rate nor its character is changed (the dependence of the character of a mutation on the properties of the genotypes is one of the most complex and as yet unresolved questions of population genetics).* This ensures the fundamentally boundless nature of the evolutionary process.

We see that the conclusions of population genetics do not contradict the brilliant principles of classic Darwinism. On the contrary, they invest Darwin's diffuse notion of "indeterminate variability" with specific content, and allow us to uncover the specific mechanisms of the beginning stages of the evolutionary process. Population genetics also gave rise to some theoretical difficulties, however. These difficulties turned out to be connected directly or indirectly with an analysis of the rates of the evolutionary process.

A mathematical analysis of the possible effectiveness of individual selection carried out by Haldane (1954) led the author to the conclusion that in the case of horotelic evolution, the substitution of one allele requires about 300 generations. Haldane supposes that his conclusion corresponds to the available data defining the "average" rate of the evolutionary process. At the same time, essentially similar conclusions were drawn on the basis of experimental studies on changes of polygenically determined traits in *Drosophila* (Buzzati-Traverso, 1955). It was shown that under the influence of selection in experimental populations, traits such as body dimensions or wing length are changed at a rate of 0.00024% per generation. This rate corresponds roughly to the maximum rate of phylogenetic reorganizations established by paleontologists (the index of skull height in the evolution of man). The authors draws far-reaching conclusions to the effect that the known genetic processes (reproducible experimentally) are adequate mechanisms of the phylogenetic reorganizations that the paleontologist

* In the experiments of Gershenzon (1965) with the ichneumon fly, *Marmoniella vitripennis*, data were obtained that show that there is an increase in the frequency with which dominant mutations arise in males produced by crossing two lines of different geographical origin.

fixes. Even though we are striving to reduce the element of discussion to a minimum in this chapter, we must note that both authors, expressing the viewpoint of a large group of researchers, regard evolution as a horotelic process. Nevertheless, there is absolutely no basis on which to consider it demonstrated that the rate of evolutionary reorganizations is constant (for more detail, see Chapter VII). Even if one regards the process of evolution as horotelic, however, a series of facts compels one still to recognize the presence of some kinds of mechanisms for the reorganization of populations that can be reduced to the concept of "natural selection" only in its very broad interpretation. Let us limit ourselves to a few examples, since the specific analysis of this question deserves special examination (Chapter V).

The sparrow that was introduced into America changed to such an extent during 100 years that according to all taxonomic rules, it should be raised to the rank of clearly expressed subspecies. Taking into account differences in the living conditions of sparrows in the new and old homelands, a similar rate of formation clearly does not go into the framework of the theory.

Other facts also attest to the great speed of evolutionary reorganizations. Lake Lanao on the Mindao Peninsula (Philippine Islands) is no more than 10,000 years old. During this time, no fewer than 18 species and 4 genera of fish have been generated there (Myers, 1960). Speciation occurred just as rapidly among fish in some North American lakes (Hubbs and Raney, 1946; R. R. Miller, 1961). In Lake Nabugato (Uganda), no fewer than 5 species of fish (Cichlidae) were formed during 4000 years. The house mouse, which was introduced into the Faeroe Islands about 300 years ago, changed so much that some authors consider it a new species (Mayr, 1963). It is very likely that such phenomena are in no way unique. We simply have not learned to observe them. Too little attention is given to the investigation of morphophysiological characteristics of acclimatized forms (Shvarts, 1963*a*; L. V. Shaposhnikov, 1958), but cytogenetic investigations on a wide geographic plane are still rare. It is therefore possible that a whole series of microevolutionary phenomena, occurring before our eyes, remain unnoticed. Testifying that they do are the observations showing that from 1940 until 1957, the genetic structure (the frequency with which different changes in chromosome structure are encountered) of *D. pseudoobscura* changed substantially over a vast territory of the southeastern United States (Dobzhansky, 1958). A contrary example can be derived from the well-studied paleontological history of the horse. There was a steady increase in the diameter of the molars of the fossil ancestors of our horse. This increase occurred in connection with an adaptation to feeding on tough food. But it occurred at an average rate of 0.2 mm per million years. Moreover, the range of variability within different populations reaches 3 mm! A comparison of these examples shows that apart from the pressure of natural selection, there should be still other factors determining the rate of the evolutionary reorganizations of populations. Hypotheses examining these factors were already formulated nearly 30 years ago in the theory of genetic automatic processes. Their development is connected with the names of S. S. Chetverikov, S. Wright, N. P. Dubinin, E. Mayr, and some other geneticists and zoologists.

The essence of the idea of genetic automatic processes can be summed up as follows: If some isolated part of the range (islands for land animals, lakes for fish, and so forth) is colonized by representatives of a certain species, the colonizers are not full-valued representatives of the original population. The gene pool of the new population is not only impoverished, but is also specific, since it is determined by the genotypes of the individual founders. Since the colonization of new land and water areas occurs accidentally, the genetic structure of the new population must also, to a significant degree, be determined by chance. It is self-evident that in a new homeland, too, a population will be subjected to the forces of selection, but since selection always works on the available gene pool, the original composition of the new population cannot help but influence the result of selection, especially if the influx of new individuals from outside ceases (isolation!). A new population rapidly arises, which changes in its own distinctive direction. Thus, for example, on an island archipelago, there arises a group of closely related forms (subspecies, species) in which isolation and morphological differentiation occur significantly more rapidly than in a continuous part of the range. It is difficult to explain the peculiarities of these forms by the action of natural selection alone, but they can be explained by the theory of genetic automatic processes.

It is easy to understand that fully analogous principles act not only in space, but also in time. After a sharp drop in the numbers of a species, caused by nonspecific environmental factors (floods, storms, spring snowfalls, and so forth), the population is reestablished by the few individuals remaining alive. Their gene pool does not coincide, for understandable reasons, with the initial gene pool of the population; therefore, the very same principle of chance founders operates during the reestablishment of numbers as during spatial isolation. Without going into details, it is useful to point out that in accord with the majority of theoretical views, the evolutionary reorganizations of populations occur especially rapidly in those cases in which the species is represented by relatively isolated medium-sized populations.

The manifestation of genetic automatic processes in space is known as *Mayr's principle*; their manifestation in time, as *Wright's principle*. In a series of works, these authors quite adequately formulate the theoretical credo of "the synthetic theory of evolution."

S. Wright (1948, 1955, 1959) emphasizes that evolution is determined by the action of a series of factors: mutation, selection, and chance factors. Wright considers the most effective "mechanism of evolution" to be the subdivision of a species into partially isolated local populations, which become differentiated under the joint action of chance processes (the founder principle) and individual and interdeme selection. Mayr (1954) is even more definite in his opinions (p. 70):

> Isolating a few individuals (the "founders") from a variable population which is situated in the midst of the stream of genes which flows ceaselessly through every widespread species will produce a sudden change of the genetic environment of most loci. This change, in fact, is the most drastic genetic change (except for polyploidy and hybridization) which may occur in a natural population, since it may affect all loci at once. Indeed, it may have the character of a veritable "genetic revolution." Furthermore, this "genetic revolution," released by the isolation of the founder population, may well have the character of a chain reaction. Changes in any locus will in turn affect the selective values at many other loci, until finally the system has reached a new state of equilibrium.

In one of his subsequent works, Mayr (1965) finds it possible to assert that under the influence of selection, the genotypes are almost automatically reorganized in a harmonious way.

Thus, geneticists—and, in their footsteps, many zoologists and botanists also—arrived at the conclusion that the evolutionary process is determined by natural selection, isolation, and the dynamics of numbers (waves of life). Different authors attach different significance to these factors of the evolutionary process, but on the whole, this conception is acknowledged by the majority of biologists. As was already pointed out, the most eminent evolutionists understand very well that the theory of evolution, in its modern form, is still far from absolute perfection. But surely it is possible to say this of any theory of natural science. And, like any theory, it should be developed. Just what are the paths of development for the theory of the evolution of the organic world? Here we abandon the firm ground of facts long ago conquered by science and take our stand on the immeasurably more unstable ground of hypothesis.

Undoubtedly, in the near future, the significance of population genetics research will not only not be diminished, but indeed will grow significantly. Attesting to this are the works of recent years that have disclosed a series of fundamentally new rules. Thus, during the last decade, it was shown that the level of the genotype's integration grows in the process of evolution; that the enrichment of the population's gene pool is accompanied by an increase in the reserve of the genetic potentials of separate individuals, and that, corresponding to this, the role of isolation depends on which populations the founding organisms come from*; that during the crossing of two populations, after several dozen generations, a new population arises, a new polygenic system, better adapted to the conditions of the external environment. It is difficult to overestimate the significance of such investigations for the perfection of the theory of evolution. But it seems to us that no degree of their development can solve the task at hand in all its incredible complexity. On what is this crucial assertion based?

Undoubtedly, the first step of the evolutionary process consists of the reorganization of the population. In this direction, population genetics has done a great deal: the basic paths of the process are disclosed and can be reproduced experimentally. Evolution, however, is not exhausted with its first step. To explain evolution means to disclose the mechanisms of species formation with that degree of detail that would allow one to control them.

* As we will attempt to show further on, this position has exceptional significance for an understanding of the ecological mechanisms of the evolutionary process. It is important, therefore, to emphasize that it is based on strict experimental facts (Carson, 1961). Populations of *D. robusta* from the center of the species distribution, which are distinguished by a high degree of genetic heterogeneity, are characterized by a great genetic storage capacity of individual karyotypes. This is demonstrated by the fact that experimental populations that arose from one pair of founding flies did not differ from populations that were founded by a large number of individuals. On the other hand, it was shown that enrichment of the gene pool of such populations by means of crossing them with flies from peripheral populations did elicit the usual effect of raising the fitness of the population. This attests, in the author's opinion, to its genetic saturation. It is the business of geneticists to analyze the mechanisms of these interesting phenomena. Here it is only important to note that they agree well with many ecologists' observations, which have exceptional significance within the limits of our subject.

Chapter II

An Ecological Estimate of Interpopulation Differences

The beginning stage of the evolutionary process consists in the irreversible reorganization of the population. It follows that the estimate of interpopulation differences gives material by which to judge how, specifically, the first step of the evolutionary differentiation of species is manifested, and gives an idea of the scale of intraspecific reorganizations and of the forms of these phenomena.

At this point, it can be considered proved that any reproductively isolated population of a species is morphophysiologically specific. Differences among populations (including neighboring populations) are established for the dimensions and form of the body, the coloring and pattern, the general level of metabolism (Slonim, 1962; Cook and Hannon, 1954), thermoregulatory mechanisms (Slonim, 1950, 1952, 1962; Kalabukhov, 1951; Kalabukhov and Ladygina, 1953; Herter, 1935, 1943), endocrinologic reactions (Svetozarov and Shtraukh, 1936; Shvarts, 1960*a*; Jocum and Huestis, 1928; Riddle, 1928; Sumner, 1932), manifestation of sexual dimorphism (Mashkovtsev, 1940; F. D. Shaposhnikov, 1955), biology of reproduction (Stein, 1956), fertility (Shvarts, 1963; Pyastolova, 1967; Lidicker, 1960), rate of development (J. King, 1961), duration of the period of lactation (I. A. King *et al.*, 1963), relative growth rates of different parts of the body (Shvarts, 1961*a,b*; Ishchenko, 1966; I. A. King and Eleftherion, 1960), specific characteristics at different stages of development (Pasteur and Bons, 1962), immunological reactions (Fujino, 1960; Syuzyumova, 1966; Zhukov, 1967*a,b*), content of blood plasma proteins and of other tissues (Mikhalev, 1966; Larina and Denisova, 1966; Dessauer *et al.*, 1962;

Cei and Bertini, 1962), amino acid content of proteins of different body tissues (Hillman, 1964), blood groups (Iversen, 1965), mutability (Gershenzon, 1965), and selection of biotopes (C. J. Cole, 1963; Serafinski, 1965*a,b*). Finally, there are often cases in which the specific adaptive feature of a population consists not in the absolute expression, but in the range of variability, of individual traits (G. V. Nikol'skii and Pikuleva, 1958; Bentvelzen, 1963). The ecological characteristics of individual populations are even more diverse than their morphophysiological characteristics. They are repeatedly described and are well known. It is important to note only that the paradoxical cases, in which the adaptive characteristic of a population consists of special mechanisms to maintain a low numerical size, have been completely described (Klomp, 1966).

It would be possible to expand many times over the list of differences that have been discovered by various authors during comparative studies of populations of different animal species. It would be possible to expand many times over the number of investigators cited above. There is scarcely any necessity for this, however, since at present, the thesis of the inexhaustible diversity of interpopulation differences does not demand detailed substantiation. The question of the ecological analysis of these differences is more complex.

When the differences among populations are manifested in such animal characteristics as fertility, reaction to a change in environmental conditions, rate of development, and so forth, their ecological significance is revealed without difficulty. In many cases, however, populations are distinguished by characteristics that have no self-evident adaptive significance. This circumstance can create a false impression of the chance nature of interpopulation differences and create the basis for false generalizations of broad significance. That this is so was recently correctly noted by Tinbergen (1965), who showed that in the majority of cases, a special analysis is needed to judge the adaptive significance of many animal traits. Traits that at first glance appear useless can in reality have substantial adaptive significance. The author justly believes that this possibility should be taken into consideration during the elaboration of evolutionary theories, for, with no clear idea of the diffusion of nonadaptive interpopulation differences, it is also difficult to form a well-founded representation of the role of different factors of the evolutionary process (selection, genetic automatic processes, isolation). Many data speak to the fact that insignificant differences among populations can have substantial adaptive significance (or, to the contrary, can signal deep physiological differences), but in order to reveal this significance, ecological and physiological analysis is needed.

Let us use for illustration the simplest case. Let us try to analyze the biological significance of differences in body dimensions. It is known that difference in size is one of the most widespread phenomena of interpopulation differences. Their adaptive significance is undoubted, since a change in the dimensions of the body is very closely linked with a change in the conditions for maintaining the energy balance (Kalabukhov, 1950; Vinberg, 1950; Kestner and Plaut, 1924; Hatfield, 1935; Kramer, 1932; Benedict, 1932; 1938; Kleiber, 1947; Krebs, 1950; and many others). It should be mentioned that the nature of the connection between body size and the general level of metabolism can be different, owing to the physiological and ecological features of the forms compared. The complexity of the question is accentuated by the circumstance that in large animals with a more economical type of metabolism, a change in the speed of movement or a change in the position of the body is connected with a relatively greater expenditure of energy than in small animals (Buddenbrock, 1934). A monograph by Rensch (1959) cites a long list of morphological (in particular, histo- and cytological) animal characteristics that are connected with body size. A change in size determines the position of the species in the biocenosis, and unavoidably leads to a change of its interrelationships with the environment. Let us note only one of the manifestations of this rule. With a change in the animal's body size, its requirement for food resources is changed: a larger animal requires relatively less food. With this change, it receives relatively fewer vitamins per unit mass of its body. In a series of cases (probably in the majority), this has a negative effect on the state of the animal (Hickman and Harris, 1946). The way out of this contradiction is either in a change in the tissues' requirement for vitamins (a very substantial physiological change) or in a change in the natural food supply. This idea is corroborated by experiments conducted on rats, which show that a deficiency of a certain vitamin leads to a change in food preferences (a survey of the data is in *Nutrition Revue* for 1954). It becomes evident that even insignificant differences in the sizes of the populations compared can lead to a change in the nature of their nutrition, i.e., to very substantial biological distinctions.

Differences in coloration can be defined as insignificant genetic features of the organisms or populations compared; they can also attest to very significant distinctions among them. The research of Dement'ev and Larionov (1944) and of Dement'ev (1951), which showed that the diversity of melanic coloring in birds could be reduced to a difference in the degree of oxidation of the single propigment melanogen, has great significance for an understanding of this question. The shades of coloring are determined by the intensity of the oxidative processes, and the pattern by its irregularity. The coloring as a whole reflects the features of the oxidative processes in the organism of birds. It should be added that the biochemical conditionality of isolated details of coloring and pattern has been demonstrated (Novikov, 1949, 1952; Novikov and Blagodatskaya, 1948). Experiments on rabbits showed that different parts of the skin contain different quantities of a specific enzyme (Danneel and Schaumann, 1938). These data indicate that insignificant differences in coloration can often be regarded as signals of those population characteristics that can be the cause of intraspecific differentiation.

Often, differences in coloration are spontaneously elicited by differences in the conditions of existence. Well-known experimental data attest to this. The character of the nutrition can frequently influence the coloration substantially, as is shown with particular clarity in insects. Thus, when the caterpillars of some geometrid moths (*Selenia bilunaria, Tehrosia bistortasa*) were fed plants poisoned with lead salts, melanistic forms of these moths were successfully obtained. In this way, it was experimentally demonstrated that changes in metabolism elicit variations in the coloring of the moths that are practically indistinguishable from characteristics of coloring in the natural subspecies of these insect species inhabiting large cities (Harrison and Garret, 1925, and others). This finding clearly shows that outwardly identical differences among populations can arise both as a result of selection of genetic variants and on the basis of environmental mechanisms.

Changes of metabolism, connected with a change of nutrition, elicit well-known changes in the coloring of birds. For example, the action that hemp exerts on birds of the family Fringillidae is generally known: crossbills, pine finches, and bullfinches darken, and the red tones are replaced with black-brown ones. In crossbills, the red color around the base of the beak is replaced by yellow-brown, and the body plumage turns black. In siskins, black patches appear on the body, and the little red caps of the linnets become orange. Feeding some weaver birds with hydrolyzed carotenoids caused a clearly noticeable yellowing of their coloration (Adlersparre, 1938). Cases in which equally marked changes in coloring were elicited by changes of nutrition in mammals are known (Svihla, 1931).

Changes in coloring can also be elicited by changes in the conditions maintaining thermal balance. This change was demonstrated by a number of authors in insects and poikilothermic vertebrates, and by Rhoad (1938) in cattle.

It is very important that direct observations show that changes in the conditions of existence elicit changes in coloring in natural populations of animals. Leraas (1938), for example, was able to ascertain a definite connection between the natural food supply of separate populations of *Peromyscus maniculatus osgoodi* and their coloring. In representatives of another subspecies of that species (*P. m. gracilis*), it was shown that in very old rodents (older than 5 years), changes in coloring that are connected with changes in metabolism are observed (Dice, 1936).

In nature, as in an experiment, changes in coloring can be elicited by very specific factors of the environment. The best example is the so-called "marsh melanism." The essence of this interesting phenomenon is that many species of mammals (water voles, meadow mice, *Microtus minutus*), reptiles, lepidopterans, beetles, rotifers, and others darken when they inhabit the marshes. The mechanism of this phenomenon is environmental. It is most probable that the darkening is elicited by the action of the marsh gases on the respiratory system of the animals (Lehmann, 1966). Finally, it is well known that features of coloring can be connected with ecologically important traits (e.g., an increased metabolic rate). Such correlations are often very specific. Thus, it was discovered that in some voles, the genetic variants of different coloring are distinguished by different fertility and different sex ratios (Lidicker, 1963).

Tomich (Tomich and Kami, 1966) correctly analyzes differences in the coloring of rats as markers of "physiological types"; more recently, the research of Keeler and co-workers (Keeler *et al.*, 1968) showed that foxes of different color variants are clearly distinguished by the dimensions of their adrenals and by their behavior. Differences that are seemingly unimportant can have substantial ecological significance or signal important differences in the conditions of existence of the forms compared. More complex cases, of course, demand even more complex analysis. These analyses can be given full weight only if they take the form of thorough investigations of populations compared in nature. These investigations, however, are very laborious. Therefore, we judge the features of populations of a predominant number of species, and will for a long time continue to judge them, on the basis of museum collections or, at best, more detailed morphological inspections. It is important, therefore, to show that morphological material (more accessible for mass collection) can also be utilized successfully for the disclosure of the true nature of interpopulation differences.

Let us avail ourselves of a specific example. Two forms of the *oeconomus* voles are compared: the southern *Microtus oeconomus oeconomus* and the northern *M. o. chachlovi*. A series of important differences between them come to light (these are indisputably "good" subspecies), including craniological differences. The latter differences are most distinctly revealed in a comparison of the interorbital distance and of the widest dimension of the skull. If one selects voles of comparable age and equal size (body length 100–120 mm)—i.e., if one fulfills the ideal conditions of comparison, which are not often fulfilled even by modern systematists—the following is revealed: the index for the widest dimension of the skull for *M. o. oeconomus* is 0.441; for *M. o. chachlovi*, 0.478; with $t = 6.6$ (the differences are highly significant). If, however, we try a comparison to lead to another principle, the results obtained are different. If we compare animals with skulls of equal dimensions, we cannot succeed in discovering any differences between *M. o. oeconomus* and *M. o. chachlovi*: allometric curves establishing the dependence between the length and the width of the skull coincide completely (according to the data of Pyastolova, 1967). We arrive at the conclusion that the highly significant differences between them turn out to be imaginary. Even this conclusion turns out to be premature, however. Construction of the allometric curves for "body length to skull length" establishes that the differences between the northern and southern *oeconomus* are very great. During the entire period of the animals' growth and development, the condylobasal length of the skull in the northern forms is certainly less than in the southern. In the size group analyzed (100–120 mm), the condylobasal length of the skull in *M. o. oeconomus* is 27.8 ± 0.186 mm; in *M. o. chachlovi*, $26.2 + 0.225$ mm ($t = 5.5$!). Since an increase in body size in rodents, as in an overwhelming number of other mammals, leads to a relative decrease in skull dimensions, then precisely because the proportions of the skull are absolutely equal in the forms compared, they will always differ in a comparison of southern and northern forms of equal size. Our initial conclusion is essentially made more precise: the forms compared differ not in the proportions of the skull, but in its dimensions relative to the body. Let us try to make even this conclusion more precise.

It is known that the relative length of the skull is determined not only by the absolute size of the animals, but also by the growth rate. In rapidly growing animals, the skull is relatively smaller. Direct observations showed that in the example we analyzed, the differences in skull dimensions are determined by differences in the growth rate of the animals. The northern form grows more rapidly under natural conditions, and the relative dimensions of its skull are correspondingly smaller. From this ensue all the differences in the proportions of the skull noted by systematists. A taxonomic character of the forms compared, which initially had no ecological meaning, is filled with new content. In correspondence with a specific condition of existence, the northern forms are distinguished by more rapid growth, and all those distinctions determined by the systematist are only a secondary effect of genetic differences that directly determine the ecologically important characteristics of the animals.

An analogous analysis was used in our laboratory by Bol'shakov (1972) in a comparison of *Clethrionomys rufocanus* from the southern and polar Urals. We will limit ourselves here to inspection of the graphs (Fig. 1), which show that in this case, the craniological differences between the forms compared are completely determined by differences in their growth rate. These examples show that a thorough morphological analysis, based on the use of general morphological rules, helps us see behind the

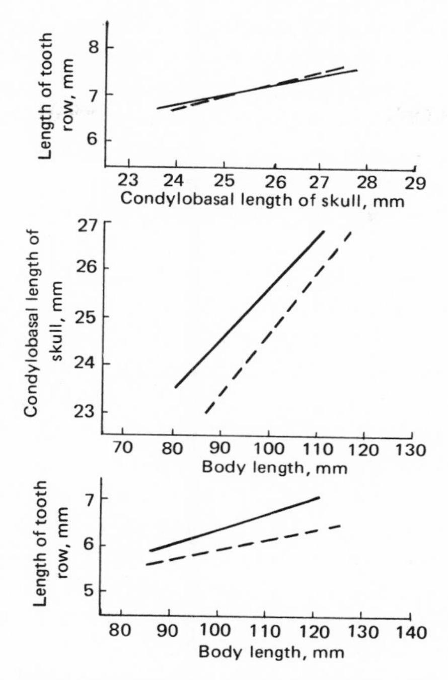

Fig. 1. Craniological differences between southern (solid lines) and northern (dashed lines) populations of *Clethrionomys rufocanus*.

immediately visible attendant animal characteristics to their principal characteristics, which have clear ecological meaning, and helps us understand the biological meaning of the taxonomic differentiation of species.

In the example analyzed, the principal (defining) feature of the populations is the growth rate. Differences in this feature turn out to be hereditary. [That they are was verified experimentally in model populations (Ovchinnikova, 1964, 1966).] The differences in growth rate elicited by environmental mechanisms, however, would also entail similar consequences to the complex of fully analogous differences between the compared populations. It is extremely important, therefore, to show the range of interpopulation differences based on various mechanisms. Of course, this problem can be solved relatively simply by experimental methods, but these methods are so laborious that their broad application at the present time is out of the question.

Earlier (more precisely, nearly 20 years ago), the matter seemed relatively simple. If populations are distinguished by "stable" traits (as opposed to hereditary), the differences are not hereditary, but environmental. Now the situation has changed substantially. It has been shown that even such "stable" traits as the craniological characteristics of mammals are susceptible to strong phenotypic variability under certain conditions (Dehnel, 1949, and others), and changes in the proportions of the body and skull of animals (one of the taxonomists' favorite traits) can be elicited no less simply than changes in the weight of reserve fat (an example of an extremely labile index). A change in growth rate elicits a change in the animals' constitutions, which is no less substantial than differences among many subspecies.

The work of Dehnel's school indicates the possibility of very substantial interpopulation differences based on environmental mechanisms. Other investigations, including some fairly old ones, also give evidence for this. Thus, Klatt (1926, cited by Hesse-Doflein, 1943) showed 50 years ago that feeding tritons mollusk meat leads to an acceleration of their growth, which is connected with the hyperfunction of the hypophysis. At the same time, the form of the body is changed: the head is shortened; the lower jaw is lengthened. Klatt's work is corroborated by recent research (Mihail and Asandei, 1961) showing that feeding tadpoles with mollusk meat activates the hypophysis and leads to a 60% increase in their weight. One may assume that "Dehnel's effect" also has an endocrinological mechanism as its basis. The conclusions of some experiments conducted on amphibians, which indicate a seasonal variability of skeletal features (Cohen, 1962), have something in common with Dehnel's research. In invertebrates, environmental conditions may appear to be the direct cause of cardinal changes in morphology and physiology. G. Kh. Shaposhnikov (1965, 1966) showed that when breeding the aphid *Dysaphis anterisci* on various food plants, one can obtain forms approximating different species. What is more, the author asserts that forms arising in this way lose the ability to cross with their own species, but prove to be fertile when crossed with other species. It is possible that the fact that the systematics of aphids are not sufficiently worked out had an influence on the author's final conclusions, but his investigations testify very clearly to the enormous influence of external conditions on the morphophysiological characteristics of insects.

In many cases, the external environment acts as a switch, determining the course of development along one of several possible alternative pathways. A change in the salt concentration in the environment elicits a profound reconstruction of intercellular organization in *Naegleria gruberi* (Bistadiidae). These protozoa have an amoebic form when developing on a dry substrate in the presence of bacteria. When developing in water, they form flagellates, and all the cells lengthen and acquire the structures characteristic of flagellates (Willmer, 1956). One finds three sharply distinct types of coloring in the freshwater snail *Theodoxus fluviatilis*. Neumann (1959) found that a simple change in external conditions (temperature, pH, and salt content in water) elicits the display of one or the other type of coloring. An insignificant change in the surrounding conditions leads to the appearance of different antigens in paramecia (Beale, 1954).

Analyzing the significance of such observations, K. Waddington (1964) writes (p. 208):

> . . . an insignificant change in the surrounding conditions leads to the appearance of one or another of the hereditary potentials of the individual. In *Theodoxus fluviatilis*, however, these potentials do not absolutely exclude one another and can appear simultaneously, so that spotted and transverse models are superimposed on one another. Besides that, the phenotypic traits in a given case are much more complex than the antigens of *Paramecium*. A very large number of interacting cellular processes, which inevitably are associated with the activity of many genes, participate in the formation of the spotted or transverse coloring. Those factors (be they genetic factors or factors of the external environment), which lead to the formation of one or the other coloring, apparently therefore, act as switches directing the development toward the formation of one of the two alternative systems, each of which is significantly more complex than the system of formation of one antigen in *Paramecium*. Perhaps the closest example is the switch between the flagellar and ameboid forms in *Naegleria*.

Similar but perhaps even more interesting observations were made on butterflies of the species *Papilio* (see the review by Sheppard, 1959). It was observed that when the caterpillars of *P. machaon* pupate, the chrysalises on the leaves acquire a green color, and those on the bark a brown color. The same thing happens in *P. polyxenea*. An analysis of these observations showed that the green and brown chrysalises were distinguished not only by their color, but also by their rate of development: the majority of brown chrysalises overwinter; the green ones produce butterflies in the autumn. In this case, we encounter a very interesting form of polymorphism, which is based not on the existence of two sharply diverse genotypes in the population, but on different displays of one genotype. This type of polymorphism can be called *environmental polymorphism* (Sheppard, 1959). The biological significance of the environmental polymorphism is understood. It is practically impossible for chrysalises to overwinter on the leaves, and therefore the cryptic (same color as the leaf) chrysalises should complete their metamorphosis as early as possible. These observations indicate very clearly that the simplest changes in external conditions (the color of the substrate) can lead to cardinal changes in the physiological characteristics of animals (developmental rate). The character of the environmental influence on an organism is determined, of course, not only by the nature of the factors operating, but also by the genetic specificities of the organism. (The assertion that the norm of reaction, not traits, is inherited is now taken as an axiom.) This means that one and the same hereditary change elicits a different phenotype, depending on what environment the animal develops in.

This appears especially clearly in those cases in which the so-called "secondary effect" of gene action emerges first and foremost. The essence of this effect is that relatively second-degree genetic differences entail diverse and sharply expressed morphophysiological consequences, which do not have a direct genetic basis. The mutation *frizzled* in chickens can serve as an example. The direct manifestation of this mutation is the frizzling of the feathers. The secondary effect (connected with a deterioration of the normal maintenance of heat balance) is the intensification of metabolism, hypertrophy of the aortic ventricles, increase in the general volume of the blood, acceleration of the heartbeat, and enlargement of the intestines, adrenals, and thyroid gland (Landauer, 1946).

One cannot call the *frizzled* characteristics environmental, since they are connected with a definite mutation, but it is also difficult to call them genetic characteristics, since the effect of the mutation leads to markedly different morphological consequences in different environments. Once again discoveries of the expression of the rules of morphophysiological correlates place before zoologists a new round of questions, these questions being linked with a general problem: not to be limited by a description of the differences among compared forms, but to elucidate their adaptive and genetic nature. It is hardly necessary to demonstrate that resolution of this problem would permit us to mobilize the data—accumulated over many decades—that characterize the morphological specificity of near forms of different taxonomic rank for the solution of modern problems of evolutionary theory.

Comparison of the data cited here shows, first, that direct environmental influence can elicit changes of a taxonomic scale in populations. Second, interpopulation differences are very rarely (probably never) determined only by either environmental or genotypic mechanisms. They are determined both by differences in the genetic constitution of the populations and by differences in the environmental conditions. Hence, it follows that only complex ecological and genetic research can be coordinated for major progress in the theory of microevolution. One cannot forget in the process that extremely complex cases are possible. Their analysis leads to the conclusion that it is necessary to study the most complex epigenetic mechanisms during an analysis of what would appear to be very simple natural situations. The difficulties that one may encounter are very well demonstrated in the very interesting research of Barnett (1965), which deserves to be analyzed in greater detail.

Mice form stable populations in cold rooms at a temperature of −10°C. Mice cannot adapt to such a low temperature because of physical thermoregulation (small body size!), and the means by which they adapt to conditions of life in refrigerators remains unexplained. In the author's experiments, laboratory mice were raised at temperatures of −3°C for many generations. It was established that in complete agreement with theory, the insignificant improvement in heat-insulating properties of their outer layers could not be the cause for their rapid adaptability. This conclusion was brilliantly supported by the fact that the ability to adapt to the cold was constitutional in hairless mutants as well. Calorimetric measurements showed that the experimental animals acquired an ability to utilize nutritional substances better. That they did received indirect support from the finding that lactating females from the experimental mice needed less feed per 10 g of offspring than did the controls. In the rearing of a genetically heterogeneous population of laboratory mice in the cold, this result could have been ascribed to the action of selection for cold-resistant genotypes (the mortality of offspring in the first two generations varied from 40 to 80%, and then fell to 10% in the 12th generation). Exactly the same result, however, was obtained by the author when he worked with inbred lines of mice: their cold resistance increased, the mortality of offspring had already fallen to 5–10% by the 7th–10th generation, body weight increased, and fat content in the organism increased. The author considers it axiomatic that selection is powerless in a strongly inbred line, and he explains his results by the physiological characteristics of the mothers, which had adapted to the low temperatures. (This view was supported by experiments with substituted litters.) On the whole, the author considers the experimental results enigmatic. In other investigations by the same author (Barnett and Coleman, 1960), it is shown that environmental conditions also determine the results of intraspecific hybridization. It was established that heterosis appeared significantly stronger during the crossing of two inbred lines of mice if the animals were crossed at low temperatures.

Apparently, Barnett is correct. The results of his experiments are indeed enigmatic, since it is difficult to explain them with orthodox genetic concepts. To explain them, it is probably necessary to take recourse to an analysis utilizing data that characterize the phenomena of specific inhibitors (Lenick, 1963). These data explain the influence of the mothers' physiological features, acquired during the process of development in a unique environment, on the morphophysiological features of their posterity. It is also necessary, apparently, to recall the recently discovered phenomenon of "maternal inheritance" (Tamsitt, 1961; Dawson, 1965; Falconer, 1965; Tenczar and Bader, 1966), which has demonstrated that under certain conditions, other conditions being equal, the traits of the mother dominate in the offspring (the so-called "maternal effect"). It is the business of geneticists to analyze the results that follow from work similar to Barnett's research. For the ecologist, it is extremely important to take into account that one and the same external effect can be elicited by diverse genetic mechanisms, including very intricate and complex mechanisms. That it can be indicates that interpopulation differences are not only themselves boundlessly diverse, but also are based on mechanisms of virtually boundless diversity. Even an insignificant modification of a population is a complex biological phenomenon. The idea that in one case it is based on genetic automatic mechanisms, in another on the action of individual selection, in a third on the action of group selection, in a fourth on environmental mechanisms, and in a fifth on the assimilation of acquired traits (Waddington's principle), and so forth, is beyond criticism. It is logical to assume that different mechanisms in various combinations participate in each act of the modification of a population. Their combined action leads to the effect that we will directly determine—the reorganization of the population.

Research of recent years has shown that environmentally conditioned traits of individual populations can be no less stable than traits fixed by heredity within very narrow ranges of variability. (This stability is connected with a relative stability in environmental conditions and a stability in the reactions of animals to changes in these conditions.) It follows that the stability, *per se,* of individual traits in populations cannot be used as a criterion of their genetic nature. At the same time, the genetic evaluation of interpopulation differences (and, in individual cases, of differences among taxonomic units as well) is one of the most important prerequisites for the further development of evolutionary theory. This is precisely why the attention of investigators has been drawn in recent years to basically new methods of work that allow them to make more definitive judgements about the genetic differences among closely related forms (immunological and karyological methods, electrophoresis of proteins, the biochemistry of enzymes and fats, the chemistry of nucleic acids, and others). These methods help in the understanding of many very complex cases (Avrekh and Kalabukhov, 1937; Larina and Kul'kova, 1959; Zhukov, 1967*a,b*; Mikhalev, 1967; Syuzyumova, 1966; Moody *et al.,* 1949; Fujino, 1960; Cei and Bertini, 1962; Salazar and Morrison, 1962; and others). In particular, it has been established with the help of these methods that even in the cases in which neighboring populations do not differ substantially in their morphophysiological characteristics, they are genetically unique (Syuzyumova, 1967, and others). On the other hand, the "new methods" helped to establish that the zone of hybridization between morphologically differentiated populations is significantly broader than could have been ascertained by morphological methods (Dessauer *et al.,* 1962). The results of these investigations make it possible to establish a connection among the morphophysiological differentiation of populations, the conditions of their existence, and their degree of isolation. It is difficult to overrate the significance of such work. The "new methods" (let us avail ourselves of this conventional expression), however, require significantly more labor than do, for example, morphological methods, and mainly they cannot be used, for obvious reasons, for the treatment of collections, which will long continue to be extremely significant for the elaboration of evolutionary problems. It seems important to us, therefore, to show that with the help of morphological analysis also, one can obtain valuable data not only about external traits, but also about the nature of interpopulation differences.

An important method of research is the study of allometric relationships, which reflect a change in the body proportions of animals during the process of growth and development of different forms under different environmental conditions. Allometry as a method of research has long been known, but in recent times, interest in it has grown. It is no coincidence that only in recent years have a great number of works been published that endeavor to use the method of allometry for the solution of taxonomic problems (for a survey of the newest research, see Hückinghaus, 1961; Gould, 1966; Ishchenko, 1967).

The most interesting result of these investigations is that even a sharp change in environmental conditions, eliciting a very substantial change in the animals' development, does not lead to a change in the ratio of the dimensions of individual parts of the animals' bodies. Thus, the conclusion follows that if two forms that do not differ substantially in their growth rates differ in the regularities of growth of individual organs or parts of the body, one can maintain with a very great degree of likelihood that the differences between them are hereditary.* In the following chapters, we will have the opportunity to dwell on some specific investigations carried out with the help of allometric methods. Here, let us emphasize that with a foundation of morphological data, the contemporary ecologist and evolutionist can form their judgments not only on the basis of information characterizing the genetic nature of interpopulation differences, but also on the basis of a comparison of the character of the animals' variability in different stages of ontogenesis. The basic theoretical concepts from which this statement of the question directly ensued were formulated by A. N. Severtsov. Matveev (1963) recently turned his attention to them.

In "Etudes on the Theory of Evolution," Severtsov (1912) wrote (p. 75):

> This struggle and its result—destruction of a certain number of individuals—begins for each individual with the appearance of the individual as such in the world, i.e., from the stage of the fertilized egg, and continues during the course of the individual's whole life: in the course of different periods of life, but, namely, in the periods of morphogenesis, growth, and sexual maturity (I enumerate only the very important periods), the unfavorable conditions of this

* Not only indirect evidence, but also special genetic investigations (Mayrat, 1965), indicate that they are.

> struggle, i.e., the circumstances threatening life such as enemies, competitors for food and space, injurious climatic conditions, and so forth, are very different. Consequently, the means by which the animal struggles should also be different,

and further:

> . . . in the course of the individual development of the animal's body, there are a whole series of developing organs that do not relate to the surrounding environment at the given time (they function in the adult animal); the struggle for existence in the given period and the changes of organization ensuing from it do not directly concern these organs.

This thesis is very important. From it follows the idea that the character of the variability of organs should be different, depending on the degree to which the organ participates in the struggle for existence of the individual and the species. For well-known reasons, a comparison of embryos with adult forms has special significance. In this case, let us avail ourselves of a specific example for analysis.

Variability in the weight of the heart, liver, and brain of the little sea-gull (*Larus minutus*) was studied on the Yamal Peninsula. The following data were obtained: the coefficient of variation (C) of the relative weight of organs in embryos of the same age was found to be equal: heart, $22.2 \pm 2.32\%$; liver, $11.2 \pm 1.24\%$; for the parents, the corresponding percentages were $8.45 \pm 1.08\%$ and $15.22 \pm 1.94\%$. Comparison of these figures leads to very interesting conclusions. The sharp decrease in the variability of heart weight in adult birds (in comparison with embryos) testifies that in the process of the birds' active life, there is a leveling off of differences among genetically different animals or animals in which the process of organogenesis deviated from the norm. It is not improbable that some portion of the individuals deviating most sharply from the norm were eliminated before reaching adulthood. In order to verify these suppositions, the coefficients of variation for embryos within individual clutches were calculated.* In this case, the comparison was made on closely related groups of animals (siblings). The variability of the index analyzed was therefore sharply reduced ($C = 11.2 \pm 1.2\%$). It is significantly more important, however, that the variability of the adult birds turned out to be lower than the variability of the embryos, even if the embryos represented an immeasurably more genetically uniform group of animals. This means that in actual natural populations, the phenotypic variability does not intensify, but masks, the genetic heterogeneity of a population in regard to such an important trait as the dimensions of the heart.

An examination of the variability of the relative weight of the liver

* A special method of mathematical analysis, which it is not necessary to dwell on here, was applied to the material.

gives an entirely different result. The livers of embryos vary less than those of adults. If, as in the previous example, we compare the variability of a genetically homogeneous group of embryos ($C = 8.82 \pm 1.18\%$), the observed difference is almost doubled. It becomes clear that in this case, the phenotypic variability reinforces the genetic heterogeneity of the population. We obtained a completely analogous result when we compared animals that were of different ages, but had already developed to an independent stage of life.

These examples show that by obtaining data that characterize different animal traits at various stages of ontogenesis, the prerequisites for determining the genetic nature of characteristics of individual populations are created.* These very data help us find the path for investigating the genetic nature of interpopulation differences. If differences among populations are significant at early stages of the animals' development and decrease somewhat with age, but remain substantial, then we can assert with high probability that the characteristics of the populations are genetically determined. We also have the right to speak of the genetic nature of differences in the case in which the differences among animals at early stages are insignificant, and not only do not intensify with age, but diminish. If, with the same initial situation, the differences intensify with age, one can assert with almost complete certainty that the characteristics of the populations are basically environmentally conditioned. If, however, the initial differences among populations are great, and they do not smooth out or intensify with age, then we become certain that the genetic differences among populations are sharply intensified by their phenotypic variability.

All that has been said is no more than a general sketch of the ways in which to utilize data on the variability of animals' morphological traits in order to determine the genetic nature of differences among populations. This sketch shows, however, that determination, by morphological methods, of the genetic nature of differences among intrapopulation groups is, in principle, possible.

The data cited lead to two important conclusions. They show that the manifestation of interpopulation differences is endlessly varied. Their genetic nature is varied and complex. Even so, this nature can be established, even in those cases in which we deal with natural, not experimental, animal populations. The conditions are thus created for the concrete analysis of those mechanisms that determine intraspecific differentiation. It is thereby revealed that the ways in which populations are reorganized are more diverse than they appeared on the basis of theoretical analysis issuing from

* This line of work seems very important to us, since direct investigations show that genetic factors often determine less than 50% of the phenotypic variability of even such stable traits as features of dentition (Bader, 1965).

general notions of the relative roles of different factors of the evolutionary process (selection, isolation, and genetic automatic processes), and that the specific action of these factors operates indirectly on characteristics of the population structure of species. These conclusions indicate the leading role of ecology in the perfection of evolutionary theory.

CHAPTER III

The Reorganization of Populations Homeostatic Alteration of the Genetic Structure of Populations and Microevolution

In modern theoretical literature, there is wide use of a contrast of the notions of microevolution and macroevolution. Microvevolution is understood to mean intraspecific reorganizations; macroevolution, the complex of processes leading to the formation of taxa of higher order. The process of speciation links micro- and macroevolution in the single process of phylogenesis.

The majority of investigators understand that the phylogenetic development of living organisms is, in essence, a single process of development, proceeding through a series of stages so closely linked that one can speak only conditionally of their independence. In the majority of cases, however, assertions of this kind remain general philosophical declarations, and are not part of a strict theory. On the other hand, most followers of the synthetic theory of evolution clearly adhere to the idea that different laws govern interspecific reorganizations and the macroevolutionary development of animals. This book is not a history of modern evolutionary thought, and it is therefore sufficient for us to cite the opinion of one of the most eminent representatives of Neo-Darwinism—Dobzhansky, who asserts that in contrast to macroevolution, microevolution is reversible, predictable, and repeatable (Dobzhansky, 1954, and others). In the view of many modern theoreticians, it is difficult to express the difference between micro- and macroevolution more exactly and clearly.

Dobzhansky's assertion is based primarily on facts drawn from work in experimental population genetics. This work showed that genetic reorganization of populations can be predicted, and that changes that arise under the influence of directional selection are reversible. The best example is the emergence of poison-resistant "races" of insects or microorganisms (see Chapter I). Under the action of poison, a new population, possessing new properties that can be foreseen beforehand, rapidly arises. Recently, works have appeared that show that the ability to predict the reorganization of populations can be extended to animals of any taxonomic group (experiments under laboratory and natural conditions). In this connection, the research of Ferguson (1963), which was mentioned in Chapter I, deserves special attention.

These investigations make it absolutely clear that if the major selection factor (in the examples selected, this factor is the selective action of the poisons) is known, the character of the reorganization of the population can be predicted. A change in the nature of selection leads, as one would expect, to a change in the direction of the population's reorganization. A property that has had major significance (resistance, in regard to poisons) loses its significance, and will be lost in a number of generations.

The impression is created that Dobzhansky's aphorism—microevolution is reversible, predictable, and repeatable—is supported by experiments and observations. This aphorism, nevertheless, seems mistaken to us. A number of experimental observations show that the genetic mechanisms of the population's adaptation to even the simplest changes of the external environment can be varied, despite an almost complete identity of its phenotypic expression. The predictability and reversibility of microevolution, even if we understand microevolution to mean any reorganization of the population, can therefore hardly be called complete. Rather, the essence of microevolution is something other than predictability and reversibility.

Dobzhansky's aphorism is based on the idea that any genetic reorganizations of a population should be regarded as manifestations of microevolution. A number of observations of samples from different groups of animals showed, however, that the genetic structure of populations (the gene pool and the system of its realization in specific genomes) is in constant dynamic equilibrium, which is maintained by ecological mechanisms. The stability of the population's genetic structure is only apparent: the genetic reorganization of populations is a necessary condition for the maintenance of their numbers under fluctuating environmental conditions.

The dynamics of the quality of the population is just as characteristic a property as the dynamics of its numbers. In Chapter V, we will try to lay the foundations of this important position with specific data. Here, we will only note that numerous observations by different authors on animals of different groups showed that a change in environmental conditions in different seasons of the year or in different years leads to a natural displacement of the genetic structure of the population. The norm does not remain constant. A new change in the conditions of existence returns the population to its "initial" norm. A fluctuation, peculiar to the population, occurs in the quality of the population over some average of many years. This fluctuation has as its basis the fluctuation of the genetic makeup of the population, which corresponds to changes in environmental conditions: under some conditions, the advantage goes to certain genotypes; under different conditions, to others. The fluctuation of the genetic makeup of a population creates the prerequisites for its flourishing under fluctuating environmental conditions: selection creates, but does not stabilize, the optimal correlation of genotypes, since any such stability would be disadvantageous in a changing environment. We are dealing with a typical homeostatic reaction, which we can call the homeostatic reorganization of the genetic structure of a population. Not every reorganization of the genetic structure of a population, therefore, can be equated with microevolution.

The genetic reorganization of the population is one of the most important mechanisms of population homeostasis, maintaining the population's viability and adaptability. Microevolution is the beginning stage of the evolutionary process. The distinction is a major one, but up to this time, there has been no serious attempt to analyze if from a broad general biological perspective.

For convenience of analysis, we will use the terminology of intraspecific systematics, even though this major division of systematics has only recently become an arena for debates on principles. It is well known that a multitude of names have been proposed at different times by various authors for the intraspecific forms of different ranks. Only one of these names—*subspecies*—has withstood the test of time. We do not need to examine in detail the numerous discussions on the nature of subspecies or the correlation of this concept with other intraspecific categories. We will stress only the most important. Modern systematics developed on a phylogenetic basis, and in the works of a great number of systematists, "subspecies" is used as a synonym for the Darwinian term *variety*. In this sense, phylogenetic and evolutionary significance is clearly attributed to subspecies. It was not without reason that the overwhelming majority of earlier systematists used the expression "the species with its subspecies." Even then, the subspecies was not regarded as a typical representative of the species (a nomenclatural relict often encountered even now is "a typical subspecies"), but as an evolutionary deviant that had not yet achieved the species level. On the lips of the early systematists, the word "subspecies" was not only taxonomic, but also evolutionary.

The later position changed substantially. Formalistic tendencies in intraspecific systematics became increasingly clearer, and any intraspecific forms that were clearly localized spatially and distinguishable morphologically from others forms of the species came to be described as subspecies. The concept of the "subspecies" began to lose its evolutionary meaning. It became clear more quickly, however, that at the same time the concept became meaningless in general. The development of population studies played a leading role in this interesting process of the development of science.

The basic conclusion of these studies is that any population that is reproductively isolated (complete isolation, not facultative) is morphologically specific (see Chapter II). Not only species and subspecies, but also any natural populations, are different. The ascertainment of these differences is only a question of technique, time, and industrious investigators.* It follows that what has become the classic notion of subspecies as morphophysiologically specific intraspecific categories is insufficient. It cannot be used to discriminate between the concepts of "subspecies" and "populations." The conclusion drawn by Dement'ev (1946) 30 years ago is confirmed: "In the works of practical systematists, any morphologically distinguishable and spatially isolated populations are usually acknowledged as subspecies. This position is, in essence, hardly acceptable." The majority of modern systematists and theoretical evolutionists nevertheless do not see the connection between the development of population studies, on the one hand, and the crisis in subspecies studies, on the other.

Much has been written and is being written about the crisis of intraspecific systematics. An ever-larger number of investigators urge that application of the concept "subspecies" be relinquished, and that intraspecific differentiation be described in terms of geographic variability (Bogert *et al.*, 1943; Wilson and Brown, 1953; Gillham, 1956; Hagmeier, 1958; R. A. Pimentel, 1959; Terent'ev, 1957; and others). Such is the reaction of the theorists. This is also reflected in the practical work of

* That this is so was very clearly shown by the excellent work of Grünberg (1961), which established that settlements of rats inhabiting different warehouses in a big city differ in roughly the same way as populations of "wild" rodents. It is important that differences were found among settlements of rats in the structure of teeth and skeleton. Such differences could fully characterize good subspecies.

zoologists: they began to describe clearly characterized intraspecific forms as species. We will limit ourselves to one example. Ladygina (1964), having ascertained that *Mus musculus hortulanus* is distinguished from other subspecies of the house mouse by essential morphological and physiological characteristics, expresses complete certainty of the specific independence of this form. The basic principles of systematics (reproductive isolation of species, presence of a morphological hiatus) appeared forgotten, and the author seemed a captive of the typological notions of "the old systematics." This example is far from unique. The danger of discrediting the polymorphic conception of species has become real. With this approach to the problems of intraspecifc systematics, the concept of "subspecies" has indeed become unnecessary.

Consistent use of a formal criterion in distinguishing and describing subspecies leads to a negative concept of subspecies. What are the possible consequences of the projected tendency, and what is its significance within the realm of problems of interest to us?

First of all, it should be noted that it is possible to find in any group of organisms a body of examples of sharp morphophysiological differentiation. This differentiation is manifested in the isolation of individual intraspecific forms having a degree of differences incommensurable with those differences that usually characterize populations. Without going into details, let us recall at least the Teleut squirrel, numerous examples of subspecies of the red deer, of the forest steppe subspecies of willow ptarmigan, of grouse, the Ussuri tiger, the grizzly, and so forth.

May not one call such forms species? In our opinion, one can only give a negative answer to this question. To distinguish such forms as species leads to a veritable theoretical catastrophe, since the principal criteria of species will be made null and void. The Teleut squirrel is distinguished from all other forms of squirrels no less than from many species of rodents (including some from the squirrel family). To distinguish the Teleut as an independent species, however, would be to ignore the most importent criterion of species—reproductive isolation. In our laboratory, Pavlinin (1966) gave special attention to the study of this question. It turned out that the Teleut has unlimited fertility in crosses with any forms of *Sciurus vulgaris,* and the nature of its variability is such that the species *S. vulgaris* represents a single whole connected by an uninterrupted series of transitional forms. The same argument is also justified in regard to other "good" subspecies. This means that to raise clearly differentiated subspecies to the rank of species leads not only to a crisis in the concept of "subspecies," but also to a still more serious crisis in the concept of "species," and this inevitably entails a crisis in systematics in general and, in the final analysis, chaos in evolutionary concepts.

It is hardly advisable to pretend that forms in which the highest degree of intraspecific differentiation is embodied do not exist. Analysis of the paths of their formation is, we must point out, one of the most reliable ways to study microevolutionary processes. These forms can, of course, be called something else, but there is no necessity to remove the term "subspecies" from the broad usage it has won. We come to the conclusion that it is advisable to retain the concept "subspecies" only if its original, evolutionary meaning is returned to it. In this way, we eliminate the contradiction—between intraspecific systematics and systematics above the species level—a contradiction that is unquestionable even if it is not sensed by everyone. Indeed, systematics constructed on a phylogenetic basis reflects the evolutionary process, expressed in the system of coordinated taxonomic units. Taxa above the species level reflect stages of phylogenesis, stages of macroevolution (how precisely the given taxonomic system corresponds to real phylogenesis is not of paramount significance; progress in this area of knowledge is regulated by natural laws just as much as in any other science). It is logical to assume that the taxa of subspecific systematics should reflect the microevolutionary process. From this point of view, the basic question of our subject can be formulated thus: What properties should a population deserving distinction as an independent subspecies possess, or—to ask essentially the same question—how are the beginning stages of the microevolutionary process manifested? Is it possible for us to make a distinction between homeostatic changes in the genetic structure of a population and those of its changes that signify the microevolutionary process?

The principal answer to the latter question is given in the literature. It is advisable to elevate to the rank of subspecies only those populations that are characterized by a specific evolutionary tendency (Lidicker, 1962). Few modern zoologists have so distinctly expressed this demand of modern systematics, a demand that also fully coincides with the modern demands of evolutionary study, but the view is transparent in the work of many modern taxonomists. It seems to us that the criterion of the independence of the evolutionary tendency is invulnerable; in its practical application, however, one encounters great difficulties. It is no coincidence that many authors see the possibility for its application in the elevation to subspecies rank of only physically isolated populations: isolation guarantees their independent development! We arrive at the same problem that population genetics comes up against. Isolation creates the best conditions for the development and formation of species, but under the conditions of isolation in small areas of the biosphere, it is theoretically impossible to expect reorganizations on a macroevolutionary scale, and morphophysiological progress under the conditions of isolation is practically excluded (see p. 9). Investigations conducted in our laboratory led us to the conclusion that the criterion for the independence of evolutionary fate (in the language of systematics: the criterion for subspecies) can be found. It consists in the irreversibility of microevolutionary reorganizations.

In contrast to the homeostatic changes in the genetic structure of a population, the microevolutionary process is, in principle, irreversible. Is it possible by studying the properties of a population to determine whether its features are reversible or irreversible? We answer this question in the affirmative. Ecological analysis allows us the possibility of determining whether we are dealing with the reversible reorganizations of a population's genetic structure or with the microevolutionary process. Investigation of this problem is the most interesting chapter of modern evolutionary studies. It falls completely within the competence of evolutionary ecology.

This question can be viewed from two aspects: practical and theoretical.

From a purely practical point of view, the given task can be resolved fairly simply in a very large number of specific cases: the forms having differences from other forms of the species that do not exceed the limits of chronographic variability cannot be elevated to subspecies (Shvarts, 1963*b*).* If the differences among the populations compared are less than those among animals of the same population in different years, elevating them to subspecies is in no way justified. It is artificial. In investigating the dynamics of the basic population characteristics, it is not difficult to establish the entire range of homeostatic variability in the genetic structure of the population. Characteristics of forms that fall within this range certainly cannot be acknowledged as the basis for distinguishing them as independent subspecies. With regard to a large number of subspecies admitted by modern systematics, this criterion may turn out to have been already decided at the contemporary level of our knowledge. To convince oneself of this, it is sufficient to look at any summary list. In the list of mammals of North America (G. S. Miller and Kellog, 1955), many "species" don't deserve even subspecific designation. In the multivolume monograph by Ognev (1928–1950), the later volumes of which are characterized by a more sober approach to the description of subspecies, imaginary subspecies are cited for almost every species. Thus, for example, the overwhelming majority of subspecies of water voles accepted by Ognev differ among

* By *chronographic variability*, we mean a reversible change in the features of populations over time. The most usual manifestation of this variability is the differences among generations of animals born at different times (Shvarts *et al.*, 1964*b*). Chronographic variability, however, does not mean simply seasonal changes. That it does not is attested to by changes in the mean of the variability of populations in different years (Pavlinin, 1959; Numerov, 1964).

themselves significantly less than do autumn and spring generations of one and the same population, to say nothing of the more protracted chronographic changes. From this stems one of the most important tasks of evolutionary ecology: study of the limits of variability of populations in their natural habitats and under experimental conditions. This task is subdivided into two: study of the phenotypic variability of populations and study of the homeostatic changes in their genetic structure.

Even very distinct subspecific characteristics are reversible in principle. We devoted special research to this question, and the results have been published (Shvarts and Pokrovskii, 1966). This research can be summarized as follows:

The nominal (*Microtus gregalis gregalis*) and northern (*M. g. major*) subspecies of the narrow-skulled vole possess clear differences in coloring. Colorimetric investigation showed the authenticity of these differences. An experiment was undertaken to try to obtain a more similar coloring in these forms by means of selection in a laboratory population of the northern subspecies. The results of selection were not long in expressing themselves. Colorimetry of 223 skins of animals of the 4th generation showed that 48 individuals (21.5%) exceeded the limits of variation in coloring of the initial laboratory colony. The distribution of deviations in colorimetric characteristics of the deviant individuals is interesting. *Microtus gregalis major* deviated 5 units past the lower limits of variation found in representatives of the color of the initial population, and 44 units in the direction of the nominal subspecies past the limits of variation in the initial population as regards whiteness. All individuals that deviated beyond the limits of variation of coloring in the initial population were indistinguishable in this trait from typical hybrids obtained by crossing the subspecies. The results of the experiment showed that selection in the course of 4 generations (less than 4 years) proved to be enough to eliminate the discontinuity between clearly expressed subspecies in one of the most important diagnostic traits. This trait proved to be reversible.

The material cited should not be treated as evidence for the possibility of transforming one subspecies into another. It is not difficult to show the irreversibility, in principle, of "real" subspecies with narrow-skulled voles as the example.* The results of the experiment described show clearly, however, that the degree of morphological differences cannot serve as the basis on which to make judgments about the reversibility or irreversibility of differences between compared forms. Differences in coloring between *M. g. major* and *M. g. gregalis* could not serve as the basis for distinguishing them as subspecies, even though a hiatus in coloring was observed between

* Vinogradov and Argiropulo (1941) considered *M. g. major* and *M. g. gregalis* independent species.

them before selection was begun. The search for the criterion of irreversibility should be directed along another path.

The irreversibility of species-level reorganizations is doubted by no one. It is based on hereditary incompatibility, manifested to different degrees in the reduction of hybrid viability (all the way to complete inability to hybridize). In other words, the irreversibility of species-level reorganizations is based on their genetic nature. Along with this, as we noted earlier (Shvarts, 1959), the irreversibility of evolution is determined by the different reactions of different species to changes in the environment and in the conditions of existence (for more details, see Chapter VI). Forms of one species are united not only genetically (unlimited fertility among any forms of the species), but also by the reaction, common to the species as a whole, to changes in environmental conditions. That they are thus doubly united emphasizes not only the genetic, but also the ecological, unity of the species. Since the origin of a new species proceeds by means of the reorganization of one (or several) populations of the old, however, it is natural that "specific" features should be observed in different intraspecific forms. Such features are observed even in the case in which the most important manifestation of specific independence—the inability to interbreed—is investigated. We have already had the opportunity to touch on this question (Shvarts, 1959), and we will therefore limit ourselves to just a few examples.

Hybrids of *Peromyscus maniculatus areas* × *P. m. gracilis* are infertile. *Peromyscus municulatus areas* does not cross with some other subspecies of the same species when contiguous with them in nature (Lin, 1954). Some experimental data (van Harris, 1954) relate to the sexual isolation of two subspecies of *P. maniculatus*. The initial stages of genetic isolation between subspecies of sagebrush vole (*Lagurus lagurus*) were determined by complex investigations (Gladkina *et al.*, 1966). Males of the 2nd generation of *P. truei truei* × *P. t. griseus* are infertile, but the females possess a lowered fertility (Dice and Liebe, 1937). Sumner (1915) did not succeed in obtaining progeny from crosses of *P. m. gambelii* × *P. m. rubidus*. The progeny from crosses of some subspecies of the house mouse have a high rate of growth that explains their viability; however, a great deal of mortality is observed among newborns, which is a characteristic trait of interspecific hybrids (Little, 1928). There is an indication of a relatively lowered fertility of some subspecies of pigeons (Withman, 1919). Hybrids of *Thomomys bottae passilis* × *T. b. meva* have lowered viability (Ingles and Wormand, 1951). *Rana pipiens* from the northeastern states of North America and from eastern Mexico behave as species (Moor, 1946). *Hyla aurea* from different regions of Australia produce abnormal progeny when crossed. Hybrids of western and eastern forms of *Crinia signifera* are absolutely inviable (Moor, 1954). When different populations of some species of amphibians are crossed, substantial disturbances of normal development are observed (syndrome of micro- or macrocephaly; underdeveloped circulatory system). This phenomenon was studied in detail by Ruibal (1962) under experimental conditions.

These examples, which are indicative of the hereditary physiological incompatibility of subspecies, should not be confused with those cases in which the infertility of different intraspecific forms is connected with pronounced differences in the size or growth rate of embryos. We will limit ourselves to one example. In some regions of Assam, domestic female buffalo are crossed with wild bulls (the domestic and wild forms belong to one species, *Bubalus bubalus*). Mating does not always lead to fertilization. Even more frequently, the female perishes because of the large size of the fetus. Finally, 75% of the calves perish in the first 8 days, since the mother cannot provide them with enough milk (Gee, 1953). It is completely possible that analogous phenomena take place when different subspecies cross in nature.

The great biological significance of the question broached and the paucity of studies about it make it useful to give some examples from invertebrates. In this connection, Rubtsov (1948) writes: "Races of trichogramma, for example, either do not cross, as with the malarial mosquito, or produce sterile progeny; i.e., they behave as good Linnaen species." Differences in the copulatory apparatus of the most strongly expressed varieties of *Carabus monilis* are so great that they exclude the possibility of successful copulation (Franz, 1929).

The number of similar examples increases yearly. Thus, it follows that if even so specific a property as the inability to interbreed is detected in subspecies, there is a basis for assuming that we may detect in subspecies another, more important, property of a species—the specific reaction to conditions of existence. That this is actually so can be shown in the narrow-skulled vole.

The northern subspecies of this species is distinguished from the southern by, aside from the coloring, a series of very important biological features: *M. g. major* is characterized by large body size and a high growth rate, high fertility, an early beginning of reproduction (under the snow), and a series of other features (Shvarts *et al.*, 1960). On the other hand, *M. g. major,* as the larger and more rapidly growing form, is more sensitive to a lack of food resources. Its requirement for wet food is also higher than that of *M. g. gregalis*. It is hardly necessary to prove that these are the kinds of characteristics that determine the originality of the reaction of animals in the compared subspecies to a change in environmental conditions.

Let us assume that *M. g. major* finds itself in environmental conditions characteristic for *M. g. gregalis*. This circumstance does not lead to its becoming more similar to the nominal form. A protracted spring, which appears to be a factor that seriously disturbs the normal tempo of reproduction in a population of *M. g. gregalis,* does not exert a strong influence on *M. g. major,* since its reproduction proceeds successfully under the snow. Even a very sharp reduction in the period propitious for reproduction does not elicit serious consequences in the latter, since its fertility is higher. An early approach of the autumn cold can lead to the death of *M. g. gregalis.* It affects *M. g. major* incomparably less, since the latter has acquired the capability of intensive growth in the autumn, when the growth of the young *M. g. gregalis* has practically stopped. Naturally, for all these reasons, under equal conditions of existence, *major* and *gregalis* will be subjected to completely different selection forces. Their divergence under equal conditions may be strengthened, not smoothed out.

Several more analogous examples can be cited. The subpolar form of the white hare utilizes trees as a food resource for a large part of the year. This has affected its morphology, particularly the morphology of the gastrointestinal tract. The cecum of the intestine in the northern hares is almost twice as large as in the hares of the central Urals (Shvarts *et al.,* 1966*b*). The subpolar hares have not been subjected to special physiological investigation, but it could scarcely be doubted that they have biochemical adaptations for better utilization of cellulose. Naturally, such features modify the character of the animals' reaction to a change in environmental conditions. In particular, it makes them less sensitive to a lack of nonarboreal resources. Hence the inevitable consequence: under equal environmental conditions, the northern and southern forms of the hare will, with the certainty of law, undergo different selection forces, with all the ensuing consequences.

Sharp ecological differences are observed in different forms of the house mouse (for a survey, see Serafinski, 1965*a,b*). Even under equal conditions of existence, *Mus musculus domesticus* almost never deserts human dwellings; *M. m. hortulanus* leads a wild form of life and stores food; *M. m. musculus* does not store food, but in favorable times of the year prefers open biotopes. It is impossible to doubt that these forms undergo different selection forces that guarantee the irreversibility of their reorganizations.

There is no need to cite further examples of this kind. Any ecologist could easily cite them from his personal experience.

It is no less important that a large number of forms are known in which the differences among them are such that even if they are subjected to equal selection forces (considering both the direction of selection and its strength), their changes would be of a different character. A good example in this regard is the narrow-skulled vole, which we have studied in detail under both field and laboratory conditions (Shvarts *et al.,* 1960). In the nominal subspecies (*M. g. gregalis*), an increase in body dimensions is negatively correlated with the width of the interorbital distance. In the northern subspecies (*M. g. major*), however, which is distinguished by significantly greater body and skull dimensions, this index is increased. Let us point out that the width of the interorbital distance is the most important diagnostic trait not only of the species, but also of subvarieties. The northern subspecies was descended from the southern (Shvarts, 1964). If selection is directed at the present time toward diminishing the size of *M. g. major,* a form will be obtained that does not differ in size from *M. g. major,* but that is distinguished by the most important craniological traits. Return to the ancestral form does not occur. In the process of the reorganization of the ancestral form, irreversible changes have arisen. The detailed investigations conducted by Ishchenko (1966) in our laboratory help in understanding the causes of this phenomenon. He established that the compared subspecies are distinguished not only by the dimensions and proportions of the body and skull, but also by the nature of the allometric growth. In the southern subspecies, an increase in skull dimensions is accompanied by a decrease in the interorbital distance. In the northern, this relationship is practically absent (see Fig. 7, p. 133). It is not difficult to understand what elicits the change in allometric growth during the process of differentiation in the compared forms. If we assume that during the origin of *M. g. major* from the nominal form, the relationship between skull length and interorbital distance characteristic for its type was maintained, then given *M. g. major's* dimensions, the interorbital distance would be so shortened that the skull could not function normally. There was selection for individuals with a less pronounced negative relationship of "skull length to skull width." The nature of the allometric dependence changed as a result of this. There would be no reason for a new change under reverse selection. A return to the ancestral form would be impossible. In this case, the irreversibility of reorganizations arising in the process of microevolution has as its basis a change in the correlated relations of the organism's morphogenesis.

A very similar example can be drawn from the wealth of literature on variability in *Peromyscus*. The relative growth of the cerebral part of the skull compared with the facial slows down more significantly in *P. maniculatus gracilis* than in *P. m. bairdi* (I. A. King and Eleftherion, 1960). Thus, selection for a change in growth rate or body dimensions will lead to different consequences in these forms.

The material cited shows that there are frequent cases of unreversed intraspecific reorganizations. Precisely such cases are, in our opinion, a manifestation of the microevolutionary process. *The microevolutionary process is the process of a population's irreversible adaptive reaction, determined by the specificity of its evolutionary fate*. It is advisable to call the forms in which the microevolutionary process materializes *subspecies*. The causes of the irreversibility of microevolutionary reorganizations can be diverse, but the origin of morphophysiological differences that change the relationship of animals to their habitat and change the system of morphological correlations are the most significant.

There is no need to talk about the fact that hardly any subspecies realizes the specificity of its evolutionary fate and leads to a new species. On the contrary, the vast majority of facts indicate that only a very few subspecies are reorganized into new species. Each species in its own development, however, passes through the subspecies stage.

Thus, we picture the process of intraspecific differentiation as follows: undirected homeostatic change of the population's structure—directed change of the population's structure—irreversible change of the population's genetic structure (microevolution = subspecies formation)—speciation.

A detailed discussion of this very interesting question of gradations of evolution's beginning stages would lead us far from our basic subject—the problems of theoretical systematics. We will therefore emphasize only the most important points.

A concordant body of facts shows that even though there is great diversity in manifestations of intraspecific differentiations, it appears in two different forms: reversible reorganizations of genetic structure, and irreversible reorganizations determined by the specificity of the intraspecific forms' evolutionary fate. We must make a few particular observations before proceeding to an analysis of the important consequences of this conclusion.

We will give major attention to internal prerequisites for the irreversibility of subspecific reorganizations. Undoubtedly, however, in a large number of cases, irreversibility of this type is determined by external, not internal, causes. We have already mentioned one of these causes—isolation. Isolation has really important evolutionary significance, however, not in small areas (islands and so forth), but in extensive territories, spatially isolated. The irreversibility of differences between *M. g. major* and *M. g. gregalis* is determined by internal causes, as was shown. Even if these internal mechanisms that make the reversibility of the polar subspecies' features theoretically unthinkable and practically unfeasible had not existed, the irreversibility of this form's features would be guaranteed by its spatial isolation. Until external causes of irreversibility in the reconstruction of a population's genetic structure elicit internal ones (in the sense indicated above), however, one cannot speak of a microevolutionary process, since it is impossible to make comparisons in reality.

One of the main tasks of evolutionary ecology ensues from this impossibility. It consists in investigation of the question: under what conditions do irreversible population changes become inevitable?

The basis of the irreversibility of microevolutionary changes, as was already mentioned, is a change in a species' reaction to change in environmental conditions. This process begins "within a species" and is completed in speciation. Let us clarify this position, to which we attach very great significance, by analysis of a specific example.

Many plant and animal species begin to reproduce (or renew growth and development) in the spring, with the advent of warm weather. If a return of cold weather follows the warm, the animal young that are still weak and the plants having their growth affected will die. This phenomenon has been termed *provocation of development* in the botanical literature.

In this case, we encounter the limited capacity of an organism's phylogenetically fixed reaction to change in environmental conditions; in certain situations, this reaction turns out to be harmful. Until the time when mistakes such as "spring provocation" substantially influence the realization of a species' biological potential, however, the reaction maintains its significance. Changes in the conditions of existence (including external changes completely unconnected with the reaction to a change in environmental conditions) that reduce the species' reproduction (e.g., the appearance of a new enemy) inevitably lead to the perfection (change) of the observed reaction. The species can no longer permit itself the luxury of placing some of its individuals under the threat of death during a return of cold weather. The reaction to environmental temperature changes is made more precise and perfect. In some species, development begins only when the average absolute temperature reaches a certain magnitude; in others, a rise in temperature stimulates development only against a background of long days (late spring); and so forth. It is important that the very same reaction lies at the basis of perfection of the observed type of adaptation: reproduction (or development) is stimulated to begin by a rise in environmental temperature. As we will see later, this circumstance has major significance: the reaction is not changed, only perfected. But attainment of absolute perfection is theoretically impossible, since the organism's reaction, which is phylogenetically fixed, corresponds, under any conditions, only to an "average over many years." Deviations from this average are not only possible, but necessary. In each specific case, therefore, development of individual animals is found to contradict in certain ways the norm fixed in the process of phylogenesis. This norm corresponds best to "average" conditions of the species' life, but can never encompass all possible deviations. That it cannot leads to genetic heterogeneity's becoming fixed in populations.

The rule of perfection of reaction described above can be observed, in a philosophical sense, as a most interesting manifestation of a very important position of Lenin (1934) on the correlation of phenomenon and law (pp. 148 and 149): "Law takes a static case, and therefore law, any law, is narrow, unfulfilled, and approximate. Phenomenon is richer than law." The phylogeny of the species determines the law of its development, but the actual form of the manifestation of this law corresponds to it only *approximately*. A clear understanding of the contradictions in the actual realization of the phylogenetically fixed reaction helps one to understand more deeply the specific mechanisms that govern the development of the organic world.

It is convenient to take one of the most important positions of modern cybernetics as the starting point for our further discourses. Norbert Wiener (1964) in his book *I Am a Mathematician*, writes (p. 327): "Thus, from the point of view of cybernetics, the world is an organism, neither so tightly jointed that it cannot be changed in some aspects without losing all of its identity in all aspects nor so loosely jointed that any one thing can happen as readily as any other thing." In living nature, a contradiction between real conditions of development and "ideal" conditions (stipulated by phylogeny) can be considered the norm, and it is only thanks to a "flexibility" of the "organism–environment" system that the normal development of a great number of individuals proves to be possible.

In our example, a phylogenetically fixed reaction is expedient, but the expediency is relative and limited. It is expedient to begin reproduction with the establishment of warm weather (in the spring). This corresponds to the biological characteristics of the majority of species (in temperate and subpolar climatic zones). That this expediency of reaction has limitations, however, is expressed in two aspects: First, a temporary warming, not connected with the final triumph of spring, is possible. Second, the later course of temperatures or of any conditions cannot correspond exactly to the phylogenetically conditioned demands of the species. Development never proceeds under absolutely optimal conditions, and it is only thanks to numerous compensatory reactions that it is satisfactorily completed. In all such situations, however, the basis of the reaction remains expedient. Other cases are possible, however, in which natural selection brings the biological significance of the very basis of the reaction into doubt. Let us analyze one such case.

The developmental rate of amphibians, as a rule almost without exceptions, is directly proportional to environmental temperature (within the limits of temperatures normally encountered in the species' range). This relationship is a consequence of the chemical bases of metabolism; moreover, in the majority of cases, it is biologically expedient, since in high temperatures (hot summer), the reservoirs in which the frogs shed their eggs quickly dry out, and the only path of salvation is to complete metamorphosis and go out onto dry land. It is not surprising, therefore, that in arid regions, the acceleration of the developmental process occurs even more markedly than demanded by the chemical law expressed in Arrhenius' well-known law. That it does emphasizes the point that developmental rate reflects the organism's reaction, fixed in the process of phylogeny, to a change in environmental conditions, and not simple chemical laws.* Under cold climatic conditions, metamorphosis is arrested; amphibians overwinter in the larval stage, and a 1-year cycle is transformed into a 2-year. Since the intensity of metabolism in cold-blooded animals decreases with the lowering of the environmental temperature, the latter phenomenon is easily explained not only from a physicochemical viewpoint, but also in correspondence with the general biological characteristics of the group of animals observed. A different situation obtains in the far north and high in the mountains. In these regions with exceedingly low temperatures, the soil and reservoirs are frozen a long time, which practically excludes the possibility of amphibians' overwintering in the larval stage. (They can overwinter only in the water.) Here, a delay of metamorphosis signifies the death of the population. Actually, the majority of amphibians go no farther north than the taiga. Only a few species—a very few—penetrate into the tundra and form stable populations there. It was found that even though their development proceeds under extremely low temperatures, it is completed in a shorter period than that of their nearest relatives in the south. It deserves attention that this discovery was made almost simultaneously in high mountain species and in amphibians of the far north (Shvarts, 1959; Toporkova and Shvarts, 1960; Toporkova and Zubareva, 1965; Moor, 1963; Ruibal, 1962).

The acquisition by northern amphibian populations of major new biological characteristics is an obvious phylogenetic change. How did it occur, or rather, how could it occur? During the amphibians' advance into the north, the phylogenetically fixed reaction, based on the same elementary chemical laws, came into sharp contradiction with the conditions of development. In this case, the reaction of separate individuals and the reaction of the population as a whole (the phylogenetic reaction) proved to be opposite in direction. The amphibians' advance into the tundra or the high mountains became possible only as a result of selection for individuals with an indistinctly expressed reaction to the reduction of environmental temperatures. (It is not important to know in this case which specific physiological characteristics of the animals this proved to be connected with.) One may assume that in years with unusually warm weather (such

* An increase in temperature stimulates the thyrotropic activity of the hypophysis (Uhlenhuth, 1919; Etkin, 1964).

years are not so rare even in the far north), individuals with a poorly expressed reaction could successfully complete metamorphosis, leave progeny, and form populations characterized by new properties. This supposition is supported by facts: even now, many amphibian species in the northern limits of their range do not reproduce each year.

Northern (but still not polar!) amphibian outposts have probably existed for a fairly long time within the limits of the sphere of life accessible to this class. But as long as the character of individual reactions to a change in environmental temperature remained essentially unchanged, not one amphibian species could penetrate beyond the limits of that region in which, even in favorable years, the warm period of the year is shorter than the normal period of metamorphosis. Within the indicated geographic limits of amphibian distribution, however, important population reorganizations occurred by degrees. Selection operated on behalf of individuals that completed metamorphosis quickly, despite low environmental temperatures. Under the conditions that had been established, this trait proved to be the primary one, determining the adaptability of different individuals. All the remaining features of the animals became secondary. This promoted the accumulation of mutations that accelerated development and the appearance of individuals with inversion reactions, and these individuals developed rapidly even under low temperatures.* Such individuals are of special value. Their appearance created the possibility for amphibians to advance into the tundra.

This example shows very clearly that in the process of the intraspecific differentiation of individual populations, species acquire features that change the animals' relationship to the habitat. This relationship determines their specific evolutionary fate and the essentially irreversible nature of the reorganizations they have achieved. It is also clear that the process of forming and solidifying such population features is immeasurably longer and more complex than the simple perfection of traits and properties already fixed in the phylogeny of the species.

It is very important to emphasize that the differentiated population, having opened a path to a new habitat, acquires a complex of distinctive characteristics. As a result, a new species arises that differs dramatically from its nearest relatives. In this case, also, the class of amphibians seems to us the best example. In recent years, the Vietnamese crab-eating frog, *Rana cancrivora,* became a distinctive zoological celebrity. Outwardly, this

* It is the manifestation of this process that interests us here, not its physiological mechanism. At low temperatures, the skin of amphibians loses its sensitivity to thyroxine (Etkin, 1964). It is possible that in northern populations of the frog, this reaction is expressed less strongly and that this leads to an increase in the rate of their development. Selection among northern populations of frogs should occur at the cytological level.

frog is quite similar to a very common species—the tiger frog (*R. tigrina*). Amphibians are typically freshwater animals. Not one amphibian species is capable of existing in water of high salinity. The limit of salinity that even species with the very greatest "fortitude" (*Bufo viridis*) can withstand is 1%. Yet the crab-eating frog develops in marine estuaries, in which the salinity of the water exceeds 3%! The ecological significance of this characteristic of crab-eating frogs is understood. The frog acquired the capability to master water sources inaccessible to other species. It escaped the pressure of its closest competitors, mastered a new sphere of life, and overcame one of the basic ecological barriers—the osmotic barrier. The biochemical features of the crab-eating frog proved to be connected with a capability completely unique for amphibians—the ability to actively regulate water–salt exchange in the tissues and liquid cavities. In an evolutionary sense, this example really justifies the popularity that this species of frog acquired among zoologists. It shows that adaptation to unique environmental conditions can lead to the origin in individual species of properties that place them in a unique position not only in relation to phylogenetically close species, but also in relation to the entire class. Detailed investigations (Gordon *et al.*, 1961) showed that the concentration of salt in the plasma of the crab-eating frog reaches a level at which some enzymes are denatured and the oxygen affinity of hemoglobin decreases. The great fortitude of the crab-eating frog's tadpoles in water of high salinity compels us to assume that they have undergone a major change in the type of nitrogen exchange.

It is perfectly obvious that the beginning stages of the process that led to the formation of so peculiar a species still took place within the bosom of the ancestral ("normal") species, since for *R. cancrivora's* unique characteristics to become refined, the ancestral form must have had a long existence in the new environment.

Comparing the results of research on polar and marine frogs, we arrive at the conclusion that in the process of microevolution, there actually do occur changes in animals that determine the specificity of their further evolutionary fate. It is self-evident that not all such phenomena are expressed as markedly as those described above, but they are the same in essence.

How, exactly, does the microevolutionary process differ from homeostatic reorganizations of population genetic structure? It is evident from the examples cited that the mechanism of population reorganization is the same, in principle, in both cases: selection of genetic variants that, under given conditions, leave more progeny. The result of this process in each case, however, is different. In the first case, irreversible phenomena arise; in the second, reversible phenomena. It is natural that irreversible changes arise only in the case when they are the most correct, inevitable path of adaptation to unique conditions of existence. The irreversibility of intraspecific reorganizations is determined by ecological mechanisms—a change in the reaction to external conditions.

The facts cited show that in the principles of modern biology, there are fully objective methods that permit us to draw a boundary between microevolution, on one hand, and, on the other, the adaptive reorganizations of a population's genetic structure, which do not change the general direction of the population's development and do not determine its evolutionary fate. Microevolution begins with the origin of irreversible reorganizations of the population, which determine the course of its further development, but do not lead to the population's reproductive isolation from other forms of the species, i.e., to its genetic isolation. In a great many cases, of course, the boundary between homeostatic reorganizations of the population's genetic structure and the microevolutionary process cannot be drawn with sufficient reliability. The boundary is justified, however, in regard to literally any biological phenomenon or process. For example, although no one doubts the reality of biocenoses, actual boundaries very often cannot be drawn between individual associations. There is a large body of literature devoted to the theory of the boundaries of biocenoses. It represents very different viewpoints (often mutually exclusive), but this does not influence, to the least degree, our confidence in the reality of cenoses and the tremendous theoretical significance of analytical studies on biocenosis. More than that, it does not influence the possibility of applying conclusions from biocenology in practice (e.g., in forestry, the ecology of cultivated communities, and so forth). Precisely the same holds true for the concept of "population." That populations exist is a fact, but it is not always possible to determine the boundaries between them. Finally, let us recall the different concepts concerned with the ontogeny of organisms: infancy, childhood, adolescence, youth, maturity, and old age. No one resolves to define exactly the boundaries between these states of the developing organism, nor does anyone resolve to deny their reality, and no one doubts that in nonborderline cases, they can be distinguished in a practical way without mistakes. To close one's eyes to the existence of stages in animal development on the basis that boundaries between them cannot always be established would mean to distort the real picture of ontogeny and to close important avenues of research. In precisely the same vein, we can say that to deny the presence of qualitatively different phases in the initial stages of phylogeny on the basis that boundaries between them cannot always be simply drawn means to distort the real position of things in nature and to deprive science of important avenues for the cognition of phylogeny. It seems to us that even in the first stages of phylogeny, its qualitatively different stages, in nonborderline cases, can be determined just as clearly as the stages of ontogeny. Purely technical difficulties do not have essential significance.

In drawing the boundary between homeostatic reorganizations of the population and microevolution, it should be especially emphasized that the phenomena of population homeostasis play an important role in the evolutionary process. The dynamics of population structure creates the preconditions for the rapid, directed reorganization of the population during a change of external conditions—during a change in the direction of selection. The greater the range of the population's homeostatic reorganizations, the higher the population's capability to master a new environment, and the greater the possible rate of microevolution. It is possible for a population to evolve immeasurably more rapidly than one might determine on the basis of an analysis of the geographic variability of organisms without an accompanying ecological analysis of the variability of habitats. We touched on this question lightly in Chapter I. Here, we must deal with it in greater detail.

As already noted, there are many experimental data on the possible rate of the reorganization of a population's genetic makeup. It is important to emphasize that this rate affects all traits and properties of the organisms.

Wilkes (1947), in the course of an insignificantly short period (4 generations), succeeded in creating an experimental population of *Diprion hercyniae* that was distinguished from the initial selected sample by a lower preferred temperature, greater fertility, a larger number of adults, and a longer life span. The parasite *Dahlbominus fuscipennis* served as a factor of selection in this author's experiments. This result shows that a new factor of elimination leads to a complex reorganization of the population's genetic makeup. Allen (1954) increased the rate of propagation in a model population of *Horigenes molestae* by a factor of 24 through 3 years of selection. The possibility of a dramatic change in the sex ratio as a result of selection was proved experimentally in *Anoplex carpocasae* (Simmonds, 1947). It was also found to be possible, during the course of a few generations, to change the degree to which a sexual dimorphism was expressed (Korkman, 1957; experiments on laboratory mice). In the experiments of D. Pimentel (1965), the resistance of flies to parasitic wasps (*Navonia vitripennis*) increased greatly over the course of 33 years. As a result, the numbers of parasites were reduced by almost half, under unchanged environmental conditions and without a change in the numbers of the host. It was also shown by D. Pimentel (1965) that the joint maintenance of two competing species (*Musca domestica* and *Phaenicia sericota*) elicits a change in the population genetic structure of both species, which leads to a change in their relative numbers. The effectiveness of selection on resistance to poison was discussed earlier. These experiments are especially interesting in regard to their clear portrayal of the reversibility of acquired features. Thus, I. G. Robertson (1957), through the action of DDT on an experimental population of *Macrocentrus ancylivorus*, raised its resistance by 12 times, but during the development of the population in a normal environment, it returned to the initial state in 13 generations.

Can the results of these and analogous experiments be extrapolated to natural populations? Observations on the acclimatization of animals give a simple answer to this question.

Acclimatization is the process of formation of a new animal population possessing a series of common specific characteristics. The most important and dominant force of this process is natural selection. It is rational to assume that among the specific characteristics of the new populations, physiological characteristics of the animals occupy one of the leading positions. Study of these features is one of the serious tasks facing researchers concerned with the theoretical bases of the acclimatization process. At present, however, we have almost no notion of the physiological specificity of acclimatized animals. We are therefore forced to illustrate the thesis of their formation in the process of the acclimatization of specific animal populations with the study of their morphological characteristics, although these characteristics also, unfortunately, are studied far too little.

Various subspecies of the American mink (*Mustela vison*) were brought to the territory of the USSR. The predominant number of wild animals were attributed to the eastern Canadian variety (*M. v. vison*). The thorough investigations of Popov (1949) showed that starting from the moment the American mink were brought to Tatar and for the next 14 years, such drastic changes occurred in their morphology (dimensions, cranial characteristics) that it is now already impossible to assign them to any one of the known subspecies.

The raccoon dog (*Nyctereutes procynoides*) brought to the Kalinin region in 1943 underwent even greater changes: the width of its cheekbone increased, body length decreased by 6% and weight by 9%, the tooth row was shortened, and its fur became thicker. Sorokin (1953, 1956) gives the following explanation of the changes that occurred in the morphology of the raccoon dog: a less available food resource caused the decrease in its dimensions. It went through an impossible hibernation, and the animal seemed under the direct influence of relatively severe winter conditions. This entailed a change in its fur coat. The shortening of the tooth row, of course, can be explained by the relatively greater quantity of animal food in its diet. Morozov (1955) cites analogous data. Summing up the raccoon dog's acclimatization in the European part of the USSR, Tserevitinov (1953) also emphasizes the increase in its dimensions in some places and the development of luxuriant fur.

Drastic changes also occurred in the morphology of the muskrat during the process of acclimatization. In an article devoted to this question,

Tserevitinov (1951) states that the fur-bearing qualities of the muskrat change in relation to the climatic conditions of different areas. This makes it possible to pick out four extremely distinct differing forms of muskrats: eastern Siberian, western Siberian, northern, and Kazakhstan. Tserevitinov investigated only those characteristics of the muskrat that are of the greatest interest from an economic point of view (weight and area of the skin, coat thickness, hair length and thickness, and coloring). It was perfectly obvious, however, that other traits also underwent changes.

The results of the research of Povetskaya (1951) are of great interest. She showed that in the process of acclimatization of valuable subspecies of squirrels, significant changes in their morphology were observed, including those traits that serve as a basis for dividing different populations of rodents into independent subspecies. Thus, the Teleut squirrel (*Sciurus vulgaris exalbidus*), introduced in 1940 from the pre-Altai pine forests into the Crimean preserves, underwent the following changes during 9 years: body weight increased an average of 23 g; the thickness of the fur diminished by 18.5%, the height by 1 mm, while breadth increased by 5.73%; hair strength increased; the thickness of the hide decreased by 12%; and the fur color changed.

Povetskaya obtained analogous data from an analysis of the results of acclimatization of the Altai squirrel (*S. v. altaicus*) in the Teberdinsk preserves: during 10 years, the fur became 20% thinner (on the average) and coarser, hair length diminished by 1 mm, and the hide became thicker. The results of these observations were confirmed recently by Meladze (1954).

The American "brown squirrel" (*S. carolinensis*) has been repeatedly brought to England. At present, the morphological characteristics of the British population of this species do not correspond to the diagnostic features of any one of the American subspecies (it is closest to *S. c. leucotis*, but is inferior in its dimensions).

Acclimatization of the hare (*Lepus europaeus*) in Siberia was accompanied by substantial changes in the morphology of its fur coat (Gerasimova, 1955). Stroganov and Yudin (1956) believed that by 19 years after the acclimatization of the hare from Bashkir to the Novosibirsk region, it had changed so much that it deserved independent subspecies status: *L. e. orientalis* subsp. *nov.* Yazan (1964) writes about the morphological and ecological changes in the beavers reacclimated in the Pechoro-Ilychskii preserves.

Very interesting changes occurred in Australian rabbit populations (Birch, 1965). A cline in color was produced in Tasmania during the course of 50 years: in the mountains, black individuals predominate; their proportion in the populations decreases regularly with a decrease in elevation (Barber, 1954). In Australia, an analogous cline was formed during 40–60 generations (Birch, 1965). Morphological changes that have occurred in the Australian populations of house mice are poorly studied, but Birch (1965) says they are huge.

The house sparrow (*Passer domesticus*), having spread throughout North America, underwent a series of changes that affected the length of its hip and shoulder. These changes are especially noticeable in the study of North American populations inhabiting localities with severe winters (see Chapter I).

The North American species of frontal finches (*Carpodacus frontalis*) that migrated to the Hawaiian Islands was reorganized into a clearly distinct subspecies after several decades (cited by Puzanov, 1954). The well-known changes in the coloring of birds with changes in humidity should also be assigned to the category of phenomena described. In individual cases, it is possible to observe that birds maintained under conditions of increased moisture become more like natural varieties from regions with a moister climate. Such cases are known, in particular, for the thrush, *Hydrocichla mustelina,* and the inca pigeon, *Scaradafella inca* (Hesse, 1924). The greatest changes were observed in the desert weaver birds of Australia, *Munia flavaprimna,* after it had been maintained for 3 years in England: the bird acquired the color and pattern characteristic of another weaver that inhabits coastal regions of Australia and is described as an independent species.

All the amphibians inhabiting the Hawaiian Islands at present have been introduced. Comparison of the Hawaiian representatives of different species with the source forms demonstrates substantial differences. Thus, the Hawaiian *Rana rugosa* are distinguished from the Japanese by greater mottling of the abdominal surface, stronger development of the flotation membranes, and diminution of the callus internus (Oliver and Shaw, 1953). The distinctions between these forms are evidently of subspecies rank. Acclimatization of fish, which possess relatively more plasticity, leads to even more substantial morphophysiological changes. For illustration, it is enough to cite the remarkable research of Kirpichnikov (1966*a,b*), which has shown that during a historically short period, the carp from Lake Balkhash acquired features that did not differ in scale from the features of "good" subspecies. When the subspecies are hybridized, their offspring, apparently, very quickly acquire the appearance of the local aboriginal variety.

In order to improve local herds of deer in Germany, marals (Siberian stags) and elk (*Cervus elaphus canadensis*) were repeatedly introduced into the range of *Cervus elaphus elaphus*. After several years, however, the "blood" of the imported deer no longer had any effect, and the German type of deer remained unchanged. A detailed analysis of the results of work conducted during the last 70 years (Beninde, 1940) shows the latter situation. Completely similar results were obtained in the acclimatization of sables from the eastern subspecies to the Urals (Pavlinin, 1959).

There is no doubt that many population characteristics that arise in the process of acclimatization are environmental in nature. But the aggregate of data cited clearly shows that the reorganization of population genetic structure plays a great role in this process. This testifies to the fact that during a change in environmental conditions, phenomena occur that can with full justification be called microevolutionary. Since the habitats of organisms change very rapidly under the influence of man, we may expect that rapid evolutionary reorganizations are occurring before our eyes on a very broad, planetary scale. Thus, the use of chemical poisons to fight pests leads to the emergence of specialized forms, and some pure bacterial cultures for growth are already needed now for antibiotics (and what will happen later?) (L. C. Cole, 1966). Thus, the question of the rate of microevolutionary reorganizations is in no way of purely academic interest.

Some direct observations also have bearing on the possible rate of population reorganizations under the influence of natural selection. One such observation is the origin of industrial melanism (see Chapter I). This example, although it enjoys well-deserved fame, is relatively simple, since it deals only with a change in the population frequency of a monogenically determined trait. More interesting are the data showing that even within the lifetime of one generation of animals, a noticeable change can occur in the average values of population variability for traits that include polygenically conditioned ones. There are not many such observations, but they do exist. Thus, Van Valen (1965–1966), using a completely biometric analysis of the material, showed that under certain conditions, tooth width (M^1) of mice in the old age groups is greater than that of young mice, despite the stabilization of this index at a very early age. The author is convinced that there is differential mortality of the different genotypes in the populations examined, which leads to a displacement of the average value of population variability during the life of one generation (the average change in the trait is not less than 0.154 SD, which corresponds to a selection intensity of 0.1–0.3). Similar research was done in our laboratory at the same time. Kopein (1964) studied the skull dimensions of ermine from the subpolar regions of the Urals and Yamal. It was found that even though the skull continues to grow during the first years of life, its average size at age 2+ is less than at age 1+, and this size is less than that of the current year's young. A very great amount of material was studied (more than 3000 carcasses); the differences are significant at the 80% level. The sources of selection to decrease the dimensions are still not firmly established. It is interesting that in the more southern regions (the northern taiga), Kopein did not succeed in detecting a decrease in skull size with age in the ermine. On the contrary, the skulls of the current year's young turned out to be substantially smaller than those of older animals. These findings also convince us that the decrease of skull length in the subpolar area is the result of selection, not a phenomenon connected with the physiology of the ermine's growth.

The research on variability in the water snake, *Natrix sipedon* (Camin and Ehrlich, 1958), should be included in work of this kind. The authors studied snakes of this species on the islands of Lake Erie. They established that in contrast to the "mainland" populations, the island snakes are characterized by a much smaller number of striped individuals. It is especially interesting that striped animals are encountered significantly less often among adult snakes than among young ones (the difference is statistically significant). Analysis showed that this can be understood only as a result of selection, the effectiveness of which is revealed during the lifetime of a single generation. It was established that the probability of survival for striped individuals is 4 times less than that for uniformly colored snakes (the agents of selection are birds). The authors correctly assume that if a constant migration of striped serpents from the "mainland" did not exist, they would long ago have ceased to be encountered in the island populations.

The material presented in this chapter testifies to the fact that the homeostatic reorganization of a population's genetic makeup is a common and very widespread phenomenon.* It ensures the possibility for individual populations of a species to exist under constantly changing (fluctuating) conditions of the external environment. In those cases in which environmental conditions change in a given direction, analogous changes also occur in the population's genetic structure: in the course of a few generations, it acquires new properties and the population oscillates around a qualitatively different mean. In the first stage of this process, the changes that arise are reversible. The population can return to the initial norm of variability. It follows that the apparent stability of the morphophysiological characteristics of different geographic forms of a species testifies, in the majority of cases, not to the constancy of their features, but to the relative constancy of environmental conditions. This circumstance has very great significance in the determination of a series of principal problems in systematics and zoogeography.

When the directional change of the population's genetic structure has gone a long way and new population properties change the very character of

* We therefore cannot agree with S. Wright (1945), who affirms that the elementary evolutionary process reduces to a change in the frequency of genes in a population.

the organism's relationship to the environment (which inevitably leads to a change in the direction of selection), the irreversibility of intraspecific reorganizations is already assured, not by external (constancy of the environment), but by internal (quality of the population) mechanisms. This stage of the population's reorganization can with complete correctness be considered microevolution. These are the external manifestations (phenomenology) of the microevolutionary process. Its basis is a change of the species' norm of reaction to a change in external conditions.

CHAPTER IV

Ecological Mechanisms for the Maintenance of the Genetic Heterogeneity of a Population

When the genetic heterogeneity of a population takes the character of an expressed polymorphism, its biological significance becomes obvious. It consists in the great range of conditions that the population as a whole can utilize for the maintenance of optimal numbers. The data cited and discussed in Chapters I and III show that modern ecology is still far from an understanding of the role of polymorphism in the life of animals. The degree of adaptation of different genotypes to different environmental conditions attains a degree of specificity about which we can only guess at the present time. Some experimental data address this point. They indicate that not only the representatives of different phases, but also the different genotypes, occupy a different position in the system "population–environment," since they occupy different ecological subniches (Lewontin, 1955). Convincing indirect data support the ideas developed by Lewontin. It was found that populations distinguished by expressed chromosomal polymorphisms are also distinguished by a broader distribution and greater diversity in the biotopes occupied (Cunha and Dobzhansky, 1954; Cunha *et al.*, 1959). Furthermore, populations with an impoverished gene pool are characteristically limited in the number of ecological niches occupied.

It would seem that complete clarity in this most important question of evolutionary theory can be brought about only by intensified ecological research, which permits us to relate the characteristics of individual populations not only to general genetic characteristics, but also to the degree of genetic heterogeneity. So long as there are essentially no such investigations, we are necessarily limited to a very general assertion: an increase in a population's genetic heterogeneity is advantageous, since it promotes a fuller utilization of environmental resources. On the other hand, enrichment of a population's gene pool greatly increases the possibility of its adaptive reorganization as a unified whole, greatly increases its viability, and, in essence, guarantees its existence in changing environmental conditions. As will be shown below, the very same mechanisms that ensure the population's adaptation to changes in environmental conditions in its home territory create the prerequisites for expansion of its range. Thus, it is understandable that there should exist in nature diverse mechanisms to maintain a population's genetic diversity and to constantly enrich its gene pool.

A general enrichment of a population's gene pool not only raises the adaptive opportunities of the population as a whole, but also increases the genetic capacity of separate individuals (Chapter I). The experiments of Carson (1961) are of special interest in this regard. It was found that a population produced by one pair of *D. robusta* from the center of its range did not differ in degree of viability and capability for further adaptive reorganizations from populations arising from a large number of progenitors (founders). Carson's experiments were conducted with great care, but, in essence, they only supported many data obtained earlier. It is known that laboratory colonies of a number of wild animal species are produced from a few founders. In our laboratory, several pairs of individuals were the founders of large colonies of Middendorf voles, *oeconomus* voles, sagebrush voles, and narrow-skulled voles. Several hundred Middendorf voles were obtained from one pair.

Populations of many acclimatized species are produced by a very small number of founder individuals. Drachevskii (1961) shows that in Kirghiz, over 4 years, a flourishing group of nutria numbering 300 individuals was obtained from 3 pairs. In England, populations of nutria arose from individual animals that had escaped from wild animal farms. There could hardly have been many such individuals. Nevertheless, in 10 years, the nutria spread over a territory of several counties, and in 1961–1962 had reached 100,000 animals (Norris, 1963). Three pairs of elk released on Newfoundland produced a flourishing population, numbering 30–40,000 head (Pimlott, 1961). Canadian populations of hares were started by 7 females and 2 males, taken from Germany in 1912 (Dean and DeVos, 1965). From 14 wild boars trapped in 1909 in Russia arose a population in the United States that now occupies a territory exceeding 1000 km^2 (Ivanov, 1962). L. V. and F. D. Shaposhnikov (1949) describe a flourishing population of beavers that arose from 3 pairs. There are many similar examples known. In a summary article, Nasimovich (1961) writes: "There are numerous examples in which the issue of a small number of animals led to rapid propagation and later distribution of the introduced species to an enormous territory (rabbits, muskrats, house sparrows, and others)."

We (Shvarts *et al.*, 1966*d*) conducted a special experiment, the aim of which was to investigate how a population's variability changes during its formation from a few chance founders. We compared the variability in coloring of rodents from natural populations and from laboratory colonies. The coloring was determined colorimetrically, which assures objectivity in its evaluation. The founders of the colonies were animals from the same populations that served as our controls: Middendorf voles (*Microtus middendorffi*), 2 males and 1 female; *M. oeconomus*, 6 males and 4 females; northern narrow-skulled voles (*M. g. major*), 5 pairs; *M. g. gregalis*, 4 males and 6 females. The animals were bred for 4–6 generations. It was established that a substantial change in the range and direction of variation did not occur in even one of the four cases. The founder principle did not work. In light of the results of the experiments of Carson (1961), this did not seem surprising to us. We had the opportunity to show that populations arising from a few chance founders fully maintain the capability for directed changes under selection. A special experiment (Shvarts and Pokrovskii, 1966) relates to this phenomenon. Its results are discussed in another part of this book.

Immunological investigations of transplantation antigens also give information about the genetic capacity of separate individuals, and consequently about the possible genetic heterogeneity of populations arising from a few founders. In our laboratory, Syuzyumova (1967) studied the time to rejection of skin transplants in different groups of the *oeconomus* voles. When the recipient and donor belong to different subspecies (*M. oeconomus oeconomus* and *M. o. chahlovi*), the time to rejection of transplants varies from 6 to 7 days. When skin patches are transplanted within the same family (i.e., when the recipient and donor are the progeny of one female), the average duration of a transplant is, of course, increased (variations from 6 to 30 days). These experiments, in themselves, indicate that within closely related groups of animals (one family), it is possible to detect antigenic and, consequently, genetic differences commensurate with differences between subspecies. The enrichment of the gene pool in the animal groups studied led to the situation that with intrafamily transplants (brother-to-brother), the duration of the transplanted skin patches is often reduced even more. Thus, within one of the hybrid families, the duration of transplants varied between 6 and 8 days. In other families, great variability in the reaction to transplants was observed (the duration of transplanted patches being from 7 to 52 days). It seems evident that populations that arise by the propagation of such families can be genetically just as heterogeneous as flourishing natural populations. Under these conditions, the founder principle can have only limited significance.

All these and, in addition, a large body of other facts show that the enrichment of a population's common gene pool has enormous biological significance. It is natural, therefore, to expect special mechanisms for the maintenance of genetic heterogeneity in populations. One of these mechanisms is the increased viability of heterozygotes.

This question has been elaborated at length in the genetic literature, and there is no need to analyze it in detail here. This fact is important: in the overwhelming majority of cases, heterozygotes possess greater viability, which promotes the maintenance of genes in the population that are harmful or even lethal in the homozygous state. The actual causes of the heterozygote's heightened viability cannot be considered as being ascertained. There are data showing that it is not heterozygosity in itself that is important, but a fully defined combination of genes. In particular, the experiments of Wallace (1955) attest to this. They show that among heterozygotes of a single population, all degrees of viability are to be found, including individuals who are distinguished by little vitality. On the other hand, there are observations that indicate specific partial causes for the increased survival of heterozygotes. The best example is the distribution among humans of sickle cell hemoglobin (Hb^S). Homozygotes (Hb^S Hb^S), as a rule, do not reach sexual maturity. Nevertheless, in some African localities, the Hb^S gene occurs with a frequency of 20%. Allison (1956) established that this is connected with the increased resistance of heterozygous individuals to malaria. It was found later that Hb^S is also encountered in other parts of the world in which malaria is an important mortality factor. These observations lead to the conclusion that deviation from the normal type of hemoglobin causes the conditions for the development of the parasite to deteriorate, and thus reduces the overall percentage of mortality among local inhabitants, despite the significant mortality of the homozygote Hb^S Hb^S. It is very probable that the distribution in populations of other types of heterozygotes can be explained in a similar way. It is clear, however, that the higher the genetic heterogeneity of a population, the greater the possibility for the origin of the most favorable variants. This justifies the search for the mechanisms to maintain high genetic variability in a population.

Undoubtedly, some of the principal mechanisms for the maintenance of heterozygosity in populations and the general enrichment of the gene pool are the very same mechanisms that promote the intermixing of populations and micropopulations. Examples of this type are very well known, and we will therefore limit ourselves to citing only a few.

In the majority of mammalian species, there is a clear seasonal change in habitats. Thus, the water vole abandons the islands and, partially, the lake shores in the autumn to settle in meadows and haystacks for the winter. In the hunting season, massive migrations of muskrats are seen. These migrations usually occur because of a decrease in water supply. Often, a consequence of these wanderings is the settlement of new reservoirs by the muskrats, and undoubtedly they lead to constant intermixing of populations. Many small species of mouselike rodents occupy very diverse biotopes in the steppe and forest steppe regions in the autumn and the first half of the summer: areas in the steppe forests, thickets of shrubs, plots of cultivated plants, ground that has long lain fallow, and so forth. With the approach of summer's warmth, the majority of them abandon these habitats and settle along the shores of reservoirs. This resettling undoubtedly leads to the mixing of individuals from different habitats.

We established the presence of obvious migrations of the common brown-tooth shrew in the forest steppes of the Transurals. In the summer, under the conditions of an arid climate, shrews of this species are encountered only along the shores of reservoirs. With the approach of autumn weather, they are less closely tied to the moist habitats, and they occupy very diverse biotopes in the steppes and forest steppes. When they are concentrated at the reservoirs, the intermixing of populations the following summer is unavoidable. In the winter, elk remain in the bogs favored during the summer, and often confine themselves to a limited section of riparian woods. The polar fox and northern deer abandon their summer habitats in the tundra in the autumn, and winter in the forest tundra or even in the northern parts of the forest zone. A colleague from our laboratory, V. S. Smirnov, banded foxes in Yamal. One of the animals was recovered in Alaska after several months. In the winter, the gazelle, *Gazella gutturosa,* appears in the steppes of the Transbaikal and northern Mongolia, and in the autumn it returns to southern Mongolia. Seals, walruses, whales, bats, and other mammals carry out yearly migrations over huge distances.

Migrations, and especially migrations for long distances, inevitably increase the probability that animals from different populations will mate. As is well known, migrations are even more characteristic of birds than of mammals. Remember, too, that migrations are known among reptiles; they have been observed in common water snakes (Terent'ev and Chernov, 1949), in *Elaphe dione* (Khozatskii and Églon, 1947), and in the lizard, *Lacerta agilis* (Shchepot'ev, 1952).

When speaking of the mechanism by which heterozygosity is increased, one must note the special significance of the resettling of young during their transition to an independent form of life. This phenomenon is well known for relatively few species, but it is probably a widespread phenomenon.

Kalabukhov and Raevskii (1935) established that in the little suslik (*Citellus pygmaeus*), the young animals are most inclined to emigrate. By means of banding, Pavlinin (1948) established that young moles have a tendency to go great distances from the place of their birth. Of 112 newly caught and banded moles, 79% of those found at a distance of 50–500 m from the point of release were adults; 58% were subadults. At a distance greater than 500 m, only subadults were encountered. Among the moles caught in the area of banding, 71% proved to be marked adults, and only 47% were subadults.

Nasimovich *et al.* (1948), in an analysis of the question of the migration of Norwegian lemmings, wrote: "Subadults and juveniles emigrate in especially large numbers from the settlements; therefore, the summer–autumn migrations of lemmings are primarily the resettlement of the young." The data of Raevskii (1947) relate to the resettlement of young sables out of the locality of the maternal nest. In some mammals, sexual maturation comes at a strictly defined age and is relatively late (in the beaver, marmot, common hamster, mole, and shrew). Shrews, for example, live 13–15 months, and do not mature sexually until 8–10 months. A relatively long period of sexual maturation favors a broader resettlement of young individuals, with all the ensuing biological consequences.

It should be noted especially that in a number of species, individuals of different sexes and ages either have different regions of hibernation or abandon them at different times. In some species, the females and males that constitute a single population are subjected to different external conditions at certain periods. In the northern deer, old males and barren females pass the winter farther south than other individuals. Among the seals, the mature males are the first to arrive at the Komandorskie Islands in the spring, and in 3–4 weeks the sexually immature males arrive, and last of all the sexually immature females. The females of the sperm whale almost never go beyond the limits of the tropics and subtropics, but in the warm months of the year, the males migrate north, to the Barents Sea and the shores of Kamchatka (Tomilin, 1938). In Tadzhikistan, the wild boar spend the winter and spring in lowland thickets. In the summer, the young animals and the middle-aged males migrate to the hills, but the mothers with nursing young and the old lone individuals remain in the low-lands. In a number of species of mountain sheep, females with young remain at comparatively low heights; adult males and juveniles inhabit the upper zones, going up to the high mountains (Tsalkin, 1945).

Analogous examples could be cited by any zoologist from his own practical experience. A very important conclusion follows from these observations. Since males and females lead different forms of life, they are inevitably subjected to different selection forces. Because they are, their genetic makeup cannot be the same, which unavoidably leads to a general increase in the population's genetic heterogeneity. From this viewpoint, new meaning attaches to the well-known facts on the marked ecological and morphophysiological differences between males and females, which is especially clearly expressed in fish.

It is known that in a great number of fish species (salmon, deepwater anglers, certain Labridae, and others), along with the normal males in the population, there are dwarf males that differ radically in both morphology and form of life. The best example is the resident dwarf males of the migratory species of whitefish. Normal reproductive functioning has been demonstrated in dwarf males; their numbers fluctuate, depending on external conditions (Evropeitseva, 1962), and are correlated with the numbers of migratory males. The mean relative numbers of the males are apparently fixed by genetic mechanisms, since they constitute a fairly significant percentage of the young reared under artificial conditions (observations on the Koura River salmon, *Salmo trutta caspius*). Of course, dwarf males and normal females undergo completely different selection forces. Their genetic makeup cannot but be different. On the strength of the presence of dwarf males in the populations, the population's genetic heterogeneity is maintained, and the richness of the store from which selection draws the resources for the continuous perfection of living organisms is increased.*

For mobile animals, migrations and displacements are undoubtedly one of the basic mechanisms for the maintenance of a population's genetic heterogeneity and an obstacle to the impoverishment of the common gene pool. A large number of almost sessile animals are known, however, for which this mechanism can have only limited significance. Here, another mechanism comes into effect, the essence of which is that pairs are formed from animals belonging to different generations. Since the genetic structure of different generations is different (for details, see Chapter V), the ultimate effect of this mechanism should be very similar to the migration mechanism for the maintenance of genetic heterogeneity.

* The presence of dwarf males in the population can have another biological meaning also, which is the economizing of the population's food resources. If this factor also has significance, however, the role of dwarf males in the maintenance of the population's heterogeneity remains undoubted. It is not so important which factor is primary and which is secondary. Both (the economy of resources and the maintenance of genetic heterogeneity) can have significance. In such a case, we encounter the unique multifunctionality of the species' ecological features.

As is known, in a whole series of mammals, fights for females take place between males during the rutting season. The sexually mature young males do not take part in reproduction until they achieve full physiological sexual development and can stand up to the males of mature age. Several examples illustrate this.

Female bison achieve sexual maturity at 3–4 years of age, males in the 4th year, but the males begin reproductive activities no earlier than the age of 6–7 years, since they are driven off by the older males. A similar situation is observed in the wild boars: males and females mature at the same time, as a rule, but the young males are allowed to reproduce only after reaching 6–7 years of age. Relative to the Siberian mountain goat, Tsalkin (1950) writes as follows: "Females achieve sexual maturity at about one-half year, and by the second year often already have young. The males reach sexual maturity at the same time, as a rule. They begin direct participation in reproduction only at a significantly later time, however, since the bigger and older males drive them off." A completely analogous phenomenon is observed in the Pinnipedia. Female fur seals reproduce in their 3rd year; the males reach sexual maturity at the same time. The male seals, however, do not enter into reproductive activity until 7 years of age, and they occupy different beaches. The struggle of males for females also takes place among carnivorous animals (wolf, bear, tiger), although it is in a less overt form. This gives us reason to assume that among them, also, the reproductive males are those that have already achieved the full development of their powers.

Everything that has been cited indicates that in a large group of mammals the younger males do not participate in reproduction, and the young females mate with older males. On the other hand, numerous observations show that old males, who still have sexual potency, are forced to yield their place to younger reproductives. This indicates that the oldest males are excluded from reproduction, just as the youngest are. Because of the higher natural mortality of mature males compared with mature females, their average life span is, as a rule, less than that of females.

A similar phenomenon was recently described in the black grouse by Chel'tsov-Bebutov (1965) under the designation "age cross." The author showed that the complex character of the black grouse's breeding grounds leads not to random pairing, but "to preferential encounters between defined and, what is especially important, physiologically heterogeneous groups in the population: the less active males that have been expelled from the breeding grounds have a better chance of impregnating the most sexually active females, since the less active females, which have a lower threshold of irritation, are more often mated by the strongest and most active black cocks, which remain the conquerors of the breeding grounds."

Comparing the "activity" of the birds with their age, the author arrives at the very important conclusion that the breeding grounds of the black grouse ensure the preferential pairing of mature females with young males and of young females with older cocks. Chel'tsov-Bebutov assumes that the basis of the age cross is the increase in vitality of offspring from parents of different ages. This supposition is not devoid of basis, since it is supported by several observations on domestic animals (Anorova, 1959, 1960, 1964; Briges, 1953; Vakhrushev and Volkov, 1945; Milovanov, 1950; Starkov, 1952; and others). Whatever the primary cause of the age cross, however, its most important genetic consequence is the increase of population genetic heterogeneity.

A more complex question concerns the rules that govern selection of pairs of animals that do not possess sexual dimorphism and for which the struggle of males for females is not characteristic. it would seem that in this case, there are no general rules, and everything is decided by chance. Our observations on rodents showed that this is not so. It is known that female rodents mature somewhat earlier than the males. We will show with a concrete example what this leads to.

Under the conditions of the forest steppe in the Transurals, where we conducted our investigations, the first litter of the water vole is born at the beginning of May or (in some years) at the end of April. The females of the latter generation mature in the first half of June. At this time, the young males are not yet sexually mature, and therefore the young females can be mated only by older males (those who have over-wintered). This actually takes place, since at the time the young males are still sexually immature, the overwhelming majority of females are found to be pregnant. Thus, for example, in 1951 in the Kurgansk region, the first sexually mature male born in the current year was obtained on the 15th of June. Pregnant females, meanwhile, had already been obtained within the first 10 days of June, and in the latter days of May, several young females in heat were found. By the time the young males had matured, some of the young females had already reared young. Thus, on the 15th of June in the Zverino-golovskii district, a young female was obtained that was both pregnant and lactating. Analogous phenomena were also observed in the Kurgansk region in 1950. We began to encounter sexually mature young males only from the 19th of June, whereas at the end of the first 10 days of June, some of the young females were already pregnant for the second time (a combination of lactation and pregnancy). There is no doubt that the males that had achieved sexual maturity quickly entered into reproductive activity, since the (old) males that had overwintered were already so few at this time that if the young males had not already entered the role of progenitors, there would have been a general state of unmated females, which did not occur.

At first, after the maturation of the young males, the females of this generation are already pregnant. Consequently, the young males mate for the most part with females that have overwintered—females that at this time, with full justification, can be called old. The water voles of the second litter are born in June. Far from all the individuals of this generation succeed in reaching sexual maturity in the year of their birth; females preferentially do. Consequently, in this case, the young females are mated by older males.

Observations showed that exactly the same picture also emerges in an analysis of the "pair makeup" of other rodent species (*M. oeconomus,* the narrow-skulled vole, the Asiatic red-backed vole, and others). According to the data of Sludskii (1948), based on a study of a very large amount of material, male muskrats never mature in the year of their birth, while at the same time, some of the females are bearing litters at no more than 4 months of age. Naturally, they can be mated only by older males. Our considerably more modest data support Sludskii's conclusions for the northern forest steppe of the Transurals.

The examples cited show that the different rates of sexual maturation lead to a situation in which pairs are formed, for the most part, between animals of different ages and different generations. A consequence of this is the continuous renewal of the population's gene pool, even in the case in which in different generations it appears to be substantially disrupted. It seems highly characteristic that the rates of sexual maturation of males and females are different in all animals. The examples we cited concern mammals. G. P. Nikol'skii (1965) noted as a well-known fact the different rates of sexual maturation of fish. He also cites a number of specific examples, clearly showing that the formation of pairs of animals from different generations becomes inevitable in a number of cases. "Thus, for example, in the common carp, under exceedingly favorable growth conditions, the difference in the time of maturation of males and females is smoothed over (under ordinary conditions, the males mature somewhat earlier than the females), and both sexes mature at the same age, 1+ (Schäperclaus, 1953). Under these conditions, the relative number of males in the younger age groups diminishes, but the number of older males grows" (p. 161). This observation emphasizes that the conditions for the formation of pairs of the same age are created only in the most favorable environment. It is possible that this principle is also prevalent in other animals. Its biological meaning is clear from the position developed here. Without doubt, the sexual maturation of animals of the same generation at different ages promotes the maintenance of the genetic heterogeneity of populations of fish.

There can scarcely be a doubt that such wide distribution of such an important biological regularity as the different rates of maturation of the sexes is not accidental. It is logical to assume that fundamental biological rules lie at the basis of this phenomenon, which is almost as general as the sexuality of almost all living creatures. The factual analysis conducted makes it appear probable that this fundamental biological rule is the mechanism of the formation of pairs, which prevents the possibility of impoverishing the gene pools of animal populations.

It is possible that the maintenance of genetic heterogeneity is also determined, in some cases, by ethological mechanisms. On this question, there is much that is still not clear, but material is gradually accumulating that indicates that selection of marriage partners is not subject to simple chance. Nor does this situation prevail only among animals for which mating battles for the females are characteristic. Selective pairing is displayed in protozoa (Jennings, 1911), insects (Tower, 1906; Petit, 1956), amphibians (Sawada, 1963), birds (O'Donald, 1959), and mammals (Frederickson and Birnbaum, 1956; Mainardi, 1964; L. Levine and Lascher, 1965; and others). The majority of observations seem to indicate that selective pairing is realized on the basis of morphological similarity. This is evidenced, in particular, in the fact that when a choice is possible, pairs are formed of animals of the same, not different, subspecies (Mainardi *et al.,* 1965). A more detailed investigation by the same group (Mainardi, 1964; Mainardi *et al.,* 1965), however, showed that within subspecies, the females prefer to pair with genetically unrelated males. A first attempt to evaluate the significance of selective pairing in the development of a population was made with the help of a computer. Data were obtained showing that if the marriage partners are similar, one of the alleles (usually the dominant) is fixed; in the opposite case, a stable polymorphism is established (Mainardi *et al.,* 1966). No less interesting are the observations that the combination of different genotypes in the formation of *Drosophila* pairs changes with a change of external conditions (Petit, 1956). It is possible that a broadening of research in this direction will help elucidate new mechanisms in the maintenance of a population's genetic heterogeneity. Males and females are distinct, not only physiologically, but also ecologically. Their relationship to the environment is different. The latter difference is especially noticeable in their different mortality. Let us note only a few of the most interesting observations along this line. On the basis of a study of the literature on the sex ratio in 54 species and 12 races of mammals, Kubantsev (1964*a,b*) arrived at the conclusion that in the majority of species, more males are born than females. It is known, however, that in the majority of populations, the sex ratio of adult animals is approximately 1:1. This in itself indicates the differential mortality of animals of different sexes. The increased mortality of males is manifested differently under different conditions. In the black-tailed deer (*Odicoilens hemionus columbianus*), it is shown that at the very earliest age, the males die with significantly greater frequency than the females (Taber and Dasmann, 1954). This greater mortality is explained by their greater activity, which, in turn, is determined by a higher metabolic rate. During worsened environmental conditions, therefore, the sex ratio changes sharply in favor of the females. The unfavorable influence of low temperature causes an increase in the mortality of sexually mature males in mice, but no differences are observed in survival among the young (Zarron and Denison, 1956). A relatively greater mortality is also observed when there is a serious intervention in the life of the population, which occurs, for example, during the control of agricultural pests. The experiments of Kalabukhov (1944*b*) showed that during chemical treatment of the land, the common vole (*Microtus arvalis*) is contaminated, 30–50% of the rodents remaining alive, principally the females, the majority of which are pregnant and nursing. Later, Kalabukhov and his co-workers (Kalabukhov *et al.*, 1950) showed that pregnant females of the little suslik (*Citellus pygmeaus*) take poisoned bait reluctantly, and their mortality during chemical treatment of contaminated territories is immeasurably less than the mortality of males and active juveniles.

In some species, the regular excess of male mortality over female leads to the origin of certain mechanisms to promote the numerical dominance of males at birth. This is observed, for example, in muskrats. According to the observations of Okolovich and Korsakov (1951), in many districts, the sex ratio at birth is approximately 60:40.

One should not necessarily think, however, that the mortality of males is always higher than that of females. A sharp increase in the mortality of females is observed, for example, during hibernation of some species of bats (*Pipistrellus subflavus*), which leads to a reduction of their relative numbers to 20% (Davis, 1959). It is even more interesting that females are sometimes found to be more sensitive to unfavorable conditions as early as the embryonic stage of development, which leads, of course, to the numerical superiority of males among the young (W. Zimmermann, 1963; observations on chinchillas).

Changes in environmental conditions can apparently elicit a change in the population's sex ratio not only by means of a change in the relative mortality of animals of different sexes, but also by means of a change in the sex ratio at birth. The influence of the environment on the sex ratio can be both indirect and direct. A special symposium devoted to the problem of "parental age and offspring" showed that the age of the parents determines, to a significant extent, the properties of the offspring (Cowdry, 1954; symposium results; see also Uda, 1957). Naturally, this also indirectly influences the sex ratio. It becomes clear that the population's age structure can exert an influence on the sex ratio of young animals. The basis of the environment's direct influence on the sex ratio is a difference in the reaction of the sex chromosomes to a change in the organism's biochemistry (Uda, 1957). In particular, it is shown that the pH of the blood exerts such a strong influence on the sex ratio that the prerequisites are created for regulation of the number of males in the offspring by means of selection (Whirter, 1956). Astaurov's well-known experiments on the regulation of sex in the silkworm (see Astaurov, 1963, for a review) also give information on the different reactions of the X and Y chromosomes to external influence.

The observations cited make clear the influence on the sex ratio of such a factor of the external environment as a change in diet regime, an influence that has been detected even in humans (Uda, 1957), and has been observed repeatedly in domestic animals (Milovanov, 1950; Lysov and Pis'mennaya, 1951; Lukina, 1953; Aver'yanov *et al.,* 1952; and others).

The origin of complex genetic mechanisms for the regulation of the sex ratio that is observed in some animals attests to the exceptionally great significance that the regulation of a definite ratio of males and females has. Research on crustaceans (Copepoda) showed (Battaglia, 1965) that in *Tisbe reticulata,* a reduction of the population's heterogeneity inevitably leads to a reduction of the relative number of females. Ecologically, this is explained naturally by the fact that at high density, the probability of females being impregnated is high, even if the number of males is low (*T. reticulata* is polygamous). When the density of the population is lowered, the danger arises that some of the females will remain unmated and that their number will increase. Experiments showed that when *T. reticulata* is inbred, selection to increase the relative number of males proved to be effective; it proved to be impossible, however, to increase the number of females under these conditions. Battaglia succeeded in establishing not only the ecological meaning of this phenomenon, but also the genetic meaning. It was found that in the copepod, sex is determined multifactorially: females are determined by several dominant genes, males by recessive genes. The number of males, therefore, is determined by the degree of homozygosity in the population. When the numbers of the population are diminished, there is a lowering of the population's genetic heterogeneity, which is accompanied by a relative increase in the number of males, and this, for the reasons given above, decreases the possibility of females remaining unmated. Thus, in the Copepoda, a clear genetic mechanism has been developed that determines the optimal correlation among population density, genetic structure, and sex ratio. It is very probable that analogous mechanisms exist among other animals also, including other vertebrates. In particular, the interesting work of Stehr (1964), which specifically studied the role of concrete mechanisms for sex determination in microevolution, addresses itself to this question. Observations showing that different sex ratios are characteristic of genetically different laboratory populations (Levy, 1965) also belong among this research.

It is natural that differential mortality and differences in the sex ratio at birth should have as consequences both differences in the dynamics of the number of animals of each sex and differences in the dynamics of the structure of their genetic makeup. The experimental investigations of Petrusewicz (1958) made this clear. He showed that given an equal sex ratio at birth and increased mortality of the males, in laboratory populations of mice the rate of changes in numbers and the amplitude of their fluctuations proved to be higher for males in 43 cases out of 47.

As a consequence of the different dynamics of numbers, the genetic makeup of the males and females should be different. Hence, the maintenance of the optimal sex ratio is, in addition, an important mechanism for the maintenance of the population's genetic heterogeneity. There is not sufficient basis to speak of a special role for males and females in this process, as Geodakyan (1965) does, attributing to the males the responsibility for the quality, and to the females the responsibility for the quantity, of offspring. It can be asserted with full justification, however, that males and females are not only physiologically, but also genetically, different groups of animals. One may therefore agree with the authors who draw an analogy between different clones of one-celled organisms, reproducing sexually, and the sexes of higher organisms (Kallmus and Smith, 1960). Hence, the necessity for an ecological analysis of the consequences of a disruption of the normal sex ratio ensues. Investigations of this question are still in the very early stages, but already there are data clearly indicating that ecological mechanisms play a leading role in the maintenance of a population's genetic heterogeneity. On the other hand, the significance of these mechanisms makes clear a number of general biological phenomena of the most widespread occurrence (sexuality in all the living world, different rates of maturation of males and females, a different ecology of males and females, including its extreme form, the dwarfism of males, and so forth).

These findings also lead to a more general conclusion concerning the significance of the genetic heterogeneity of a population. Division of a species into two sexes lowers the general productivity of the population, since it reduces the number of individuals that bear offspring. This makes it understandable why, even among higher animals, from time to time parthenogenetic populations are observed. The observations of Darevskii, (Darevskii, 1964; Darevskii and Kulikova, 1961), who had studied the ecology of parthenogenetic populations of the rock lizard (*Lacerta saxicola*) in detail, made this completely clear.

On the whole, however, parthenogenesis is not widely found among higher forms of life. The causes of this are well known. Only cross-fertilization creates the prerequisites for the formation of diverse genetic variants, on the basis of which the optimal genotypes for the given conditions are formed, and reduces to a minimum the possibility of the birth of inviable organisms.* This process, however, could be provided by a hermaphroditic population consisting of individuals incapable of self-fertilization. In this case, the advantage of sexual reproduction would be maintained, but the potential productivity of the population would be doubled. Moreover, hermaphroditism removes a number of difficulties connected with the meeting of mates in populations of small size. Nevertheless, nature took a different path—the path of sexual reproduction. The sole explanation of this admittedly surprising wastefulness of nature is that division of the species into two genetically different groups (males and females) inevitably leads to the origin of physiological differences as well.† In turn, these differences lead just as inevitably to ecological differences that are, as we attempted to show, a guarantee that the genetic heterogeneity of the population will be maintained, even under extremely unfavorable environmental conditions accompanied by a severe reduction in population size. It follows that the importance of maintaining the population's genetic heterogeneity is so great that in the final analysis, it compensates for the reduction of the population's reproductive potential by one-half.

The genetic heterogeneity of populations is the prerequisite for their evolutionary reorganizations. Natural selection, however, cannot work on credit. This means that the genetic heterogeneity of populations not only is

* At present, this question seems more complex than in the recent past. In an asexually reproducing population, two beneficial mutations can be combined only if one of them arises in the offspring of individuals who had earlier undergone the other mutation. Among organisms reproducing sexually, both mutations can be combined as a result of recombination. Mathematical analysis (Crow and Kimura, 1965) showed that the sexual process is advantageous in those cases in which the joint action of mutations strengthens the beneficial effect of each of them, the mutants reproduce at a high rate, and the population size is great. On the other hand, in those cases in which the individual action of the mutations is negative and the joint action is positive, recombination can turn out to be harmful. The authors cited doubt the unconditional usefulness of the sexual process, and allow the possibility that diploidy should be examined as a defense mechanism against somatic mutations. The authors' general conclusions are supported by the mathematical analysis of Tomlinson (1966). He showed that when a small number of individuals is scattered about a large territory, parthenogenesis and hermaphroditism are avantageous. If the doubts expressed by Crow and Kimura contain a grain of truth, this only gives impetus to the necessity to attempt an explanation of the cause of sexuality in all the living world.

† It is highly significant that physiological differences between the sexes are immeasurably greater in higher animals than in lower.

a prerequisite of their reorganizations, but also increases the viability of the populations at the current moment in their history. The judiciousness of the viewpoint developed is supported by the phenomena accompanying sexuality, to which we turned our attention.

We arrive at the conclusion that the primary cause of the division of the sexes is the necessity to maintain, by all means possible, the maximal heterogeneity of the population. On this basis, there arose special anatomical and physiological adaptations that completed the division of labor between males and females. This division achieves its greatest development among the mammals. The major mechanisms for maintaining the genetic heterogeneity of populations are ecological. In an overwhelming number of cases, they exclude the possibility of the loss to the population of evolutionary plasticity and ensure the possibility of rapid adaptive changes of the population's genetic structure under changes in the environmental conditions—changes in the direction of selection. The ecological mechanisms for maintaining the population's genetic heterogeneity, therefore, probably play no less substantial a role in the evolution of the species than the mechanisms ensuring the direct reorganization of populations.

CHAPTER V

Ecological Mechanisms for the Reorganization of the Genetic Structure of Populations

---

The main task of evolutionary ecology is to establish how the population structure of a species influences the course of the evolutionary process. This task can be formulated differently, more concretely: what is the interrelationship between the ecological and genetic structure of a population; how does a change in the population's ecological structure affect its genetic makeup? In the process of resolving this task, the ecological mechanisms of microevolution can be completely uncovered.

1. Role of the Dynamics of Population Age Structure in the Reorganization of Its Genetic Makeup

The maintenance of an optimal population age structure is one of the basic mechanisms in the adaptation of animals to the specific conditions of their habitats. Different aspects of this question are being intensively studied at the present time. An enormous literature is devoted to them. Study of the population age structure has no less interest for an understanding of the mechanisms of the evolutionary process, however. This problem is studied to a significantly lesser degree.

Seasonal changes in the morphophysiological properties of living animals (the life cycles of which span several years at least) have long been well studied. Changes in the heat-insulating properties of the skin or coat, in the level of energy exchange, in the quantity and chemical composition of reserve nutrient substances and vitamins, in the need for food resources, in the activity of the most important organs of the endocrine system, and in the general reactions (including behavioral) to a change in environmental conditions—this is a far from complete list of the substantive physiological changes that are easily observed in a comparison of animals at different stages in the seasonal cycle of their activity.

The same thing is also observed during a study of small animals with short life cycles. In this case, however, the changes testify not only to changes of the physiological features of the animals in the process of their development, but also to the morphophysiological specificity of animals of different generations. Small rodents, Insectivora, and a few others are the most poorly studied group of animals in the full sense of the word. They are ephemeral. In an overwhelming number of species within these groups, the overwintering individuals produce progeny in the spring and die in the middle of the summer. The autumn population consists of other animals, born in the second half of the summer. (That some individuals last longer because of their size and live a full year does not change the general picture.) The autumn population is highly specialized. Animals of this generation possess an even deeper complex of morphophysiological features than autumn animals of long-lived species. And this is understandable, for, from the moment of their birth, they develop under peculiar environmental conditions and fulfill a completely determined ecological function: they must live through the winter, produce offspring in the spring, and pass on the torch of life to the next generations. It is not even possible to enumerate all the features that distinguish autumn populations of, for example, the vole, from spring populations. Indeed, there is perhaps no need to do so, since material on this species is widely published. The results obtained in our laboratory regarding this question were recently published in a collective article (Shvarts, *et al.*, 1964*b*). The replacement of biologically specific generations ensures a fuller adaptation of the population as a whole to seasonal changes of environmental conditions than is possible in a species possessing a longer life span. The diagrams presented (Figs. 2 and 3) show the complexity of the replacement of seasonal generations of rodents in different biomes.

We do not always know what specific causes determine the morphophysiological features of specific generations of rodents, but we do know that they can all be reduced to two principal differences:

1. The morphophysiological specificity of seasonal generations is the result of the organism's direct reaction to a change in environmental conditions. A partial manifestation of this rule is the influence of the mother's physiological state on the organism of the offspring (investigations in a number of laboratories, including ours, allow us to assume that certain

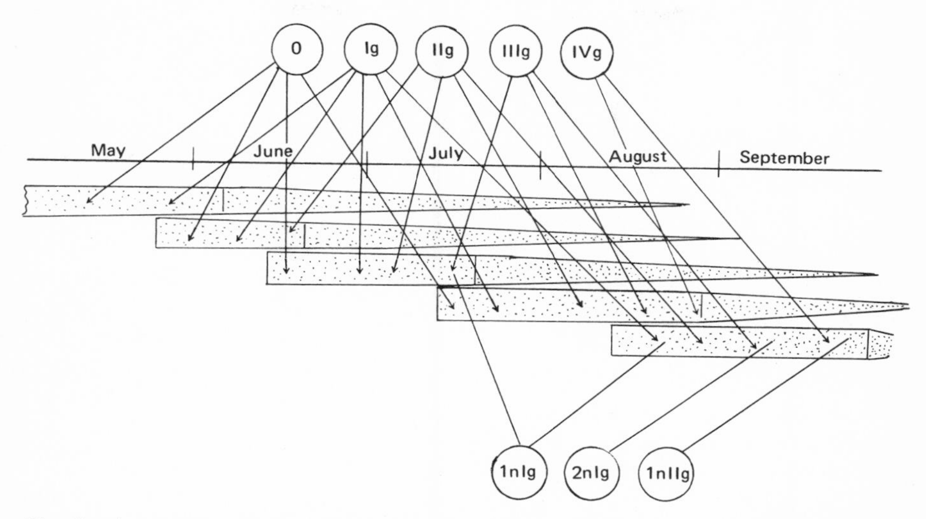

Fig. 2. Age structure of populations of the vole, *M. oeconomus*, in the far north: (O) Overwintering; (Ig–IVg) 1st–4th generation born to the overwintering voles; (1nIg) 1st litters born to the first generation; (2nIg) second litters born to the first generation; (1nIIg) first litters born to the second generation.

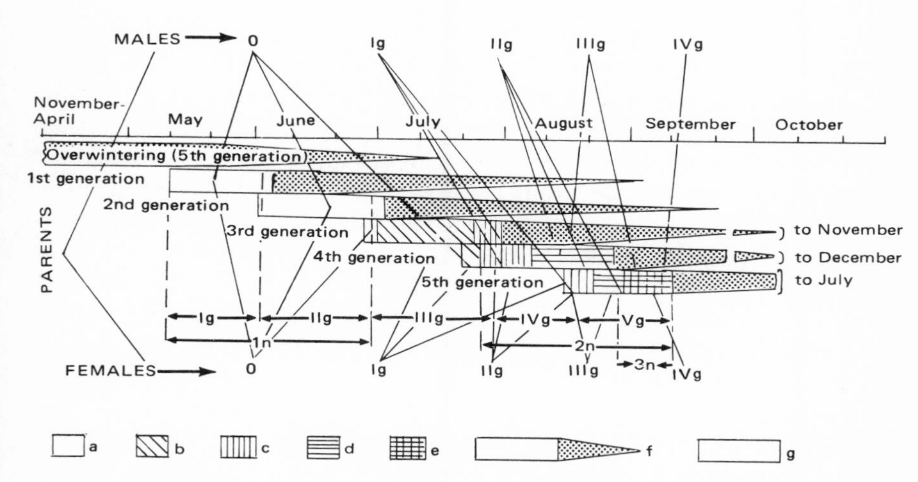

Fig. 3. Population structure and origin of generations of field mice: (O) Overwintering; (Ig–IVg) 1st–4th generation born to overwintering voles; (1n–3n) 1st–3rd litters; (a) offspring from overwintering females; (b–e) offspring from females; (b) 1st generation; (c) 2nd generation; (d) 3rd generation; (e) 4th generation; (f) life span of the generation; (g) usual time for birth of litter.

physiological features of the autumn generations can be explained by features of their mothers, which are animals born in the early spring).

2. The morphophysiological specificity of seasonal generations is the result of the restructuring of the population's genetic structure, and it is connected with changes in the direction of selection at different stages in the species' life cycle.

There is no doubt of the significance of the first factor. It is demonstrated by a large series of experiments showing that changes in the conditions of existence can elicit an "imitation" of seasonal changes in the morphophysiological features of a species at any time of the year. It should be noted, however, that such "untimely changes" are nevertheless usually less markedly expressed than true seasonal changes.

Having established the large role that the direct influence of the conditions of existence plays in the formation of the specificity of seasonal generations, we cannot consider our task fulfilled, since there is no doubt that in any natural situations, the specific features of the animal are determined not only by environmental conditions, but also by features of the genotype. That the features of seasonal generations reflect the specific conditions of their development does not exclude the possibility of a parallel restructuring of the population genetic structure.

It has been shown that the seasonal cyclicity of the vital activity of the population as a whole can be accompanied by a change in its genetic structure. Naturally, this rule can be demonstrated most clearly in polymorphic populations. Different authors have convincingly shown by the representatives of different classes of animals that the population genetic structure changes from spring to autumn and from autumn to spring: individuals that can endure the winter better predominate in the spring; the more fertile animals predominate in the autumn. Naturally, the prevalence in the population of certain genetic variants changes not only with the season, but also with the year ("chronographic variability," Shvarts, 1963*b*). Kirikov (1934) seems first to have ascertained that the "black hamster" (the black phase of the common hamster, *Cricetus cricetus*) is distributed predominantly in the hilly regions of Bashkir and in the northern limit of the European part of the species range. The author considers it probable that the black hamster is more adapted to a cold, dry climate. Gershenzon (1945*a*) established that the melanistic hamsters reach higher numbers in autumn, and die in greater numbers in winter. V. N. Pavlinin (personal communication) collected interesting material on the incidence of different phases of the hamster, according to data from the Sarapul fur base. An analysis of these data that we conducted allows us to draw some interesting conclusions. In different years, the ratio of black to variegated phases changes very abruptly. Thus, for example, in March 1953, in Chelyabinsk province, 253 variegated and

74 black hamsters were found; in December, 77 variegated and 185 black. In Bashkir, where black hamsters predominate in general, during the summers of particular years, the numbers of variegated exceed the numbers of black, but in the autumn, the black again begin to dominate.

Many such facts could be cited, but there are others still more interesting that show that the ratio of phases changes not only with the season, but also with the year. In the autumn of 1952 in the territory of Bashkir, ASSR, there were found 1357 variegated hamsters and 1724 (56%) black, but in the autumn of 1951, there had been, respectively, 7248 and 3640 (34%). Dmitrieva (1949) showed that the ratio of grey long-tailed and yellow short-tailed house mice changes according to the year. The author suggests that the yellow "phase" is more susceptible to rabbit fever and to unfavorable climatic conditions. According to our observations, the relative numbers of the melanists among the water rats in the forest steppe of the Transurals changes from year to year. In particular years, they constitute the majority of the population. Melanistic forms are encountered in almost the entire range of the vulpine phalanger, *Trichosurus vulpecula,* but they are few in number, with the exception of the Tasmanian population, in which the black phase predominates. Pearson (1938) established that the melanistic forms can bear the cold and higher humidity better. In the mollusks (*Cepaea hortensis* and *C. nemoralis*), it was shown (Schnetter, 1950; LaMotte, 1959) that different-colored variants possess different sensitivity to changes in humidity. Physiological differences between the blue and green caterpillars of some lepidopterans determine their different sensitivities to poisons (McEwen and Splittstosser, 1964). Analogous causes explain the different mortality of black and striped tritons under different conditions (Test, 1954).

Among amphibians, the lake frog (*R. ridibunda*) may be considered a dimorphic species, but the polymorphism appears only at an early age in this species. Only those frogs having just completed metamorphosis are clearly divided into two groups: in some, the back is all one color; in others, a narrow white stripe runs along the spine (forma striata). A clear dimorphism is also observed within isolated populations in the young tadpoles, which gives reason to assume that it is connected with individuals produced by the same parents. We have studied some internal indices of the young lake frogs from the Stepnyi region of Aktyubinsk province. It was found that two "phases" of the frogs are noticeably distinguishable by the important trait of the relative size of the liver. In frogs of the same age weighing 2.2–3.0 g, the relative weight of the liver is, on the average, 50.2% in frogs with a stripe, and 43.0% in frogs without.

Eisentraut (1929) established that the melanistic forms of *Lacerta lilfordi* are distinguished by relatively longer intestines. The author expresses the supposition that thanks to this, they possess a greater capability for utilizing plant food. It is natural that a change in environmental conditions elicits a change in the numerical ratio of biologically different phases. Seasonal changes in the genetic structure of populations have also been noted by other authors: Dubinin and Tinyakov (1947) in *Drosophila funebris*; S. Wright and Dobzhansky (1946) in *D. pseudoobscura*; Lukin (1966, and others) in *Pyrrhocoris apterus*; Timofeev-Resovskii (Timofeev-Resovskii and Svirezhev, 1966) in *Adalia bipunctata*; and others. Summing up theoretical investigations along this line, Timofeev-Resovskii (1964) wrote:

> The necessary basis for any form of polymorphism is a prolonged state of dynamic selective equilibrium between two or several genotypes. This equilibrium, in turn, is always based on different and competitive selection pressures on the three mutant forms existing for the same gene locus or chromosome in the population (the heterozygote and the two different homozygotes); or competitive and differently directed selection pressures on two or several different genotypes (from the general heterogeneous mass of individuals in the population) in different (in space or time) microconditions, available within the limits of the territory occupied by the population.

Thus, the study of polymorphism showed that a change in the conditions of existence and a corresponding change in the direction of selection led to a change in the genetic makeup of populations. In one season of the year, some genotypes predominate; in a different season, others. The question arises, is this an exclusive phenomenon, peculiar only to clearly polymorphic populations, in which the genetic differences among different individuals are manifested especially sharply, or is it characteristic of any population and not detected only because it is technically difficult to detect?

Thus, the first half of our task is to demonstrate the very fact of genetic reconstruction of the population as a general phenomenon.

In this, theoretical proofs have special significance, since even dozens of examples cannot in principle exclude the possibility that the observed phenomenon is unique.

Our system of proofs can be summarized as follows: Study of polymorphic populations showed that seasonal changes in the conditions of existence are really connected with a change in the direction of selection. This is the first premise. On the other hand, it has been demonstrated that the genetic heterogeneity of a population embraces every trait of the organism, including those such as fertility, rate of sexual maturation, growth rate, utilization of different nutritive substances, and so forth, the significance of which differs sharply in different seasons of the year.

Thus, for example, Maslennikova and Khromach (1954) showed that among rats, a strongly expressed individual variability in the need for vitamin B_2 is observed. It is easily assumed that under conditions of an insufficiency of foods containing this vitamin, individuals with a less markedly expressed need for the vitamin will predominate. Individual variants in the need for vitamin D are also interesting. Their hereditary perseverance is demonstrated (Harris, 1954). In populations of birds, genetic variability in regard to the capability for utilization of thiamine has been noted (Howes and Hutt, 1956).

It was demonstrated with precise experiments that the activity of choline esterase in the sensory regions of the cortex in rats is susceptible to individual variability: rats possessing great activity of the enzyme have a more clear-cut reaction to illumination (Krech *et al.*, 1954). In different genetic variants of the house mouse, the reaction of the mammary glands to estrone and progesterone proved to be different (Mixner and Turner, 1957); the receptivity of the tissues to the growth hormone of the hypophysis is also different (S. C. King, 1965). Moreover, even such traits as expressivity of sexual dimorphism (Korkman, 1957), the rate of sexual maturation at the end of the reproductive season (Pokrovskii, 1962), the preference for different foods, the selection of habitats (Wecker, 1964), wariness (Crowcroft, 1961), sensitivity to radiation (Bartlett *et al.*, 1966), and the intensity of biosynthesis of hormones (Badr and Spickett, 1965) are subject to individual variation and are determined genetically.

The heterogeneity of a population is a biological law, knowing no exceptions. Every trait of every organism is subordinate to it. A change in the direction of selection, therefore, inevitably elicits a change in the genetic structure of the population, and each generation becomes not only physiologically, but also genetically, specific. A change in the population's age structure leads, consequently, to a change in its genetic structure.

For an example, let us use the most stable traits—the cranial ones fixed hereditarily within a relatively narrow range of variability. In recent years, it has become clear that the most exact cranial features of animals are not characterized by absolute values of the individual traits or even of their proportions, but by the nature of a dependence between the general dimensions of the skull and its individual parts. This dependence is written as the equation $y = bx^a$, in which a is the allometric exponent. Recently, a great number of works have been published demonstrating the possibility of using a for taxonomic purposes (Hückinghaus, 1961, 1965; Röhrs, 1961).

Some of the results of these investigations are presented in the tables. It was shown that the allometric index is fixed by heredity within a significantly narrower range of variability than the absolute dimensions of the organs or the parts of the body, and it is practically unchanged by changes in environmental conditions. Frick (1961) divided a colony of white mice into two groups. One group developed under conditions demanding a sharp increase in physical stress; the other served as a control. As one would expect, in the experimental mice, the dimensions of the heart and kidneys were significantly increased, but the nature of the change in the dimensions of the organs concommitant with a change in body dimensions remained invariable. Analogous results were obtained by Ishchenko (1966, 1967), when he compared the dimensions of the heart in two species of the *oeconomus* vole from natural populations and laboratory populations (Fig. 4).

Under laboratory conditions, the dimensions of the organs are decreased (a result of the lowering of the metabolic level), but a is not changed. These and some other works show that the nature of the relative growth, expressed in a can serve as a very good index of genetic distinctions among populations, and can be used in a study of the dynamics of the population genetic structure.

Let us see whether the allometric exponent remains constant in the process of the replacement of seasonal generations of rats. For this, we will take advantage of data from the laboratory of V. V. Kucheruk, with their kind permission (an analysis of V. G. Ishchenko's material). This material is a magnificent series of skulls, collected in the Volga-Akhtubinsk floodlands in different years and at different times. The data are presented in Table 1.

The data show that within a species, the variability of a significantly exceeds the interspecific differences. In order to be convinced of this, it is

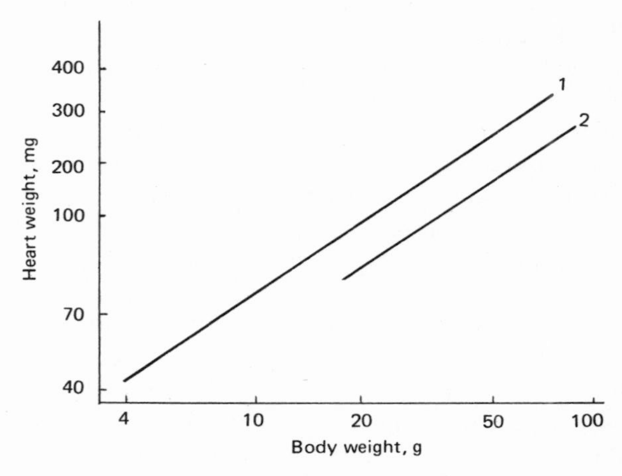

Fig. 4. Influence of external factors on the nature of the allometric growth of the heart of the *oeconomus* vole: (1) Sasykul'; (2) vivarium.

TABLE 1. Changes in the Allometric Index (mm) in a Population of *Arvicola terrestris*

Trait	Year	Spring	Autumn
Skull length	1952	0.480±0.008	0.585±0.014
	1953	0.371±0.011	0.460±0.007
	1954	0.508±0.010	0.486±0.016
Length of tooth row	1952	0.354±0.012	0.371+0.013
	1953	0.785±0.014	0.684±0.009
	1954	0.629±0.016	0.618±0.025
Width of interorbital distance	1952	−0.179±0.006	0.115±0.022
	1953	0.013±0.029	−0.475+0.037
	1954	−0.192±0.029	−0.394±0.026
Cheek width	1952	0.760±0.017	0.885±0.012
	1953	0.793±0.018	1.155±0.011
	1954	0.793±0.016	1.050±0.013
Height of the cerebrum	1952	0.768±0.019	0.515±0.017
	1953	0.695±0.054	0.640±0.009
	1954	0.568±0.034	0.571±0.011

sufficient to compare the data presented in Tables 1 and 2. Although this situation is not directly related to our theme, we will turn our attention to it. Once more, it is shown that without taking into account intrapopulation variability, it is often impossible to evaluate the differences among populations and even among species.

Within the scope of our theme, however, something else is more interesting. Experiments showed that the allometric exponent is very firmly fixed by heredity. Hereditary changes in *a* (changes in the proportions of the skull) are possible only under the most drastic changes in the conditions of the animals' existence, conditions influencing the rate of their growth (Shvarts, 1961*b*). It is completely clear, therefore, that if the observed

TABLE 2. Allometric Indices of the Interorbital Width of the Skulls of Different Species of Cats[1]

Species	*a*	*b*	Number of skulls
Felis ocreata	1.052	0.176	14
F. lynx	1.238	0.083	17
Panthera pardus	1.294	0.048	12
P. leo	0.885	0.441	11

[1] According to Röhrs, (1959).

changes in *a* are determined by predominantly nonhereditary mechanisms, especially marked differences would always be detected between spring and autumn populations, since replacement of the generations of animals born and raised under completely different conditions occurs between spring and autumn. Such changes do not occur from autumn to spring. During this time, the voles do not reproduce; consequently, these are the same animals, but their makeup has changed due to the dying out of certain groups of individuals. All cases in which the changes from autumn to spring are more significant than those from spring to autumn therefore indisputably reflect differential mortality, which leads to a change in the genetic makeup of a population. The data presented in Table 1 show that changes of this type are often manifested very clearly.

We have no basis to assume that the trait under investigation, which reflects some sort of change in the relative growth rate of different parts of the skull and of the skull as a whole, is unique in the observed situation. On the contrary, it may be asserted that precisely such data would be obtained during an analysis of other traits, more essential from an ecological viewpoint. Numerous examples of changes, according to the year, from the population's norm of variability testify to this. These examples cannot be fully explained by the direct reactions of animals to specific environmental conditions (Timofeev-Resovskii, 1940, 1964; Schmalhausen, 1946; Shvarts, 1959).

Thus, we arrive at the conclusion that the morphophysiological specificity of age and seasonal generations is determined not only by the specific conditions of their development, but also by the reorganization of the genetic structure of the populations. Fluctuations in the "quality" of a population are just as characteristic a property as are fluctuations in its numbers. These phenomena are linked: changes in the numbers (especially drastic changes) are accompanied by a change in the genetic makeup of the population. This leads us to important theoretical conclusions, which comprise the essential part of this chapter.

A change in a population's age structure is accompanied by a change in its genetic structure. As was indicated, numerous data from the literature and some of our own material testify to this. This means that if for some reason or other the mortality of animals in different age groups is substantially different, there will be a substantial change in the population genetic structure, with all the ensuing consequences. Let us conditionally call this process *age selection*. It is easy to imagine what kinds of interrelationships exist between the age structure of a population and a change in its genetic makeup. Let us use a hypothetical example of the selection of rodents for "cold resistance."

During the course of the winter, there occurs a shift in the population's genetic structure in the direction of "cold-resistant" individuals. (By what kinds of physiological causes this "cold resistance" is determined is not important to us in the given case, but in individual cases, it would be possible to say something concrete about this.) In the old group, therefore, such "cold-resistant" individuals will be relatively more numerous than in the young group. It is logical to assume that this predominance will be expressed especially sharply during more severe winters, when mainly "cold-resistant" individuals will appear stronger. Under these conditions, individual selection will work to the benefit of "cold-resistant" individuals, but age selection should act in the reverse direction, since during severe winters, mortality of the old animals will be especially marked. The final effect will depend on the relative strength of these two forms of selection. It is important, however, that they can be in opposing directions, and that situations are possible in which a mild winter will lead to a marked increase in the population of animals of the cold-resistant type. This result is not due to that form of selection to which the greatest (if not exclusive) significance is attached, but to a change in the population age structure. Experimental investigations of different aspects of this question, which we are just beginning, may be of great interest not only for theory (in particular, for the theory of microevolution), but also for certain practical fields. We will speak about this later. It is important to emphasize that a theoretical analysis leads us to the conclusion that a change in the ecological (in this case, age) structure of a population is inseparably linked with a change in its genetic structure, with all the consequences that ensue. A change in a population's ecological structure should, consequently, be regarded as the most important factor of the microevolutionary process.

The facts cited show that seasonal change in a population's genetic structure is a widespread, if not universal, phenomenon. In many animal species, it is manifested in the genetic distinctiveness of different generations. Hence, it follows that a change in the population's age structure inevitably leads to a change in the general genetic makeup of the population, and to its evolving more rapidly than it would under the influence of individual natural selection. Under relatively stable conditions, a change in the population's genetic structure, which is linked with the dynamics of its age constitution, has the character of fluctuations about a certain average over many years (analogous to the fluctuations in numbers). During a change in environmental conditions, age selection may appear to be a factor of rapid evolutionary reorganizations. A climatic change, for example, may lead to the falling out of individual generations and to a corresponding change in its genetic structure, which will not be reconstructed in the following years. As the examples cited above show, genetic differences among generations can exceed the average differences among populations in the course of thousands of generations reproducing independently, so it is not difficult to imagine what significance the phenomenon of "age selection" can have for microevolutionary rates. One must not forget, however, that "age selection" works against a background of individuals. It is just this circumstance that creates the genetic differences among generations. That individual selection during a change in the genetic structure of seasonal generations works with greater effectiveness than on differentiated populations is well known, since the developmental conditions of, for example, the summer generations of animals in the steppe and the tundra (to take an extreme example) are more similar than the developmental conditions of tundra animals of early spring and late summer. If one takes for comparison populations from closer biomes, the chorological and chronological differences in developmental conditions for the animals are clearly revealed. *Age selection (in the sense indicated above) creates the conditions for the mobilization of those changes that arise in populations under the action of individual selection.* Their combined action, in the final analysis, determines the rates of reorganization of populations. During lengthy (or irreversible) changes of climate, this reorganization is solidified and leads to evolutionary changes. Even a temporary change in the population's genetic structure, however, can have great evolutionary significance during a change in the population's spatial structure. We will analyze this question in detail in the third part of this chapter.

In order to get an idea of how widespread the phenomenon of age selection is, it is necessary, first, to consider the possible limits of the "usual" dynamics of a population's age structure and, second, the physiological distinctions among animals of different ages.

The first question is studied well enough that it is sufficient here to present two tables characterizing the dynamics of age makeup for animals with a long life span (foxes, Table 3) and for short-lived animals (voles,

TABLE 3. Number of Foxes at the Beginning of the Productive Season in the Yamal–Nenetskii National Region in Different Years[a]

Index	1955	1956	1957	1958	1959	1960	1961	1962[b]
Total number	42,861	35,893	54,895	34,025	26,094	37,480	41,743	19,026
Number of young in the given generation	31,082	27,409	43,334	6,936	18,961	29,886	28,453	259
Percentage of young	72.5	76.4	79	20.4	67.4	79.8	68.2	1.4
Young per adult pair	5.28	6.46	7.51	0.51	5.32	7.88	4.28	0.03

[a] According to the data of V. S. Smirnov.
[b] The accuracy of the data for 1962 is not high.

TABLE 4. Dependence of the Age Structure of Vole Populations on Weather Conditions in the Southern Transurals[a]

Species	Quantity of young (% of total weight)		Remarks
	1950	1951	
Water vole	1.2	22	In June 1950, only 13% of the young water vole females participated in reproduction. In 1951, at the same time of year, 57% of the young females bred, and 21% participated in reproduction twice.
Voles of the *oeconomus* group	none	2.8	
Narrow-skulled vole	12.0	23.0	
Asiatic red-backed vole	none	9.0	

[a] 1950: Spring was late; the snow had completely melted by the end of April. 1951: Spring was early; the snow had completely melted by the beginning of April. The percentage increase of individuals in the second half of May is shown.

Table 4). It is hardly necessary to comment on these tables. They clearly express the scale of the dynamics of population age structure. Analogous data have now been obtained for many species of ungulates (Skalon, 1960; Pimlott, 1961), pinnipeds (Laws, 1956), predators (Wood, 1959), rodents (Polyakov, 1964; Lavrova and Karaseva, 1956) birds (Pinowski, 1965), and other animals. Investigations in this direction are developing energetically, and factual material is accumulating rapidly. It is important, therefore, to emphasize that population age structure in individual species is often determined by the influence of predators (Lavrova and Karaseva, 1956; Baziev, 1967; Knight-Jones and Moyse, 1961; and others). It follows that a change in the numbers of one species influences not only the numbers of another, but also the structure of its populations, and consequently its genetic makeup as well. If differences in the genetic structure of different age groups are substantial, then, in the situations described above, they cannot but influence the dynamics of the genetic structure of the population as a whole.

The second question demands an ecological evaluation of those physiological differences that are detected during a comparison of animals of different ages. Since the solution to this problem creates the preconditions for the evaluation of the probable prevalence of age selection, we will dwell on it in considerable detail.

It is well known that many factors of the external environment that do not have any substantial effect on the life of adult animals lead to the death of the young. The drying up of reservoirs leads to the mass death of amphibian larvae, but does not substantially affect the numbers of adults. Spring frosts can sharply reduce the numbers of fledglings in very different bird species, while the numbers of adult individuals remain close to normal. Spring floods and autumn downpours have an especially disastrous effect on helpless mammalian young. Such facts are well known and are described in necessary detail for different species. They show that selective death is a common phenomenon. Differences in the mortality of young and adults with which we are concerned here in the evolutionary scheme, however, hold relatively less interest, since the genetic structure of the young group up to the transition to independent forms of life should be very close to the genetic structure of their parents. Probably, differences in the mortality of animals of different ages that have already passed on to independent forms of life have greater significance. There are significantly fewer data on this question, but the physiological differences among different age groups of animals are so significant that they should give rise to selective mortality.

Despite the absence of substantial changes in the capacity of the protoplasm to consume oxygen with age, at least within those ages that interest us here, the basal metabolism decreases with age, because of the decreasing number of functioning cells and a series of other causes. Accordingly, the capacity for maximal consumption of oxygen also decreases. The latter has special significance during an increase in physical stress.

This drop becomes noticeable at a relatively young age. In man, the maximal consumption of oxygen at age 40 is 20% less than at age 25. In middle age, substantial changes are observed in a number of other physiological functions: the oxygen saturation of the arterial blood diminishes; blood pressure increases; the content of lactic acid increases and the alkaline reserve of the blood is lowered (especially under conditions of increased physical stress); the elasticity of the blood vessels decreases; the maximum pulse decreases; the capacity of the intestines to absorb certain elements necessary for the normal functioning of the organism falls. With age, not only the metabolic intensity of the organism as a whole, but also the metabolic activity of the tissues and cells, declines. It is difficult to say at which age (it is different in different species, of course) this process takes on appreciable dimensions. In any case, the correspondence of the data obtained by different researchers using different methods shows that in rats, for example, in 2 years this drop was very substantial, even in comparison with fully adult animals of more than 1 year of age (Weinbach and Garbus, 1956; Barrows *et al.*, 1957).

It is especially important to emphasize that with age, those physiological features of the organism that determine its reaction to unfavorable or simply changing conditions of the external environment change. With age, the central nervous system is disturbed, and the rapidity of nervous impulses falls. A functional change occurs in the activity of the endocrine system, which is reflected in a progressive diminution of the dimensions of the cellular nuclei in the internal secretory glands, in a drop of mitotic activity, in the spread of connective tissue, and, as a consequence of the latter, in a drop in the secretion of certain hormones (Symposium of the Institute of Biology, 1956). Older animals possess a lowered thermoregulatory capacity and a relatively diminished capacity to create physiological reserves. All this cannot but lead, and actually does lead, to substantive differences in animals of different ages in their reactions to a change in external conditions.

This generalization can be very conveniently illustrated with the relationship of animals of different ages to three of the most important factors of the external environment: temperature, oxygen regime, and nutrition.

A sharp drop in the capability of older animals to adapt to lowered temperatures has been ascertained by a number of investigators. It was shown (Grad and Kral, 1957), for example, that mortality in mice of the C57B lines was much higher at the age of 16–22 months than at 4–9 months. The capability of rats to adapt to lower temperatures also decreases with age. In the authors' experiments, 60% of "adapted" older rats died in the course of weeks at a temperature at which not one of the "adapted" young rats died.

Similar results are obtained in a comparison of animals of different ages with regard to their reaction to a lowering of oxygen concentration in the atmosphere. It is established, for example, that guinea pigs of about 300 g weight are significantly less affected by a shortage of oxygen than are older animals of about 500 g weight. The different sensitivity of animals of different ages to the quality of food is illustrated by the increased requirement of older animals for vitamin B_1, which is one of the most important catalysts of the cellular oxidation–reduction systems (for a review of the data, see Shvarts, 1960*a,b*).

The data obtained allow us to suppose that the mortality of older age groups exceeds the mortality of younger animals. This was shown, for example, in a very general form in field mice by work utilizing an extremely refined methodology for the biometrical analysis of material (Hacker and Pearson, 1944). The differential extermination of animals of different ages by predators has been noted by several authors (Folitarek, 1948; Lavrova and Karaseva, 1956). There is reason to suppose that younger animals are more susceptible to disease than older (Polyakov and Pegel'man, 1950). This, apparently, is also true in regard to infection by some parasitic worms. Thus, for example, in different countries of western Europe, almost exclusively young hares are susceptible to coccidiosis.

Of special interest are data showing that the diverse relationships of animals of different ages to the habitat may be the basis of differential mortality.

Polyakov and Pegel'man (1950) showed that at a temperature of 35°C, when the subadults of the common vole die, younger animals grow energetically and do not noticeably lose viability. In correspondence to this, during the droughts common in Azerbaidzhan, there is heavy mortality at the older ages, and a general "rejuvenation" of the population occurs.

Older animals, however, can bear the unfavorable combination of winter conditions easily (Polyakov, 1956).

It is entirely natural that biological distinctions among animals of different ages inevitably lead to their different mortality rates. Unfortunately, the data on this question at the disposal of ecology are still very modest. If the dependence between age and the genetic structure of the population is established and the causes of selective mortality among the animals are clarified with the necessary exactness, this in itself will make it possible not only to see the genetic consequences of certain combinations of external conditions, but also to intervene in the beginning stages of the microevolutionary process. We believe that investigation of these processes is one of the most urgent tasks of evolutionary ecology.

In our investigations, conducted jointly with V. G. Ishchenko, we selected as our subject the sharp-nosed frog (*Rana terrestris*). Over almost its entire range, one encounters two genetic variants, *striata* and *maculata,* in populations of this frog. *Striata* has a well-marked dorsal stripe. *Maculata* does not have this stripe; spotted coloration of the back is characteristic for it. The investigations were conducted in the southern Urals, in the Il'menskii preserves, over the course of 2 years, in the same place and at the same time (beginning of August). We investigated a relatively isolated population of frogs in a small bog contiguous to Lake Miasovo. The subdivision of frogs into age groups was based on an analysis of the plotted distribution of body length. This method is not infallible, but, when working with massive amounts of material, it gives results that are completely satisfactory in their accuracy. The data obtained are presented in Table 5. Analysis of these data leads to the following conclusions:

In 1966, *striata* constituted a little less than 50% in all age groups. Differences in the genetic makeup of animals born in different years are not significant. A different situation developed in 1967. There were demonstrably fewer *striata* among the individuals born in that year than among the older age groups (the criteria are statistically significant, $t = 2.07$). In contrast, *striata* predominated heavily among the older frogs. The difference between age group 3+ and older groups is statistically absolutely significant ($t = 3.05$). These data are of interest in themselves. They underscore once more the situation that the genetic makeup of different generations of animals in one population varies, and this difference is substantial. On the other hand, these same data show that the genetic makeup of indi-

TABLE 5. Change (%) by Year in the Relative Incidence of the *striata* Variant in a Population of *Rana terrestris* (Il'menskii Preserve, Chelyabinskaya Region)

Age group	1966	1967
0+ (born this year)	39.1±2.43 (402)[a]	28.6±6.04 (56)
1+	49.2±4.35 (132)	42.6±6.33 (61)
2+	44.1±3.07 (261)	43.3±4.04 (150)
3+	44.4±3.70 (180)	37.1±3.31 (213)
4+	42.1±8.1 (38)	61.1±5.74 (72)
5+	100 (1)	64.9±7.84 (37)

[a] The number of individuals sampled is in parentheses.

vidual generations does not remain constant, but changes, and does so in a completely regular manner. This was clearly apparent in our material when we compared older frogs. In 1966, *striata* constituted 44.4% of the frogs of age 3+; in 1967, they constituted the majority (t = 2.29) of this generation (now the age 4+ group). The genetic makeup of the generation of the previous year changed in the same direction (the 4-year-old group of 1966 is compared with the 5-year-old group in 1967; the differences are significant, t = 2.24).

These data indicate that the intensity of selection under natural conditions is very significant, immeasurably more than the effectiveness that we judge on the basis of displacements of the mean norm of variability of the population as a whole. The cause is not only the regular change in the direction of selection under changes in environmental conditions, but also the different reaction of animals of different ages to analogous changes in the conditions of existence. The material presented allows us to assert that the relative mortality of *striata* and *maculata* is different in different age groups. In the older age groups, *maculata* are characterized by a higher mortality (possibly, only under those conditions that existed in the years of our work). *Striata,* therefore, significantly predominate in groups of frogs older than 4 years.

At present, we cannot conduct a full genetic analysis of our material, since we do not have data characterizing the genetic makeup of the population during reproduction and data characterizing the mortality of heterozygotes. For an analysis of the significance of the dynamics of population age structure in the microevolutionary process, however, we have sufficient information at our disposal. Let us use in the given case a well-known example of a theoretical experiment. Let us suppose that for some reason, the younger groups of frogs die out and the population begins to reestablish itself with animals of age 5+. This means that the cadre of reproductives will consist of almost 70% *striata,* which exceeds by 20% on the average (over all generations) the number of this form in the 1966 population. Such a displacement in the population genetic structure could occur through the action of individual selection, even of very high intensity, only after many years. As will be shown in detail below, we do not counterpose individual selection to age selection. Age selection (a change in the population age structure) mobilizes the genetic potential created by individual selection, and as a result increases its effectiveness manyfold. It can be said that age selection removes the contradiction between the intensity and effectiveness of natural selection.

The data we obtained also showed that sharp changes in the age structure of a polymorphic population, such as those we assumed in our theoretical experiment, are real. It is sufficient to turn one's attention to the marked reduction in the relative portion of the current year's offspring in 1967 (by comparison with 1966) and the no less marked increase in the number of oldest animals. Change in population age structure can be elicited not only by a different number of young successfully completing metamorphosis (in 1967 a large number of tadpoles died as a result of drought and attendant phenomena), but also by a change in environmental conditions in different parts of the territory occupied by the population. Observations showed that frogs of different ages inhabited different parts. Any local changes in environmental conditions (drought, early ice cover on the ground, and so forth) can therefore elicit a sharp change in the population age structure, aside from the possible differentiation of mortality occasioned by physiological features of animals of different ages. With this differentiation, a pronounced restructuring of the population's genetic makeup will turn out to be inevitable.

Naturally, change in the roles of different age groups in the maintenance of the population's numbers is determined not only by selective mortality, but also by a change in the nature of reproduction.

The dependence of the intensity of reproduction on different combinations of internal conditions is one of the most fully analyzed chapters of ecology. There is therefore no need to cite examples of the dependence of change in the structure of animal populations of different species on the number of young animals recruited into the population. The connection between the intensity of reproduction and the population genetic structure, however, has not yet been studied, and the lack of such study probably does not allow us the possibility of evaluating fully the significance of the ecological mechanisms of the beginning stages of a population's divergence.

One must keep in mind that intensification of reproduction, even from a purely ecological point of view (the dynamics of numbers), does not represent a simple increase in the relative abundance of young animals in the population. Ecological and endocrinological investigations clearly showed that a sharp intensification of reproduction is closely correlated with population density (Christian, 1961, 1963; Wynne-Edwards, 1962; and many others). After reduction in numbers, not only the fecundity and number of females participating in reproduction, but also the rate of sexual maturation of young animals, are increased. Dependence of the intensity of reproduction on population density is best studied in mammals, but there are observations showing that the dependence appears in other vertebrates (Fehringen, 1962) and in insects (Pajunen, 1966). There is already a significant body of literature on this question. The accumulated data clearly show that the change in intensity of reproduction connected with a change in density leads to a marked change in the population age structure. It is noteworthy that the specific manifestation of this rule even in very close species can be fundamentally different (Lidicker, 1965).

In some cases, a change in the age makeup of the reproducing animals exceeds the limits of the normal ecological characteristic of the species. We will limit ourselves to a single example. As is known, shrews in the central and southern latitudes usually do not enter into reproduction in the year of their birth. Stein (1961), however, showed that the number of females entering into reproduction in the year of their birth depends on population density. If in ordinary years the reproductive animals born that year constitute 1–2%, after a sharp drop in population size, the number of young shrews (*Sorex araneus*) and water shrews (*Neomys fodiens*) reaches 35%! The significance of such phenomena in the dynamics of the numbers of animals was evaluated long ago and is well understood, but no one has yet attempted to evaluate their significance in the dynamics of population quality. Meanwhile, it is clear that as the restructuring of the population's genetic makeup occurs during the course of the year, the early generations are not genetically identical, and the "illegitimate" entry of young shrews into reproduction cannot but invoke a disturbance, typical for the species, of the cyclicity of the population's genetic makeup. Examples cited earlier show that in this case, we encounter new and very interesting divisions of evolutionary ecology.

Age selection not only explains the possible causes of a change in evolutionary rates, but also creates the preconditions for the creation of a theory of control of a population's qualitative makeup. All factors that change the age structure of a population automatically lead to a change in its genetic structure. If the dependence between the ecological and genetic structure of the population is known, the elaboration of methods for the control of the genetic reorganization of populations encounters only technical, not fundamental, difficulties.

The views we have developed on the significance of a population's ecological structure in the microevolutionary process compel us to look with a new point of view on the significance of so-called nonselective elimination as well.

2. On the Significance of Nonselective Elimination

The theory of evolution prevailing at the present time accepts the following basic postulates:

1. Only individual natural selection within a population plays a creative role.
2. Especially intensive elimination can have a selective character only in exceptional cases.
3. The more active the destructive factors of the environment, the more general is the character that elimination takes, losing its selective character in the face of spontaneous forces of nature.
4. Powerful environmental factors that lead to a sharp reduction in numbers cannot elicit directed changes in the structure of populations; the changes that arise are random.

Attempts have been made, on the basis of mathematical modeling, to determine the rate of genetic reorganizations in populations. The conclusions drawn can be summarized as follows: In the beginning stages of the directed genetic reorganizations of populations, hundreds of thousands of generations are required to achieve an effect at the species level. Then the reorganization goes more rapidly, but even in this case, the period of noticeable displacements is measured in thousands of generations.

Let us attempt to analyze these conditions, taking into account the data on the genetic specificity of seasonal and age generations of animals. Let us emphasize one of the basic postulates of the theory of genetic–automatic processes: nonselective elimination is powerless in a creative sense; it creates only random changes in the population genetic structure, with all the ensuing consequences (Chapter 1). Is this so? Certainly, nonselective elimination always occurs at a certain time of year and against a background of a certain age structure and seasonal change in the genetic structure, elicited by directed selection. If elimination is nonselective, the genotypes present in the population are eliminated in proportion to their relative abundance. It would seem to follow that nonselective elimination is therefore incapable of creating any kind of directed change.

Let us attempt, however, to analyze this question more deeply. Let us assume that we are dealing with a dimorphic population, represented by the genetic variants (phases) A and B. Phase A reproduces more intensively in the summertime; accordingly, a relative increase in its numbers occurs from spring to autumn. The reality of such a supposition could be corroborated by a large body of facts.

Let us assume that the spring structure of the population looks like this: 50A + 150B (we are not assigning absolute values to the coefficients of significance; these values are used only to represent the relative abundance of the genetic variants in the population). During the period of reproduction, A increases 10-fold and B 2-fold. In autumn, then, the genetic structure of the population will have this form: 500A + 300B.

Let us see what nonselective elimination, acting at a different time of year, will lead to. Let us assume that as a result of elimination, the numbers of animals are diminished 50-fold. Since the elimination is nonselective, each phase is equally reduced in numbers (taking statistical error into account). In the spring: (50A + 150B)/50 = 1A + 3B. There is a greater probability that A will disappear completely than that B will, and with repeated eliminations, the dying out of homozygous A is virtually inevitable. In the autumn: (500A + 300B)/50 = 10A + 6B. In this case, the probability that B will disappear is greater than that A will. This simple model alone shows that since nonselective elimination proceeds against a background of regular seasonal fluctuations in the population's genetic structure, it will in the final analysis change the structure of the population as a whole in a directed way. In nature, of course, everything is significantly more complex. Populations are represented, not by two genotypes, but by many; differences in their reproductive potentials are probably much less significant; at very low numbers, the probability of the death of individual variants decreases. But the general rule expressed by our model is correct. It is constructed on precisely demonstrated regularities. We have already noted that the population genetic structure is subject to seasonal variability. It is perhaps simply repeating a truism to note that "nonselective elimination" nearly always occurs because of a seasonal calamity. The return of cold, frosts, floods, heavy rains, epidemics—all are seasonal phenomena, some of which are of primary significance to some species and others secondary. It may be difficult to name even one form of nonselective elimination (except for earthquakes and volcanic eruptions) that is not seasonal. It follows that, in a great many cases at least, nonselective elimination may turn out to have a directional effect on the development of a population.

What we have said, of course, is not meant as a negation or derogation of the notably important investigations in population genetics, but at present, they cannot be limited to purely theoretical or laboratory investigations. They should proceed from realistic ideas on the ecology of populations. Let us therefore try to bring our model closer to nature.

Let us assume that phase A is distinguished from B by a somewhat faster rate of sexual maturation. Accordingly, animals of phase A succeed in producing two litters during the normal breeding season, while animals of phase B produce only one. The fertility of animals of both phases and their mortality in the course of the summer period are identical. Given these assumptions, the normal life cycle of the population will look like this: spring, 50 A + 150 B (1:3); autumn, 50 A (reproductives) + 150 A (1st litter) + 150 A (2nd litter) + 150 B (reproductives) + 450 B (1st litter). The numbers of A increased 7-fold during the breeding season (from 50 to 350); the numbers of B, 4-fold (from 150 to 600). In order that the genetic makeup of the population remain constant, one must assume that differential mortality occurs during the winter: the numbers of A are reduced 7-fold; the numbers of B, 4-fold. In autumn, the ratio is 350 A + 600 B (7:12); in spring, 50 A + 150 B (1:3).

Now let us assume that in early autumn, nonselective elimination occurs, the numbers of animals are sharply reduced, and the entire 2nd litter dies. This is a completely realistic assumption: during the early severe frosts, for example, nonselective elimination of animals leading an independent form of life occurs, but the dependent young animals die out completely. Under these conditions, the advantage of the A genetic variant cannot be expressed, and the dynamics of the population's genetic structure are changed markedly: in the spring, 50 A + 150 B; in the autumn, 50 A + 150 A + 150 B + 450 B = 200 A + 600 B. (May we remind you that the coefficients indicate only the ratios of different forms in the population, not their absolute numbers.)

In the winter, the reduction in numbers occurs in the usual way: A is decreased 7-fold, B 4-fold. Rounding off the numbers, we get for the autumn population makeup: 30 A + 150 B (1:5).

Let us assume that in the following autumn, elimination occurs with the same attendant phenomena. At the end of the breeding season, then, we will have: 30 A + 90 A + 150 B + 450 B = 120 A + 600 B. Accordingly, in the following autumn, we will have: 17 A + 150 B. With a third repetition of the analogous situation, the genetic makeup of our population will have the form: 6 A + 150 B. Now assume that the situation causing elimination occurs at the beginning of reproduction and reduces the general numbers of the population 10-fold. According to probability theory, the A variant will in effect disappear from the population.

Under actual, natural conditions, however, A does not disappear, but is maintained in the population in the heterozygous state. Our example does show, though, how rapidly the directed genetic restructuring of the population can occur under the influence of a nondirected factor (nonselective elimination). Within the limits of our theme, it is especially important for us to emphasize that the direct cause of the change in population genetic structure is the change in its ecological (in our example, age) structure.

Let us introduce yet one more ecological refinement: during the reduction of population density, the fertility of animals increases. Let us assume that nonselective elimination occurred in the early spring (at the beginning of reproduction). The ratio of genetic variants was unchanged (50 A + 150 B), but the total numbers of animals were decreased. In accordance with the effect of density-dependent factors, the fertility of the animals increased, and in the autumn, each pair of adults produced, not 6, but 12, young. In the autumn, then, the population structure can be expressed thus: 50 A + 300 A + 300 A + 150 B + 900 B = 650 A + 1050 B (13:21). And, in this case, the change in the population's genetic structure was elicited by a change in its ecological structure.

Let us try to bring our model even closer to the real ecological situation. The effect of density-dependent factors is directly expressed with particular force after the population has been thinned out. In our example, therefore, the 1st litter should be greatly increased. The 2nd litter will then be less numerous, since the population density will increase after replenishment by its current crop of young (to say nothing of the regular reduction of fertility in the autumn). The autumn population, therefore, should take this form: 50 A + 300 A (1st litter) + 150 A (2nd litter) + 150 B + 900 B (1st litter) = 500 A + 1050 B (10:21). The population structure is even more markedly changed.

The question examined here has not yet strongly attracted the attention of ecologists. It is important to note, therefore, that the views we have developed are supported not only by general theoretical considerations based on firmly established ecological rules, but also by some direct observations (as yet not numerous) showing that a marked change in numbers is accompanied by regular changes, not random changes, in the genetic makeup of populations. In this regard, the work of B. K. Pavlov is of special interest. On our advice, he studied the dynamics of a polymorphism in squirrels and was able to establish a connection between changes in the frequency of individual genetic variants and their morphophysiological features. The author sums up this work in his paper (Pavlov, 1965):

> The squirrels of eastern Siberia are represented by a few color variants: red-tailed, black-tailed, and an intermediate group—brown-tailed. Each population has its own ratio of color variants.
>
> In the southern mountain taiga forests with their prevalence of the Siberian cedar, the black-tailed individuals dominate (up to 80%); the red-tails constitute 5–6%, the brown-tails 10–15%. In the northern larch forests, the respective percentages are 30–40%, 18–20%, and 40–50%. In the territory occupied by a single population, the black squirrels settle in the dark conifer forests, the light ones in the light conifer forests.
>
> The ratio of color variants characteristic for each population does not remain constant, but changes with the years. The general rule is that at high

population size, the black-tailed squirrels predominate; after a reduction in size, the relative number of the brown-tails increases. In certain years, the quantity of black-tails reaches 50% even in northern "light" populations.

Individuals of different color variants are distinguished by a number of morphophysiological indices. Red-tails can more easily bear a shortage of basic foods; the length of the intestines is greater than in the black-tails or brown-tails. As a consequence of this greater length, they possess a greater capability for utilizing coarse nutritive vegetation. In northern populations, the relative weight of the liver is greater in the red-tails than in the black-tails. In the southern populations, these differences are not significant. The relative weight of the adrenals is significantly lower in the brown-tails than in the red- or black-tails. Under optimal density, there is no difference in the relative weight of the adrenals betweeen the red-tailed and black-tailed squirrels. During a marked worsening of the conditions of existence, the different color variants differ in their fertility; it is higher among the red-tails. Different color variants possess different biological features.

It may be supposed that the ratio of different color variants is under the control of natural selection. Geographic variability in the phenotypic structure of a population may serve as some evidence in favor of this position. A structure with predominance of brown-tailed and red-tailed individuals is peculiar to populations with an unstable type of size dynamics. The predominance of black-tailed individuals is peculiar to populations with a stable type of size dynamics (the quantity of brown-tailed individuals is completely insignificant; the red-tails are almost always about 5–6%).

Pavlov's data characterizing the genetic makeup of different age groups of squirrels, which he generalized in the manuscript he gave me in 1967, are of special interest. The relative abundance of color variants in the various age groups in the Togodinskaya population in 1962 is as follows (in %):

	Black-tails	Brown-tails	Red-tails
Current year	85.5±2.9	12.4± 2.7	2.2±1.2
1st year	65.8±4.1	27.6± 4.5	6.6±2.5
2nd year	14.8±9.9	71.4±12.5	14.8±9.9

These data make it clear that if for some reason the age structure of a population changes, there is a marked change in the genetic structure of the population as a whole. In order to envisage the actual course of these changes, it is necessary to know the nature of the determinants of the genetic variants analyzed (these determinants are as yet unknown, but it is difficult to question their inevitability). That age selection can lead to the rapid restructuring of populations is made indisputable by Pavlov's investigations.

It is important to note that a change in the genetic structure of populations also occurs under the influence of trapping, since trapping has different effects on animals of different ages.

As was shown earlier (see p. 97), a change in the population's genetic structure appears as a change in the allometric exponent—a trait that characterizes the genetic makeup of the population as a whole. An experiment conducted by Pavlov (1969) on an experimental plot showed that during 10 days of trapping, the genetic makeup of a population of squirrels changed. The allometric exponent of the basic indices changed very substantially during this period (Table 6). No changes were observed on control plots. No displacement of squirrels occurred in the territory.

Pavlov also studied the change in the population's genetic structure, utilizing the presence of the interparietal bone as a genetic marker (for an analysis of the significance of such traits, see Berry, 1963). Before trapping, squirrels with this trait constituted 14.9% of the population; after trapping, 23.8% (the difference is statistically significant). Especially marked changes occurred in the young squirrel group: before trapping, 3.2%; after trapping, 18.2% ($t = 6.7$). Pavlov confirmed the reliability of these data not only by statistical analysis of the material, but also by shooting all the squirrels in the experimental plot. Analyzing his material, he notes that in some cases, the selective action of trapping coincides with the direction of natural selection. The reorganization of a population can occur very rapidly.

It is difficult to say what causes the selectivity of trapping. Undoubtedly, the genetic markers are linked to certain features of the squirrels' biology that determine their behavior. Change in the allometric exponent leads to the idea that these features are determined by the different activity of the animals, since a change may be connected with a change in the growth rate of the animals (Shvarts, 1961*b*). The indubitable

TABLE 6. Changes in the Allometric Index *a* in Experimental Populations of Squirrels[a]

Period of investigation	Width of interorbital distance		Length of tooth row		Width between upper molars	
	Males	Females	Males	Females	Males	Females
Before trapping	−0.066 ±0.019	+0.265 ±0.0019	−0.022 ±0.017	−0.252 ±0.012	+0.002 ±0.25	−0.386 ±0.06
After trapping	+0.926 ±0.085	−0.175 ±0.021	+0.375 ±0.014	−0.165 ±0.015	+0.531 ±0.09	+0.151 ±0.002

[a] According to the data of Pavlov (1969). All indices are calculated in relation to the condylobasal length of the skull.

fact shown by the work of Pavlov is important, however: the dynamics of numbers, determined both by natural causes and by trapping, elicit changes in the population genetic structure, which occur especially abruptly in those cases in which the process occurs through age selection (in the sense mentioned above).

It is important to take into consideration that the disproportional reduction in the numbers of one of the genetic variants not only directly changes the population genetic structure, but also may have far-reaching consequences.

In this connection, Pavlov (1969) writes further:

> It was recently shown (Shvarts, 1965) that a change in the age structure leads to a directed reorganization of the genetic makeup of the population. We succeeded in experimentally detecting the age selectivity of trapping. It appears to be especially severe among the group born in the current year. If the intensity of trapping is high (the population drops by 50% during trapping), individuals in which the dry weight of the crystalline lens of the eye is 25–26 mg (which corresponds to birth at the end of March or beginning of April) remain in the population in higher numbers. The age selectivity in this case elicits a change in the population genetic structure. These changes are found to be even more significant in connection with one of the features of squirrels of this group: in the following year, they enter into reproduction earlier than squirrels of other ages. Among them, the number of individuals that produce a second litter is significantly greater than in other groups. Certain data allow us to assert that this feature is linked with hereditary properties. The specific outcome of the reorganization of the populations will be determined jointly by the pressure of natural selection and trapping. Study of the rules of the reorganization of populations during changes in the conditions of existence and with the effect of trapping opens the path for control of natural populations.

Abrupt changes in numbers do not automatically involve a change in the population genetic structure. Ford's interesting observations and experiments (Ford, 1963) speak to this point very clearly.

In our laboratory, the investigations of Beregovoi (1966, 1967*a,b*) were devoted to the same problem. Seven populations of a polymorphic species of the common spittlebug (*Philenus spumarius* L., Homoptera: Cercopidae) were studied. It was established that in the Urals, 11 color types, known by the following names, are encountered: *typica, populi, trilineata, flavicollis, gibba, leucocephala, quadrimaculata, albomaculata, leucophtalma, marginella,* and *lateralis*. All the populations studied differ in the ratios of these chromomorphs; the differences are statistically significant. The available material does not allow us to establish a direct connection between the differences in habitat and features of the populations' makeup. Populations inhabiting a forest are distinguished from one another just as clearly as are forest populations from meadow populations. Population differences are observed, however, to clearly depend on the degree of isolation and the mutual distribution of populations. This distribution is related to the migration of individuals among neighboring populations.

The author summed up the results of his investigations as follows (Beregovoi, 1967*b*):

> The makeup of four populations was traced from 1964 to 1966. Analysis of the data obtained showed that the degree of differences between the populations was not exceeded by the differences within any one of these populations from year to year. Each population maintained its characteristic features over the course of three years. The observed stability of interpopulation differences especially underlines the established fact of drastic and noncoincident fluctuations in numbers, which frequently hamper the collection of material in individual populations. These fluctuations in numbers do not even coincide in neighboring populations within 400 m of one another.
>
> All the available material testifies to the fact that the gene pool of each population is formed in a complex dependence on environmental factors and manifests stability over time. The nature of interpopulation differences in our material reveals a great similarity to differences often observed among subspecies.

These observations show that populations possess the capability of maintaining relative constancy in their makeup, despite very marked fluctuations in numbers. A change in the genetic makeup of a population, connected with a change in numbers, can be used by natural selection for rapid reorganizations corresponding to changes in the environment. In other cases, selection over a short period of time repairs disturbances that have arisen. Populations do not find themselves at the mercy of a blind determination of their genetic specificity. Variability of a population's phenotypic structure allows it to exist in sharply differing environmental conditions.

All cases of "catastrophic," nonselective elimination, including such rare cases as volcanic eruptions and so forth, are repeated from time to time, so that they occur in different seasons of the year, and thereby exert a balanced effect on the local populations. In this scheme, furthermore, one should not expect permanent populations to exhibit changes similar to those described. They have already undergone such changes hundreds and thousands of years ago. If a population finds itself under new climatic conditions or (see below) is subjected to additional seasonally "oriented" extermination measures that are repetitive from year to year, we have a right to expect such a picture of changes in population structure.

Mathematical modeling of the rules analyzed here can be even more precise. Thus, for example, one can calculate the increase in the rate of sexual maturation of animals when the population density is lowered, the increase in sexual activization of males, the change in mortality of animals that have overwintered (reproductives), and so forth. Making such precise definitions does not enter into our task, however. We aimed to show that progress in population ecology created the prerequisites for the mathematical modeling of microevolutionary processes, a modeling significantly closer to the real natural situation than that which lay at the basis of studies on genetic drift or genetic automatic processes. It becomes clear during this analysis that a change in the ecological structure of a population, regardless of which direct causes elicited it (including nonselective elimination), has as a consequence a change in its genetic structure. That it does justifies our inquiry into the ecological rules of the evolutionary process.

We have broached only a particular instance of the problem and noted the significance of the age makeup of populations and of nonselective elimination, which changes the intrapopulation structure. In doing this, we have consciously used only an elementary mathematical apparatus, and have not striven to express the rules revealed thereby in generalized formulas. It would be no more complicated to do this, however, than to educe the formulas expressing the dependence of the rate of reorganization of populations on the selective advantage (Selektionswert) of individual genetic variants. It was important to us to show that a theoretical analysis leads to the conclusion that there is a diversity of mechanisms for the directed reorganization of population genetic structure. Individual selection, to which an almost monopolistic role in the directed change of populations has until now been attributed, is only one such mechanism.

The theoretical analysis of the problem broached should be used, first of all, as a program of correlated experimental work (both in the laboratory and under field conditions)—a program that lays the foundations of experimental evolutionary ecology. The results of this work, in turn, will influence the development of theoretical ecology.

This will help to make modern evolutionary theory genuinely synthetic.

The theoretical significance of such investigations is evident. They have no less practical significance. At present, we have no possibility at all of influencing the course of natural individual selection in nature. But we can govern "group selection." Knowing the course of seasonal variability of population genetic structure, we can with relative ease effect a directional change in population genetic structure; i.e., we can in fact govern the microevolutionary process. Furthermore, we often do this even now. Extermination methods usually take the form of selective elimination, and are always timed for a definite season. For a number of reasons, extermination measures are especially effective in the autumn, and during the process of extermination, changes in population genetic structure are inevitable. It may be assumed that extermination leads to the predominance of less fertile, but more viable and resistant, individuals. One must reflect: Is this advantageous? The prospect of changing the quality of natural populations is raised. It is difficult to foresee what sorts of consequences in the practical struggle against pests and the utilization of useful animals the making of such changes will entail.

Moreover, the very form in which extermination measures are applied leads to a disturbance of the population ecological structure, which inevitably leads to its genetic reorganization. In fighting rodents, arsenic compounds are often used. It has been found that arsenic acts selectively on rodents. Females and young animals die in relatively fewer numbers than adult males (Junkins, 1963). It is not difficult to imagine the results to which the prolonged use of arsenic bait will lead. Under these conditions, the selective advantage will go to the genetic variants that are distinguished by a higher rate of sexual maturation, since they are the chief contributors to the rebuilding of the population. There occurs to one the conclusion that the use of arsenic in the battle with rodents can lead, within a short time, to the creation of a population distinguished by the exceptional speed of the reproductive stage.

In this case, attempts at extermination lead to a qualitative reorganization of the population to man's disadvantage. If the theoretical investigations are sufficiently well developed, conditions will be created for the elaboration of extermination methods that will not only decrease the numbers of the pests, but also lower their reproductive potential. It is possible that a change in the quality of a population will turn out to be a more effective method of decreasing the numbers of a species than direct extermination measures.

Of course, this principle can be correctly applied not only to extermination, but also to trapping. The method of trapping determines both the quantitative and the qualitative makeup of the population. Let us cite a few examples.

Trapping small predators leads to a displacement of the normal sex ratio in favor of females. In some cases, this displacement can be of such magnitude as to give rise to the threat of a huge number of unmated females. Under these conditions, males distinguished by early sexual maturity should have a clear selective advantage, since when males are in short supply in the population, participation in reproduction by young reproductives should have special significance. Investigations directed toward this purpose (comparison of rates of maturation in males of trapped and untrapped regions) could thus have great importance for both theory and practice. Trapping is always selective. Its effect on the qualitative makeup of the population is determined by its selectivity. Hunting that approximates nonselective elimination should have a different consequence. Hunting squirrels with rifles, for example, has a high degree of correspondence to nonselective elimination. The more intensively such hunting is conducted, the more closely it corresponds to nonselective elimination, since when the intensity of hunting is low, the more active animals that leave more tracks will be fired on in greater numbers. In intensively hunted lands, therefore, selection should go to fertility, thereby creating a population distinguished by increased fertility.

The data cited indicate that analysis of phenomena united by the concepts of "population age structure" and "nonselective elimination" can be used for the development of a general theory of evolution. The study of the spatial structure of populations has no less significance.

3. The Spatial Structure of Populations—A Factor in Microevolution

By the spatial structure of a population we mean the features of distribution of the animals in the territory, including the origin of local micropopulations inhabiting biotopes with distinctive environmental conditions. Spatial structure as a factor in the microevolutionary process long ago attracted the attention of investigators. It is sufficient to recall that one of the chapters of the synthetic theory of evolution gave special attention to the problem of isolation (including incomplete isolation) of the structural subdivisions of the species.

Recently, interest in the study of the spatial structure of populations in the evolutionary scheme was considerably stimulated by Lewontin's theory (Lewontin, 1965) on population selection, the essence of which is as follows: If for any reason a population or micropopulation dies out, its place is occupied by others. When this happens, the new genic complex in the new environment does not simply copy the old. The population is reorganized, and an evolutionary displacement occurs.

In investigating this process, one encounters a series of difficulties, fundamental as well as technical. The reason is that when one speaks of a population dying out, one has in mind not a population in the strict sense of the word (an ecological unit, capable of the self-regulation of its numbers; see the Introduction), but the dying out of an intrapopulation group, a part of a whole population. The dying out of such a group has a complex and contradictory effect on the population as a whole. The lack of a strict approach to the concept of a "population" in this case, as in many others, can lead to mistaken ideas.

Nevertheless, investigation of phenomena falling within the concept of interdeme selection has already led to a better understanding of the mechanisms of the evolutionary process. In particular, work in this direction leads to an analysis of the possible contradictions between intrademe selection (the usual form of natural selection) and interdeme selection.

As a theoretical example, Lewontin (1965) points out the possibility of selecting animals capable of especially efficient consumption of plant food.

Such animals would leave more progeny, and this process would quickly lead to the formation of a population of "superherbivorous" animals. The vegetation would be destroyed, and the population would die out, yielding its place to another that used the available stores of food more economically.

Lewontin seeks to reinforce his supposition with examples from the laboratory and from field work (Lewontin and Dunn, 1960; Lewontin, 1962). In natural populations of house mice, the *t* mutation is often encountered. As a result of intrademe selection, the frequency of the *t* gene is sharply increased. Males homozygous for *t* are sterile, however. As a result, the productivity of the population falls, and it is replaced by another. Lewontin considers this an excellent example confirming his theory of interdeme selection.

Independently of Lewontin, the idea of group selection was expressed in the interesting work of Chel'tsov-Bebutov (1965), who investigated the evolution of the breeding grounds of the black grouse. The author believes that in order to explain the origin of the mating ritual of the black grouse, one must leave the suppositions of individual selection and think at the level of group selection, at the level of the micropopulation, by which he means the group of birds that migrate to the same breeding grounds. The complexity and intensification of the mating ritual stimulate the sexual activity of females and increase the productivity of the population. "As a result, micropopulations of the black grouse with longer periods of mating may prove to be in a more advantageous position; some of the females could lay repeated (compensatory) clutches, thus increasing the general fertility of the population" (p. 395).

It is easy to see that in the examples analyzed, interdeme selection enters as a factor in eliminating populations with unsatisfactory genetic structures, not as a creative factor. It is not casually that we define these examples as representing unsuccessful selection, but rather to demonstrate a rule. It is possible to speak of interdeme selection only conditionally, since the significance of the extinction of local "populations" (which can never be complete) can be evaluated only on the basis of an investigation of the processes occurring in the population as a whole, the population in the strict sense of the word. It is more correct to speak not of interdeme selection, but of the role of the dynamics of the population's spatial structure in microevolution.

Analysis of the question we are examining should be based on two fundamental facts. The first is: fluctuation in the numbers of species is a widespread phenomenon. It is unnecessary to cite examples, since mass reproduction of many species, belonging to different taxonomic groups, has been described repeatedly and comprehensively investigated. A peak (or a low point) in numbers often occurs not only in the whole population, but also in groups of the population. During a population explosion, there is a settlement of territories in which the species is either absent or present only in insignificant numbers in ordinary years. During such an explosion, there inevitably occurs a shift in the population or a sharp increase in the territory within which panmixia is actually realized.

The territory within the limits of which a reduction in the numbers of the species is observed is just as broad. Often, it embraces a whole biome. Thus, a drop in the numbers of lemmings or polar foxes is observed in the entire enormous territory of polar Eurasia in certain years. During a reduction in numbers, animals are maintained only in survival stations (in the sense used by N. P. Naumov, 1963) or in discrete parts of the species range in which for certain reasons more favorable conditions of existence have arisen. All that has been said is a truism of modern animal population ecology, but these truths are not yet used in full measure for the analysis of the ecological mechanisms of the microevolutionary process.

Another group of facts having special significance for an evaluation of the possible role of interdeme selection can no longer be regarded as the ABC's of ecology. We speak here of a process that is in total contrast to the one just described. If the peak of numbers and its sharp reduction embrace significant territories, in usual ("average") years, the fluctuations in numbers may differ markedly even between neighboring micropopulations. Populations of voles in forest clearings, shrub-brush areas, long-fallow fields, meadows, and fields are subjected to different environmental forces, even if they are directly adjacent and in the same geographical environment. The dynamics of their numbers, therefore, should inevitably be different. That they are is shown by numerous investigations, but the work of Hayne (Hayne and Thomson, 1965) is especially clear.

Over a period of 10 years, the author studied the dynamics of the numbers of *Microtus pennsylvanicus* in the states of Michigan and Wisconsin. Observations were conducted at 197 stationary points. The average distance between points of each pair was 10 km. Mathematical analysis of the material showed that correlations between changes in the numbers of neighboring populations either were not statistically reliable or, even if real, were so weak ($r = 0.28$) that there was no reason to consider them when carrying out eradication (control) procedures. The author draws the valid conclusion that within the limits of a single region, the fluctuations in numbers of a species are subordinate to similar rules, but local numbers (the numbers of individual populations or micropopulations) can differ sharply both from the mean of the region and from the numbers of neighboring settlements.

In our laboratory, Sosin (1967) conducted similar investigations on the muskrat. Over the course of 3 years, the dynamics of numbers in the muskrats was studied in several lakes situated 0.6–10 km apart in the forest steppe of the Transurals.

It has been established that fluctuations in the numbers of muskrats, even in reservoirs of similar nature, are asynchronous. While the number of animals grew from approximately 250 to 450 in one reservoir, a drop from 350 to 140 muskrats took place in another. In a third, the numbers during all the years of observations changed within significantly smaller limits, from approximately 120 to 160 muskrats. Change in numbers is accompanied by changes in the age structure. In settlements in which the numbers of muskrats increased in comparison with preceding years, there was an increase in the number of young borne by each reproductive female (from 5.4 to 9.4 in one lake, from 6 to 20 in another). In reservoirs in which the numbers changed little in different years, the age makeup varied within the limits of 12–16 young for the current year for each reproductive female. Comparison of the number of reproductives with the intensity of reproduction and the density of the autumn population leads to the conclusion that in this case, the dynamics of numbers are determined by the viability of young (migrations were not noted in the period of observations).

The data cited show that settlements of muskrats on closely situated lakes are governed by rules for the dynamics of numbers and can be viewed as micropopulations with characteristic features of shifts in population and age structure.

Small mouselike rodents are excellent swimmers and can swim across even fairly broad and rapid rivers. Nevertheless, there is no doubt that the exchange of individuals among populations of voles (with the exception, naturally, of species leading a semiaquatic life) is limited and probably by no means occurs every year. There is an interesting example in this regard. We investigated a population of *M. oeconomus* on the shore of a large steppe lake. In the lake, there are many islands that in the summertime represent vole habitats protected from enemies and rich in food. The islands are situated at a maximum distance of about 100 m from the shore, and the distance between individual islands is even less. In the wintertime the conditions of existence on the islands are worse than on the mainland, since the ground freezes. In winter, the majority of island populations die out. This situation allows us to get an idea of the degree of isolation of vole populations separated by a water barrier of the order of tens and hundreds of meters. It was found that although *M. oeconomus* could reach any island in 10 minutes, the islands are by no means settled every year. Special experiments were also set up. Marked *oeconomus* voles were taken to the islands and released. It was found that the overwhelming majority remain at the places of release and do not swim from island to island or to the shore.

Sosin (1967), from our laboratory, conducted essentially analogous experiments. The marking he carried out showed that muskrats do not often go from lake to lake, even if the distance between them is measured in only dozens of meters. Many similar examples could also be cited from the literature. They show that what appears to be a single population is broken down into groups of partially isolated populations, among which the exchange of genes is significantly limited.

Completely analogous data can be cited for other species as well. For illustration, let us use the data on the procurement of a number of species of fur-bearing animals in Yamal and adjacent regions of the Urals. From the wealth of material presented by Rakhmanin (1959), let us cite only a few examples. In 1954, throughout the area as a whole, the number of muskrats, in comparison with preceding years, scarcely rose (164,813 prepared skins against 155,902), but in the Yamal region, it rose 3-fold, in the Nadym region by more than twice, and in the Krasnosel'kup region it was reduced. The number of white hares in 1956–1957 in the Nadym region was reduced by half, but in the Yamal region it was increased by 20 (!) times. In these same years, the number of ermine in the Krasnosel'kup region increased 2-fold, but in the Shuryshkar region decreased 3-fold. In preceding years, the opposite occurred: in the Shuryshkar region, the number increased a little; in the Krasnosel'kup region, it decreased by more than 4 times. The fluctuations in numbers in different regions of Yamal and in other species were just as asynchronous.

Even though one cannot consider data from commercially procured stocks as completely reliable indices of changes in numbers, it is absolutely indisputable that they show that neighboring settlements of animals are often characterized by different rules for the dynamics of numbers.

A change in numbers has as a consequence an inevitable change in the genetic makeup of a population. This change can be substantial or insignificantly small, but it must occur. The firmly fixed laws of population genetics demand it. It follows that the actual effect of gene exchange among populations is determined, to a significant degree, by the comparative course of the dynamics of the animals' numbers. We will consider the significance of this phenomenon in more detail below.

Another aspect of the same problem is the extinction of micropopulations. The intensity of reproduction of a species corresponds to the conditions of its existence. The higher the death rate, the higher the birth rate. It is not by chance that the intensity of reproduction in elephants is a thousand times weaker than the intensity of reproduction in mice! Again, there follows from this a fundamental ecological truth: it is not at all obligatory for there to be a catastrophic combination of external factors for a population to become extinct. If for any reason the death rate begins to exceed the birth rate, the extinction of the population is only a matter of time. When one speaks of ephemeral animals, it is a question of an insignificantly short period of time. In the majority of cases, nevertheless, extinction does not occur, since, during a sharp curtailment in the numbers of a population, an increase in the number of immigrants from neighboring populations occurs. When a population is isolated, however, it is easily discovered that an insignificant reduction in the normal tempo of reproduction in the population dooms it to destruction. This was clearly shown in island populations of the house mouse (Lidicker, 1966). It was found that during joint habitation with *Microtus californicus*, normal pregnancy in the mice is often disturbed. This disturbance leads to a decrease in the tempo of reproduction of the population and to its more rapid and complete extinction—despite the "normal" death rate and, on the whole, a high intensity of reproduction.

Since such a change in the tempo of reproduction is undoubtedly encountered very often in nature, it follows that immigration of individuals from one population to another is a phenomenon that occurs far more frequently than is determined by direct observations. The intensity of the exchange of individuals, however, is subject to fairly strict rules, about which we have begun to obtain exact information only in recent times. On the other hand, more and more often, data are forthcoming that show that even neighboring populations are fairly isolated ecological systems. As long as a population completely maintains its viability, it actively limits the penetration of individuals from outside. The actual ecological mechanisms that hamper the exchange of individuals among populations and intrapopulation groups of animals are well known. These mechanisms have as their basis a complex intrapopulational system of dominance–submission. In this system, it is only in rare cases that the "strangers" can enter into the group of dominant individuals and compete with the "hosts" in intensive reproduction. It is possible that other mechanisms of isolation also have definite significance: the distinctiveness of the selection process of the pairs (for details, see below), disturbance of pregnancy in females with the sudden appearance of strange males (Bronson and Eleftherion, 1963), inbreeding (Scossiroli, 1962), and the urge of animals to avoid territories not occupied by the species (Haggerty, 1966). It is possible that in higher animals, an attachment to the colony and an attachment of different individuals to one another also play a role (Penney, 1964; observations on penguins). Investigation of such mechanisms of genetic isolation is only beginning, but one cannot doubt their existence. It is difficult to select a better example than the investigations of the Polish ecologists who studied micropopulations of mice in the attic and in the basement of a two-story house (Adamczyk and Petrusewicz, 1966). It was established that different types of dynamics of numbers reliably isolate these groups of animals, and this leads, as a final result, to the origin of genetic differences between them (results of experiments on transplantational immunity). It would be possible to cite quite a few analogous examples that indicate the relative isolation of groups of animals. The influx of individuals from neighboring populations increases sharply in periods of local reductions in numbers. According to well-understood causes, under these conditions, the role of immigrants in the reorganization of the population's genetic structure grows substantially, since their relative abundance in the new population that is forming increases. It was found, moreover, that the mortality of immigrants is inversely proportional to the numbers of the population (Andrzeijewski *et al.*, 1963). The mixing of populations during a reduction in numbers therefore has especially important significance. The results of Petrusewicz' experiments are corroborated by the investigations of Anderson (1966), which show that no exchange of genetic material occurred between several neighboring settlements of mice for a long period of time.

What has been said does not exhaust those most important ecological rules that one must have a knowledge of to correctly evaluate the role of population spatial structure in microevolution. A large number of works, to which we took occasion to refer in Chapter IV, clearly show that the displacement of animals of different sexes and ages is different. New facts in this regard were obtained by Petrusewicz and co-workers (for a review of the data, see Petrusewicz and Andrzeijewski, 1962). These data make it possible to consider assigning "migrants" and "residents" to ecologically different groups of individuals.

A drop in the numbers of animals has different causes, but in a relatively large number of cases, during a sharp reduction in the numbers of a population, the number of young recruited into the population is curtailed first of all (this being the consequence of a drop in the intensity of reproduction during an increase in juvenile mortality). This means that during the substitution of one population for another, there occurs primarily a displacement of old individuals of one population by young newcomers. An important task of evolutionary ecology is to establish the actual manifestations of this rule, but one can hardly doubt its broad, if not universal, distribution.

After these preliminary remarks, we will continue with an analysis of the role of a population's spatial structure in the reorganization of its genetic structure. As was already mentioned, this influence is contradictory. Logical analysis shows that the complexity of population spatial structure results in different courses of the dynamics of numbers in individual micropopulations. These dynamics promote the maintenance of the unity of the population and stabilization of its genetic makeup (given constancy in the direction and intensity of natural selection). This conclusion seems paradoxical, but it ensues directly from well-known positions (see Chapter I) on the connection between the dynamics of numbers and the change in a population's genetic makeup. A sharp reduction in numbers can lead to loss of a certain gene within individual micropopulations, but it may be assumed to be completely unlikely that the nature of impoverishment of the gene pool in two neighboring populations will be found to be the same by chance. The mixing of micropopulations and populations, therefore, inevitably leads to the establishment of a unique gene pool, and under the influence of selection, a unique relationship among the different genetic variants will be established in the course of several generations. The experimental data we cited in Chapter I testify to this.

As a specific example, we can use the investigations of the dynamic polymorphism in the sharp-nosed frog, which we have already described (see p. 106). Two micropopulations in the Kurgan region were investigated in 1966. The genetic makeup of younger animals in the spatially isolated populations was found to be different: the percentage of *striata* in one was 43.9 ± 3.7; in the other, 55.3 ± 3.46. Older frogs, however, did not appear to be different (the percentage of *striata* was 66.7 ± 7.4 and 66.0 ± 6.9). The change in genetic makeup of individual micropopulations, which was elicited by environmental changes unknown to us, did not lead to their differentiation, since, in the process of migration of adult animals, there occurred a consolidation of the population, in the strict sense of that word (see the Introduction), into a single genetic whole.

Thus, the complex spatial structure of populations does not entail the inevitable genetic or morphophysiological differentiation of the species.

A great number of small species in biomes form innumerable numbers of populations, but this does not lead to intensive intraspecific formations. Once again, let us recall the Siberian lemming. It was so poorly differentiated that Sidorowicz (1960) considered it possible to combine all the mainland lemmings of Asia into one subspecies. Krivosheev and Rossolimo (1966) objected and designated two subspecies in Eurasia, of which one (*Lemmus sibiricus chrisogaster*) is more than dubious.

The region of distribution of the Asiatic red-backed vole (*Clethrionomys rutilus*) is huge. In the northern and southern limits of the range, the species is represented by semiisolated populations adapted to islands of woods among the steppe or tundra vegetation. Nevertheless, as shown by the painstaking investigations of Bol'shakov (1962), only four subspecies in all, which are distinguished predominantly by their coloring (Bol'shakov and Shvarts, 1962), deserve taxonomic consolidation. This example is especially significant, since the range of the Asiatic red-backed vole embraces a number of biomes, from the steppes to the tundra.

A multitude of subspecies are described for many species of mouselike rodents, but the majority are assigned on the basis of insignificant (often imaginary) differences in coloring or proportions of the body and skull. It is not happenstance that each revision of the subspecific systematics of mice or voles again reduces the number of subspecies. On the other hand, subspecies of many large mammals are in fact sharply differentiated forms. A comparison of the structure of the species of the same Asiatic red-backed vole with the structure of the species of the majority of ungulates (for a review, see Geptner *et al.*, 1961) is sufficient to make clear the differences in the groups formed by large and small mammals.

It is difficult to explain these differences on the basis of the ideas of the synthetic theory of evolution that have become traditional, since small species form populations that are relatively far more isolated than those of large species. If the subspecific taxonomy of rodents is approached on the same scale as the taxonomy of ungulates, many species with a huge range (muskrat, *oeconomus* vole, field vole, water vole, Asiatic red-backed vole, and many others) should be recognized as monomorphic; the elk, roe deer, stag deer, and other species of ungulates, on the other hand, form subspecies that continue to give rise to disputes over their possible specific independence.

It follows that the rate of intraspecific differentiation is determined, in the first place, by the conditions of existence, in the process of adaptation to which the divergence of different forms of the species occurs. For small mammals in the lives of which the microclimate has major significance, the conditions of existence in the tundra forest and in the steppe forest can turn out to be quite similar; for the elk, however, they are quite different. Subspecies of voles are therefore differentiated more weakly than subspecies of elk, even though genetic automatic processes should lead to diametrically opposite results.

During a change in the conditions of existence for small mammals, their morphophysiological features also change, even if the populations compared are situated in proximity to each other and there is little isolation between them. A good example is the so-called "biotopic variability," the essence of which is that often animals (the majority of investigations were conducted on rodents) from different biotopes of one geographic region are distinguished from each other more significantly than animals of different subspecies (Polyakov *et al.*, 1958, and many others). The fact of biotopic variability is undoubted, but its analysis should be approached with caution.

In a great number of cases, the differences among animals of different biotopes are naturally explained by environmental variability. The material cited in Chapter II shows that a change in the system of correlated relationships in the development of individuals can lead to a change, under the direct influence of the environment, in, it would seem, even the most stable traits. In individual cases, one may assume that biotopic variability has as its basis differences in the genetic structure of micropopulations. These differences can be the result of the reorganization of populations in the course of one generation, but they are in no way the result of the independent evolution of populations. That the results of selection in one generation can lead to substantive results is shown both by special investigations (see Section 1) and, in particular, by the numerous observations bearing on the problem of "seasonal selection." Thus, "biotopic variability" does not contradict the assertion that subspecific differentiation due to the complex spatial structure of animal populations is not inevitable.

Only in special cases does the complex topographic structure of populations lead to the differentiation of populations and to subsequent evolutionary forms. It seems to us, however, that it is, at least in many cases, a powerful factor in the evolution of a population as a single whole. The complex topographic structure of a population enters not only as a stabilizing and uniting factor, but also as a creative factor. A synthesis of the data on the ecology of populations set forth above with the data from population genetics leads us to this conclusion.

The essence of one of the most important conclusions of population genetics is that the majority of interpopulation differences are polygenic. In this case, one and the same trait can have a diverse genetic basis, but the joint action of different genes has a mutually reinforcing effect (additive gene action).

We have already cited the most important material on the question of diverse genetic basis in our mention of the investigations on the effect of poisons on insects. It was shown (for a summary of the data, see Crow, 1957; Milani, 1957; A. W. Brown, 1958; Benett, 1960; and others) that in different insect species, the genetic mechanisms for the origin of poison-resistant populations are different. More than that, in different laboratories, the origin of poison-resistant "races" has a different genetic basis. In some lines of *Drosophila,* resistance appeared to be controlled by one dominant gene (Ogaki and Tsukamoto, 1957); in others, by the additive effect of many genes (Crow, 1957). Benett (1960) shows that in many cases, these differences among *Drosophila* lines are determined by differences in the manner in which DDT is applied to the experimental populations (aerosols, DDT on paper, as a poison added to the food, to the larvae, and so forth). It is extremely significant, however, that even in identical conditions within the lines of one laboratory colony, resistance to poisons often arises on different genetic bases.

These observations make it evident that under natural conditions as well, adaptation to identical conditions within the boundaries of different micropopulations can occur, and undoubtedly does occur, on different genetic bases. That it can is evidenced both by direct observations, on the basis of which the idea arose that identical phenotypes may be determined by different genotypes (Espinasse, 1964), and by many indirect indications. The experimental work conducted on birds, which shows that there is no strict correspondence between the phenotypic and genotypic proximity of animals (B. P. Hall, 1963), is especially interesting. This conclusion acquires special significance in the light of the general rules of additive gene action. Genes with additive effect work as a unified system, but each intensifies the expression of a certain trait, and its effect does not depend on the presence or absence of other genes (Dobzhansky, 1955).

It is very significant that the additive nature of genetic differences is observed even in such biologically important traits as growth rate and body weight (Kirpichnikov, 1967).

At the All-Union conference on the hybridization of remote plants and animals in Moscow in 1968, it was especially noted that the majority of differences among crossed forms are inherited additively (see Kuz'minykh, 1968).

The additive effect of genes is only one particular case of the manifestation of their joint action, but in the scheme analyzed here, it is of special interest. It is not only that on the basis of the ecological mechanisms described, the additive gene action can lead to a sharp increase in the rates of evolutionary reorganizations; it can also lead to the coincidence of the direction of selection and of variability. This question has occupied the minds of investigators since Darwin's time, but it seems to us that satisfactory decisions have not yet been reached. If one assumes that phenotypically similar but genetically different variants are distributed in neighboring micropopulations as the result of a given selection pressure, then when the micropopulations are united, the conditions arise for the combination in a single genotype of genes possessing similar phenotypic expressions. If the action of these genes is summed as a result of the union of micropopulations into a single population, new genotypes arise that possess a stronger phenotypic expression, even if the selection pressure is reduced or stopped. It is difficult to overestimate the significance of such phenomena, but at present the views developed may be based predominantly on theory. Confirmation of these views on the basis of field and laboratory investigations represents to us an important task of evolutionary ecology. As was indicated, it is by no means always that the union of two genes of similar phenotypic expression into a single genotype leads to a strengthening of the effect. It is probably with significantly greater frequency that their joint action leads to the manifestation of new properties of the organisms. In such cases, the mixing of micropopulations leads to other, but no less important, consequences. Enrichment of the common gene pool of the population occurs, and the possibility for its fuller and more rapid adaptation to the environment increases.

This important conclusion can now be reinforced by experimental data. Lewontin and Birch (1966) studied the process of reorganization of experimental populations of Australian fruit flies of the species *Dacus*. Preliminary observations in nature showed (Birch, 1961) that during the last 100 years, *D. tryoni* extended its range by virtue of its adaptation to high temperatures. It was established that extension of the range preceded hybridization of *D. tryoni* with a very close species ("pochti vidom"—near-species) of *D. neohumoralis* (violation of reproductive isolation occurred as a result of the enrichment of the food base of both species with new species of fruit). The authors conjectured that the rapid adaptation of *D. tryoni* to new conditions of existence was connected with the enrichment of the gene pool. This hypothesis was verified experimentally. In the course of 2 years, pure cultures of *D. tryoni* and hybrids of *D. tryoni* × *D. neohumoralis* were maintained at temperatures of 20, 25, and 31.5°C. During the first year, the pure culture turned out to be more productive than the hybrid, but at the end of the second year, at a temperature of 31.5°C, the hybrids were more viable and productive than *D. tryoni.* The authors rightly concluded that the results of the experiments indicate that the introduction into a population of genes that in themselves are not adaptive leads to rapid adaptive evolution.

This finding allows one to think that the timely separation of populations (or micropopulations) of species, during which they are genetically differentiated under the influence of ecological mechanisms, and their subsequent reunion is a powerful factor in evolution.

Furthermore, this important position can be supported by direct experiments conducted by a group of investigators in our laboratory. A. V. Pokrovskii devoted a long period of study to the ecological and morphophysiological features of two subspecies of the narrow-skulled vole and their hybrids. N. A. Ovchinnikova did analogous work with the *oeconomus* voles. Both species were studied in detail in nature by O. A. Pyastolova, K. I. Kopein, V. N. Boikov, and the author. The results of this work were published in a series of articles (Shvarts *et al.,* 1960; Kopein, 1958; Pokrovskii, 1967*a,b*; Ovchinnikova, 1967). The patterns of inheritance of individual traits were established, and it was shown that the most important morphological and ecological features of the form of species are hereditary. Ishchenko (1967) analyzed the data obtained with the help of the allometric method and showed that when two subspecies are crossed, the hybrids often possess new traits and, which is especially important, these "new traits" lead to the differentiation of the species. Let us cite a few examples.

One of the most essential differences of the northern form of *oeconomus* (*M. o. chachlovi*) from the southern is the different relative growth of the interorbital distance (relative to skull length). The allometric index of the southern subspecies is 0.040±0.006; of the northern, 0.118±0.006. In other words, when the skull length changes, the interorbital distance increases more rapidly in the northern subspecies than in the southern. In first-generation hybrids, an increase in skull length, for all practical purposes, does not lead to an increase in its width (Fig. 5). A study of the inheritance of skull length gives still more interesting material. The allometric indices of the two subspecies are identical (0.461±0.006 and 0.468±0.003, respectively; it is substantially different for the hybrids: 0.405±0.012). Experiments on the hybridization of subspecies of the narrow-skulled vole gave analogous results (Figs. 6 and 7). Dependence of the dimensions of the kidneys on body weight in the northern (*M. g. major*) and southern (*M. g. gregalis*) subspecies is identical (0.536 and 0.511, respectively). The allometric index in the hybrids is 0.779 (!). The skull length of the northern subspecies, as a measure of the growth of the animals, increases more rapidly than in the southern (0.431 and 0.394, respectively; 0.501 for the hybrids). This example is especially interesting. It may be considered demonstrated that the northern subspecies descended from the southern (Shvarts, 1961*a*). Among its differences is the distinctive growth rate of the skull. Hybridization, however, leads to an even more marked increase in its growth rate.

Essentially similar observations were made on the wagtail, *Motacilla flava*. It was found that hybrid males of *M. f. flava* × *M. f. thunbergi* were distinguished by an increased resistance to cold (Sammalisto, 1968).

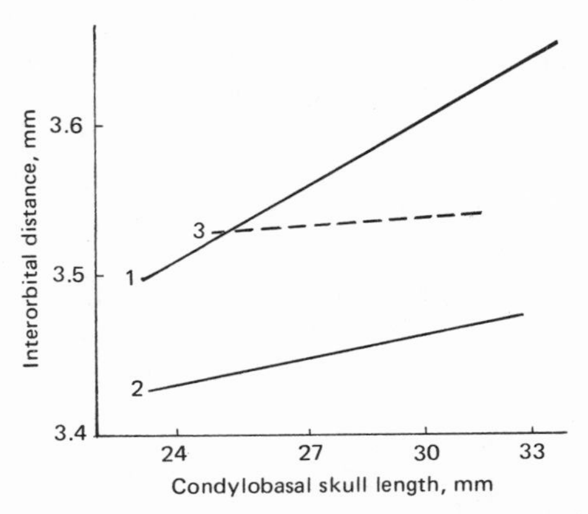

Fig. 5. Interorbital distance in southern and northern subspecies of the *oeconomus* voles and their hybrids: (1) Northern subspecies; (2) southern; (3) hybrids.

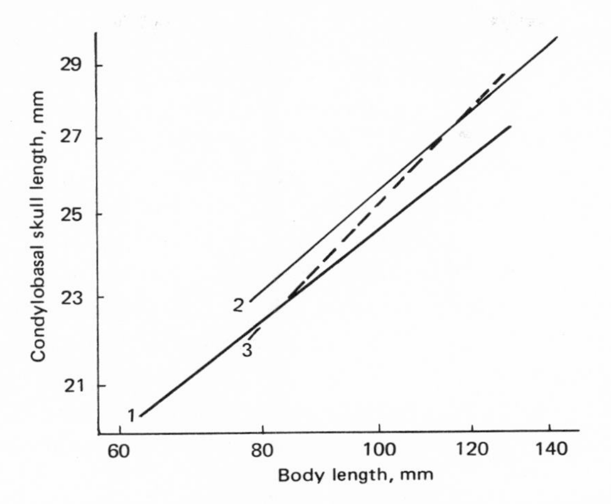

Fig. 6. Allometric growth of skull length in subspecies of the narrow-skulled vole: (1) Southern subspecies; (2) northern; (3) hybrids.

From the point of view of a geneticist, the facts cited are not at all remarkable, since it is well known that hybrids often possess new properties in comparison with the initial forms (see, for example, Dubinin and Glembotskii, 1967). The genetic mechanism lying at the basis of the phenomenon is also known (heterozygosity in many genes). The examples

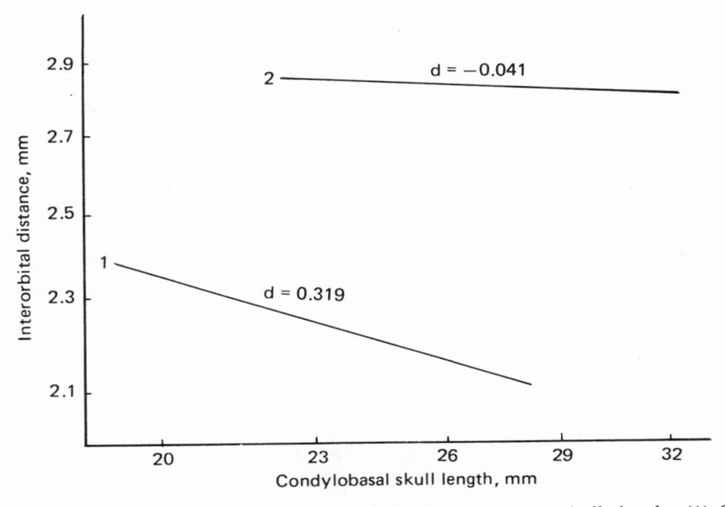

Fig. 7. Allometric growth of interorbital distance in the narrow-skulled vole: (1) Southern subspecies; (2) northern.

cited deserve attention in the evolutionary scheme, however. They show that the genetic differences among forms of one species are such that when the forms are crossed, there arise genotypes that are distinguished by a stronger expression of individual traits. It is even more important that "new" traits are detected even when externally identical forms are crossed. And in the case under consideration, the genetic nature of the phenomenon is clear: phenotypically identical forms can develop on the bases of different genotypes (see p. 129). This shows, however, that the mixing of different populations can lead to sharp morphophysiological shifts. It is self-evident that the final evolutionary result of these shifts is determined by selection in a hybrid population against a background of the segregation of these traits. Nevertheless, it can scarcely be doubted that in many cases, if not in all, the separation and subsequent reunion of populations is a powerful factor in the acceleration of the beginning stages of microevolution. For the characteristics of those processes that the zootechnologist encounters when crossing breeds of domestic animals let us avail ourselves of a quotation from the book by Dubinin and Glembotskii (1967 p. 386):

> Since the breeds have a complex and polyheterozygous genotypic structure, when crosses are made between breeds, a very complex polyhybrid segregation arises, while the majority of genotypes coming out bear a more or less intermediary character. The reproductive cross is built on this phenomenon. It is precisely on the latter that Professor Ivanov (1935) based his work on the detection of new breeds of sheep and swine. He wrote: ". . . the greater the number of genes controlling a given trait, the finer, i.e., the less noticeable, becomes segregation in the F_2 and the rarer are segregants that are close to the initial forms. The geneticist Lang calculated that if, for example, a given trait is controlled by the action of five genes, then among the 1,024 animals of the F_2 generation, only one will represent segregation to the parental forms and 912 will be of a more or less intermediary character. During the segregation of ten pairs of genes that do not have a completely dominant character the number of different phenotypes will be 59,049 and, among 1,043,776 individuals, two individuals in all will be obtained that are homozygous for all ten pairs of dominant or recessive genes. It should be noted, however, that individuals that are relatively intermediary in genotype will far from always have an intermediary phenotype, due to the phenomena of dominance and the interaction of genes. During segregation, completely new phenotypes can very often be obtained. Simultaneously, both the major difficulty and the most creative possibilities are hidden in this method."

It may be objected that in the example I borrowed from Lewontin and Birch (1966), hybridization took place between "near-species," whereas in our examples, it was between clearly expressed subspecies, which had traversed a long path of independent development. Are those distinctions that characterize neighboring populations or micropopulations sufficient? We will attempt to answer this question on the basis of material accumulated in our laboratory. And in this case we will use indices of allometric growth, since they characterize genetic distinctions among populations to a significantly greater degree than do average values (Frick, 1961; Ishchenko, 1967).

A study of *oeconomus* and two species of *Sterna* showed that even between neighboring populations that had really mixed according to a number of indices, there are such substantial differences that there is every reason to think that their hybridization would lead to just as marked an effect as the crossing of subspecies. The graphs presented (see Figs. 5, 6, and 8) can serve as illustrations of what we have said. Let us pay special attention to Fig. 8. It shows that principal differences are observed between populations of *Sterna* in northern and southern Yamal in the nature of the growth of an important organ—the heart. These differences signal distinctions in the interactions of these populations with the environment (Ishchenko and Dobrinskii, 1965; Dobrinskii, 1966). These populations are separated by considerable distances, but a change in the conditions of existence (e.g., a warming of the climate) can lead to a weakening of the conservative clustering distribution and to the intermixing of these popula-

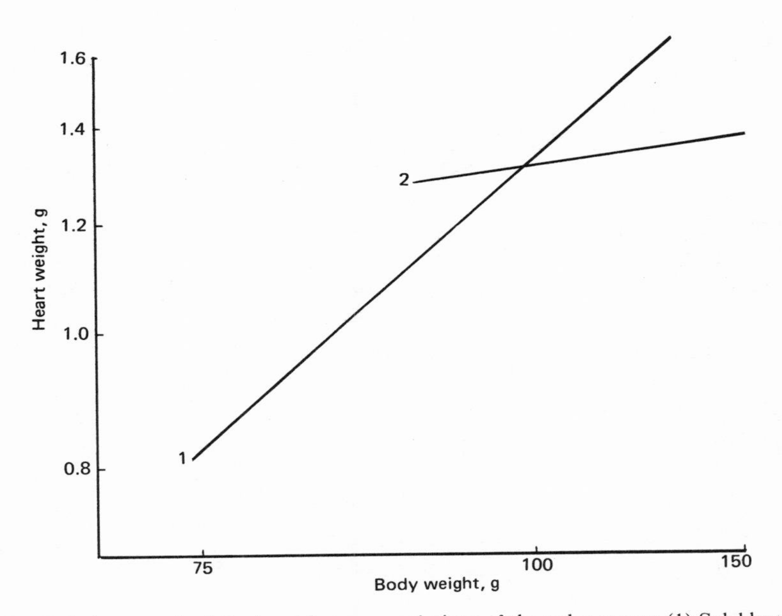

Fig. 8. Relative growth of the heart in two populations of the polar *sterna*: (1) Salekhard; (2) Northern Yamal.

tions, which would entail the same consequences as occur during the hybridization of subspecies.

It seems to me that we have a basis for speaking of an inadequately valued factor of evolution: *the dynamics of population structure*. The dynamics of population structure as an evolutionary factor is very complex. It is composed of at least the following elements:

(a) A change in the age structure of populations leads to a change in their genetic structure, which can be strengthened by eliminations that appear to be nonselective (ecological mechanisms of the reorganization of populations).

(b) The dynamics of the spatial structure of populations leads to a temporary isolation of micropopulations. This period of temporary isolation can prove to be sufficient for the origin of genetic differences among them. During the reunion and intermixing of populations, new balanced genetic systems arise under the action of natural selection. Under certain conditions, this leads to the rapid progressive development of the population and to a marked acceleration of microevolution.

In connection with the study of the role of a population's spatial structure in the reorganization of its genetic makeup, it is necessary to broach the question of the influence of predators on the evolution of prey populations. A comprehensive analysis of this vast problem is not part of our task, especially since it is investigated sufficiently fully in the general scheme. "The adaptations of predators entail the counteradaptations of prey"—this position has become axiomatic. Nevertheless, predators can play a special role in the reorganization of prey populations. It seems to us that this role has thus far gone unnoticed.

When the influence of a predator on the numbers of its prey is evaluated, the number of animals destroyed is compared with the total number in the region examined. It is observed during this evaluation that with extremely few exceptions, the predator takes only an insignificant part of the "crop" of prey. This conclusion is especially well substantiated by observations on bird predators. A generalization made recently by Galushin (1966), based on numerous data, showed that the "extent of removal" of game birds by bird predators is measured in units of or even fractions of percentages of their livestock. It is possible that predators on myomorphs eat a relatively larger number of rodents, but it is well known that even an exceptional abundance of predators cannot effect any substantial change in the numbers of flourishing populations of mice or voles. The same thing can be said in regard to insectivorous birds and insects.

A different picture is painted in the case in which the influence of the predator is evaluated, not on the basis of the total numbers of the prey, but on its numbers within the hunting grounds of the dominating species.

Galushin (1966) showed that the magnitude of the predatory birds' hunting grounds is directly proportional to the abundance of food and fluctuates, correspondingly, as much as 50–100 fold. This finding indicates that there is not a great reserve of food within the hunting grounds. The predator sharply reduces the numbers of the prey in individual areas, even in those cases in which the total number is not influenced in any way. This finding can be corroborated by the exceedingly precise observations of Zverev (1930), who established that within the limits of its hunting grounds, the kestrel almost completely annihilates the vole (results of experiments with the placing of a pole for predatory birds). We know of no such observations concerning insects, but certain indirect data make it probable that insectivorous birds (similar to predators) also significantly reduce the numbers of their prey within the limits of nesting areas. The change of hunting grounds by individual families shortly after the fledglings are capable of abandoning the nest is indicative that they do so. Similar observations have been conducted on tundra birds by colleagues in our laboratory. Thus, for example, the meadow and red-throated larks (*Anthus cervinus*) take their fledglings, which cannot yet fly, to new grounds, where they continue to feed them.

Thus, it may be considered established that a predator creates a definite vacuum within the limits of the prey territory it masters. In the process of settlement of such areas with their sharply reduced numbers, rules are observed that have a direct relationship to the problem we are analyzing. The inclination and capability for migration differ in different intrademe groups. If, for example, we avail ourselves of the example of the rodents, it may be asserted that settlement of freed territories will occur in this order: young males, young females, adult males, adult females. Subdominants will settle in the "ecological vacuum" in greater numbers than will animals occupying a dominant position in the population.

Thus, as a result of the activity of predators, even in an ideally homogeneous situation, micropopulations arise with a change in the age–sex structure. If you accept our initial premise, according to which ecologically different groups of animals also differ genetically, it follows that at the end of the season, the reproduction of such micropopulations will be distinguished by their genetic structure, and during their reunion with neighboring micropopulations, all the phenomena that accompany the mixing of genetically different groups of animals will occur.

Comparison of the conclusions of population genetics with observations characteristic of the dynamics of a population's topographic structure leads to certain essential conclusions affecting the general mechanisms of the reorganization of populations.

One of the possible variants is that despite differences in the actual conditions of existence in different biotopes, different micropopulations are subjected to similar selection forces in correspondence with the general direction of selection in a given geographical environment. The conditions in which tundra rodents live on the banks of rivers, on hill slopes, on the tundra plains, and in the shrub-brush areas are different, but they all should possess adaptations to the major factor of the macroenvironment—the cold summer. This is equally true in regard to any biome and its chorological subdivisions. The conditions of existence of the green toad in steppe reservoirs of different types (an excellent example of micropopulations) are different, but any populations of this species in an arid climate should possess a high capability for migration, which is elicited by the partial drying up of reservoirs. Populations of the green toad could not exist in the steppes and even less in semidesert if they did not possess the capability to utilize temporary reservoirs for reproduction. A great many such examples could be cited.

Different specific manifestations of the dynamics of numbers in different micropopulations, under a common selectional force for the population as a whole, should lead to the situation that during the temporary reunion of isolated micropopulations or populations of the species of one geographical region, their "successes" are summed, strengthened, and made the property of the population as a whole. In other cases, the "acquisitions" of the micropopulations are not summed, but create the basis for an integration of new genetic variants, more fully adapted to the conditions of the geographic environment. Age selection can play a large role in this process. We analyzed its rules in Section 1. The possibility of the rapid reorganization of a population's genetic structure under the influence of age selection strengthens its temporary genetic divergence and, precisely for this reason, maintains the genetic structure of the population as a whole, furthering its rapid directional reorganization, to say nothing of the continuous enrichment of the common gene pool.

It seems to us that the views developed here are in just as good agreement with the data of experimental and population genetics as are the basic postulates of the synthetic theory of evolution, but they take into account, to a significantly greater degree, the achievements of modern ecology. We may therefore consider the most important task of evolutionary ecology to be the study of the dynamics of numbers and the dynamics of the genetic structure of micropopulations, which are part of a biological macrosystem of a higher rank. The general conclusions of the views developed here differ essentially from the main conclusions of the synthetic theory of evolution.

Rapid evolutionary reorganization of populations (microevolution) does not demand the subdivision of species populations into partially isolated populations of moderate size (one of the most important postulates of the Mayr-Wright theory; see Chapter 1). No other isolation of micropopulations except that which is determined by ethological mechanisms and different dynamics of numerical size is demanded. On the other hand, a constant exchange of genes speeds up the process of evolutionary development. Biotopic diversity of the territories occupied by populations speeds up the microevolutionary process, but even this diversity is not necessary, since even an insignificant change in environmental conditions or a chance disturbance of population structure at the beginning of the animals' reproductive season leads to a change in the type of dynamics of numbers, with all the ensuing consequences.

Thus, our conclusions remove a series of difficulties, as well as one of the main contradictions of the synthetic theory of evolution, which is that isolated populations evolve most rapidly. It is clear to any impartial investigator, however, that the magistral progress of evolution takes place, not under the conditions of isolation, but in the large areas of the arena of life, exposed to all winds of the struggle for existence. It cannot be forgotten that the whole history of the animal world clearly shows that evolution is that process which led from the amoeba to man and that the magistral line of reorganizations consists of the adaptation to the macroenvironment—to life on dry land, in the forests, the steppes, the mountains. . . . It may be objected that especially rapid reorganizations of species are actually accomplished on islands. The island forms, however, are the monsters of evolution. The truth of the matter is that evolution in the large areas of the arena of life does not occur more slowly than under the conditions of island isolation, but that it does not admit of deviations from the magistral path, and therefore does not produce forms that catch the eye, such as *Raphus cucullatus,* Darwinian finches, the moa, the Galapagos iguana, and so forth. It should be noted, however, that the main factor of evolution on islands is also natural selection, and not genetic automatic processes based on the founder principle, although it is precisely here that these processes have relatively great significance. Ideas on the dominant role of chance in the evolution of island forms arose simply because "monsters of evolution" attracted more attention. Careful investigations conducted in recent times (Berry, 1964; Corbet, 1961, 1964; Eisentraut, 1965; Foster, 1964; Radovanovic, 1959, 1961; Soulé, 1966; and others) showed that, as a rule, evolution on islands is subject to the same rules as on the mainland, and the general features of island forms are explained by the general features of islands as the arena of life (impoverishment of the biocenosis, small numbers of predators, and so forth). The impoverishment of island biocenoses and the consequent weakening of the struggle for existence admit of well-known deviations from the magistral line of evolution, and the increased role of chance factors can lead to an acceleration of the tempo of evolutionary reorganizations. It cannot be doubted, meanwhile, that any increase in the evolutionary rate can also be connected with the enrichment of the biocenosis. The simplest case is the introduction into the biocenosis of a new predator, which increases the intensity of natural selection, with all the ensuing consequences. The investigations of Pimentel (1965) illustrate a more complex situation. He showed that when two close species live in the same habitat, this leads not only to their rapid evolutionary reorganization, but also to the evolution of their food plants.

The views expressed flow from a theoretical analysis based on a comparison of the data of population genetics with the data of population ecology. These views find factual documentation in the general course of the evolutionary process.

If the rates of intraspecific formations in the continuous parts of a species range were determined by subdivision of the species populations into partially isolated local populations, in which differentiation occurred under the joint influence of chance processes (the founder principle) and individual and interdeme selection, as Wright's scheme would have it and as adherents of the synthetic theory of evolution unanimously insist, then essential differences should be observed among different groups of animals within the limits of individual biomes. Small species, for which a river is in itself a barrier, form a great number of local populations, which large, widely migrating species almost never form. The rates of evolutionary reorganization of large forms nevertheless do not differ essentially from the rates of reorganization of small ones. This contradiction to Wright's scheme, it seems to us, has remained unnoticed.

Let us make use of the tundra zone for analysis. In its immense territory can be counted hundreds and thousands of local populations of lemmings, narrow-skulled voles, Middendorf voles, the *oeconomus* voles, the Asiatic red-backed voles, and shrews. Even so, small mammals of the tundra are strikingly monotonous. They almost never form groups that would serve to divide them into independent intraspecific taxa (Rausch, 1953; Shvarts, 1963*b*). The large mammals of the tundra (the northern deer, the polar fox) actually form local populations almost nowhere (with the exception of islands). Migrations over enormous distances hinder this. The intraspecific differentiation of these species is nevertheless no less than in the arctic voles. In contrast to the voles, the large arctic species form a few clearly marked subspecies (Druri, 1949; Rausch, 1953; Herre, 1964). What we have said does not mean that evolutionary forms arise more slowly in the arctic mammals. No, there is reason to assert that they occur at a greater rate, but they are not accompanied by the divergence of populations: a species develops as a single whole over a vast space. The example of the lemmings (*Lemmus, Dicrostonyx*) is especially significant in this regard.

They are one of the youngest genera, but their degree of morphophysiological originality is exceptional.

On the other hand, species that literally personify Wright's scheme evolve extremely slowly. The best example is the Urals warbler. N. S. Gashev, of our laboratory, investigated populations of warblers (*Ochotona hyperborea*) from five sites in the polar Urals. The distance between populations is from 10 to 60 km. Taking into account the mountainous relief of the region, the investigated populations can be considered to have developed independently, although an insignificant exchange of individuals should occur among them, since the intermixing of warblers over several kilometers (during one season) has been shown (Gashev, 1966). These are typical small, partially isolated populations. Study of them therefore helps to form an idea of how Wright's scheme (see Chapter I) works in nature. N. S. Gashev studied 16 craniological indices (body weight, proportions of the body, and a series of morphological indices) of these populations. They proved to be identical. (This was confirmed by statistical analysis of the material.) We will cite only the indices of individual populations for the most important taxonomic traits. The condylobasal length of the skull in the populations compared proved to be equal: 36.3, 35.4, 35.4, 35.3, 35.3, 35.4 mm; length of the tooth row: 7.6, 7.3, 7.3, 7.1, 7.2 mm; malar width: 19.6, 19.6, 19.5, 19.5, 19.6 mm; length of the nasal bones: 11.4, 11.3, 11.4, 11.1, 11.1 mm; index of skull length (in relation to body length): 106.9, 106.2, 106.3, 106.4, 106.6‰; index of the length of the tooth row (in relation to the condylobasal length): 20.7, 20.9, 20.9, 20.3, 20.3; index of malar width: 54.1, 55.3, 55.1, 55.3, 55.3%. All the taxonomically essential traits of the compared populations proved, without exception, to be identical (which makes one astonished that not even by chance during the collection of material from different populations are more than a very few even unessential differences detected).

Again, only insignificant differences in morphophysiological indices are detected among populations. This is illustrated by the following examples: heart index: 5.6, 4.1, 4.6, 4.7, 4.5‰; liver index: 49.3, 42.1, 48.4, 50.6, 49.1‰; kidney index: 7.3, 7.5, 6.9, 6.9, 7.2‰; adrenal index: 0.11, 0.17, 0.16, 0.15, 0.16‰; lungs: no data, 9.1, 9.6, 9.1, 7.7‰.

Developing under conditions of isolation for a long period of time, populations of warblers did not undergo divergence, since they developed under identical conditions of existence.

Another example: differentiation of populations of the northern *oeconomus* voles (*Microtus oeconomus*), which are distinguished from the southern groups in the relative length of the tail, relative length of the rear foot, index of the condylobasal length of the skull, width of the interorbital distance, and the malar width. Pyastolova (1967) made a comparison of these indices in three isolated populations of the voles (Yamal, polar Urals, and islands in the lower reaches of the Ob'). Animals of identical age and dimensions were compared, significantly increasing the reliability of the results. The following data were obtained: Tail length in corresponding populations proved to be equal (average values are indicated): 38.5, 40.0, 38.2 mm; foot length: 19.1, 18.5, 19.4 mm; condylobasal length of the skull: 26.2, 26.4, 26.6 mm; malar width: 14.2, 14.2, 14.0 mm; width of the interorbital distance: 3.6, 3.5, 3.5 mm. And, in this case, populations that were isolated but developing under identical conditions proved to be virtually identical.

In our laboratory, V. N. Bol'shakov studied populations of the Promtheomys voles (*Promtheomys schaposhnikovi*) in the main Caucasus range (18 individuals) and in the Adzharo-Imeretinskii range (72 individuals). The isolation of these populations was complete. Investigations showed practically complete resemblance of animals in these populations in body dimensions and craniological indices. Body length is equal, being 130.6 and 126.9 mm, respectively (animals of the same age are compared); the condylobasal length of the skull is 30.62 and 30.53; the length of the tooth row is 7.31 and 7.32; the malar width is 17.45 and 17.22; the height of the cranium is 11.9 and 11.33; the length of the facial structure is 18.43 and 18.51; the interorbital distance is 4.02 and 4.29; the length of the diastema is 9.03 and 8.99. The populations compared proved to be very similar in internal traits as well. Since even more significant differences were expected here, we will cite the average values obtained with the statistical error. The relative heart weight of voles from the main Caucasus range is 4.5 ± 0.17‰; from Adzhar, 4.1 ± 0.10‰; the liver is, respectively, 49.3 ± 1.71 and 50.4 ± 0.95; the kidneys, 5.4 ± 0.18 and 4.5 ± 0.11; the relative length of the intestines (%) is 480 ± 11 and 450 ± 6; of the caecum, 37 ± 1 and 39 ± 0.7.

Citing numerous examples of rapid evolution on islands, adherents of the synthetic theory of evolution often forget to point out facts of a completely different nature. Thus, the careful investigations of Fleming (1962) showed that the age of endemic genera of New Zealand is Neocene; of species, Pleistocene. But surely new genera of obviously mainland origin are not a bit older (*Lemmus, Dicrostonyx,* possibly *Rangifer, Ovibos, Alces, Saiga, Passerina, Lagopus, Nyctea,* and others).

These facts already clearly show that in the continuous areas of the species range, conditions often form that promote just as rapid evolution as on islands, but, in contrast to the island forms, the mainland forms are characterized by biologically essential adaptive features. They are found in the mainstream of the evolutionary process. Nor is it difficult to understand the primary causes of their rapid evolution. It is a change in environmental conditions. All the young genera of vertebrates—inhabitants of the tundra and the taiga—are from the youngest biomes. New environmental conditions elicited a burst of new formations and speciations.* Disadvantageously, from the viewpoint of the synthetic theory of evolution, the structure of species (weakly expressed splitting of the species into local populations) does not reduce the rate of phylogenetic reorganizations of the northern forms. We see the cause of this in the ecological mechanisms of the evolutionary process.

Finally, it is possible to cite still one more general line of evidence to support the views developed here. If the presence of local, genetically isolated populations were the main cause determining the rates of evolutionary reorganizations, then, other conditions being equal, the evolution of small settled forms would occur more rapidly than that of large forms, capable of distant migrations, during which a wide exchange of genes is inevitable.

The rate of evolution of the horse does not differ essentially from the rate of evolution of small steppe rodents. The rapid replacement of species is characteristic of the evolution of elephants (Zeuner, 1955). Hecht (1965) recently turned his attention to this question. Remembering that according to the calculations of Haldane, no fewer than 300,000 generations are needed for the establishment of a new species, Hecht points out the paleontological data showing that this number is overestimated by almost 10 times. For the evolutionary series *Archidiscodon subplanifrons—A. planifrons—A. meridonalis—Elephas* species, the replacement of 100,000 generations in all was required. Some species of whales are 15–45,000 years old (Davies, 1963). Let us remember, finally, the evolution of man: over several hundreds of thousands of years, the path from pithecanthropus to modern man was traversed; at least a few evolutionary blind alleys of species rank arose. It is not likely that our ancestors formed local populations, but their capability for migrations seems demonstrated now. Such facts, and they are by no means rare, compel us to accept the possibility of rapid evolution in continuous areas of a species range. The scheme advanced by us gives them a natural explanation. Analysis of the probable significance of the dynamics of a population's spatial structure leads us to the conclusion that in a continuous area of the species range, evolution can proceed with great speed, not swerving from the magistral path determined by environmental conditions and corresponding to the direction of natural selection. The main motive force of this process is the ecological mechanisms of the reorganization of populations.

* And in plants in the higher latitudes, evolution proceeds at a faster rate (Vasil'ev, 1965).

Chapter VI

Speciation

Speciation is the central event of evolution. With the isolation of a new species, two closely related forms acquire evolutionary independence. The basic criterion of species, which systematics has elaborated in its long development, is still in force. This is the so-called "triple criterion": species are genetically isolated (i.e., unable to interbreed); a hiatus is always found between very near species; species possess independent ranges. In the majority of cases, this criterion works in the resolution of the most complex taxonomic problems. Difficult cases do not compromise the triple criterion, since practical difficulties are unavoidable in the application of any theory to the boundlessly diverse phenomena of the living world.

It is not difficult to see that the triple criterion is actually twofold. The most important property of a species—its independent range—is a direct consequence of its genetic isolation. One subspecies cannot exist in the range of another, since, as a result of crossing, they become a single whole. Only in very exceptional cases, when the biological differences between subspecies is extraordinarily great, can the exclusion of one subspecies by another occur. But in such cases, the exclusion of one subspecies by another can hardly be complete; it is probably accompanied by their mutual absorption. The genetic isolation of a species allows it to penetrate into the range of another, even the closest. The competition engendered by this penetration is overcome with the help of special mechanisms that have been thoroughly studied.

The basic type of speciation is geographic speciation. Although the author's arguments pointing out the possibility of sympatric speciation are not without foundation, the general structure of the living world as a whole clearly indicates that speciation occurs in the process of the adaptation of closely related forms to different geographic environments, to different parts of an originally united range. It is not by chance that the vast majority of species can be given a clear geographic characterization. It is not by chance, either, that similarly to subspecies, many species represent both a circle of races (Rassenkreis) and a circle of forms (Formenkreis) or of species (Artenkreis), which outwardly are fully analogous phenomena (Kleinschmidt, 1926; Rensch, 1929).* After the isolation of a species, its geographic specification may be blurred: the species expands the original province of its distribution and may populate a number of geographic zones. Even in this case, however, the origin of a species makes itself very clearly known. In the steppe forest or the tundra forest, the fauna of which is composed of elements of different origin, the species of forest origin are confined to areas of the forest, those of steppe origin to the steppes, those of subarctic origin to the tundra, and so forth.

A flourishing species conquers a range that contains areas with different conditions of existence. Differentiated geographic forms arise; in extreme cases, these are subspecies. We attempted to delineate the course of these processes in the previous chapters. What events should occur in order for the geographic form to be changed into a species?

There are two possible answers to this question. The first is that changes arise by chance in one of the forms, making it impossible for this form to cross with other forms of the species. As will be shown below, some investigators even now defend this point of view.

Another approach to the problem is also possible. Speciation is the result of progressive reorganizations of the geographic form, accompanied by perfection of its adaptations to the specific environmental conditions. The final result of this process is to lead to those changes in the organism that we assess to be specific. This viewpoint on the process of speciation is ecological in essence. Before proceeding to an ecological analysis of the problem, however, we must assess the possibility of solving the problem of speciation on the basis of formal criteria, on the basis of immunological, hybridization, and karyological data.

It is well known that one of the most important distinctions between species and subspecies is their different behavior when crossed. Subspecies of the same species cross and give completely fertile offspring. Crossed species are either infertile or produce offspring that are not completely fertile. Despite a large number of exceptions to this rule, its significance remains indisputable. It is not limited to diagnostic value, but has a deep and fundamental nature.

* As an example, one may point out the monographic investigations of Kratochvil (1965) on the systematics of water voles. In the author's opinion, the genus *Arvicola* corresponds to Kleinschmidt's circle of forms.

A number of modern biologists view fertile crossing as the main characteristic of a species. Thus, Dobzhansky (1941) understands by species that stage of the evolutionary process "in which, for the first time, the actual or potential crossing of forms that turn out to be divided into two or more different formations that are not in a physiological state to cross with one another is regulated." Bobrinskii (1944) writes: "Only on the basis of the criterion of fertile crossing can we substantiate the concept of species, contrasting it with all higher-order systematic categories. For, in contrast to all higher-order systematic categories, only the species represents an aggregation of individuals bound into a single whole by the capability of normal crossing, as a result of which the traits of different individuals are constantly intermixed in their offspring and the species can evolve as a whole." It is well known, however, that in captivity, a large number of species bear fertile offspring when crossed and are not distinguished in this sense from subspecies.

Among mammals, different species of the canine family give fertile hybrids (*Canis lupus* × *C. aureus*, *C. latranus* × *C. aureus*, *C. familiaris* × *C. aureus*, *C. familiaris* × *C. lupus*), as do species of the genus of goats (at least seven interspecific hybrids are known), of the genus of sheep (three interspecific hybrids are known), of the llama genus (four interspecific hybrids), of the ox family (at least eight interspecific and even intergeneric hybrids), and so forth. It is not necessary to continue this list, since a compendium of mammalian hybrids is available (Gray, 1954). It remains only to emphasize that among mammals (just as among birds and lower vertebrates) hybrid forms are known that it would be difficult to consider closely related. Thus, for example, a hybrid of the brown bear with the white (*Ursus arctos* × *Thalassarctos maritimus*) is known, and the offspring were found to be fertile when breeding "among themselves." In the London zoo, a fertile female was obtained from a cross of *Elaphurus davidianus* × *Cervus elaphus*, which when pairing always chose a male of the red deer (Kelham, 1956). On the whole, one can say that there are interspecific hybrids in all orders of mammals, and no small number of them are fertile.

There are many fully fertile interspecific hybrids among birds. Hybrids of domestic chickens with different species of wild fowl prove to be fertile (*Gallus varius*, *G. bankiva*, *G. lafaietti*, *G. sonnerati*, *G. gallus*). The fertility of different hybrids of pigeons has also been demonstrated (*Streptopelia orientalis* × *S. decaocto*, *S. risoria* × *Onopopelia humilis*, *Columba oenas* × *C. livia*, *C. palumbus* × *C. oenas*), as has that of representatives of some other groups (*Fringillidae*, *Laridae*, *Tetraonidae*, and others).

Among the reptiles, there are fewer interspecific hybrids. As a result of crosses conducted among 17 species of lizards of the genus *Lacerta*, Lantz

(1926) succeeded in obtaining only one interspecific hybrid: *Lacerta fuimana* × *L. muralis*. Nevertheless, fertile hybrids are known even among reptiles. There are, for example, fertile hybrids of two species of runners: *Elaphe guttata* × *E. quadrivittata* (Lederer, 1950). Mertens (1956) cites data on hybrids of tortoises, lizards, and snakes. He described 14 interspecific hybrids, many of which are fertile. Specially designed experiments (Montalenti, 1938) showed that at least 7% of the known interspecific amphibian hybrids are fertile. Many species of toads (genus *Bufo*) give fertile hybrids. Hybrid larvae of *Hyla microcephala* × *H. phlebodus*, distinguished by good viability and rapid growth (Fouguette, 1960), have been obtained by means of artificial fertilization. The obstacle to natural hybridization of these species is minute but clear differences in the character of the mating cells, which were established sonographically. Peculiarities of the mating calls also serve as an obstacle to the crossing of other species of hylids (W. F. Blair, 1958), a genus most fully studied in this regard (Kennedy, 1964). Interspecific hybrids of tritons of the genus *Taricha* (Twitty, 1964) have also been described. It is interesting that they are distinguished by a higher rate of metamorphosis.

All these data are not new. They relate to the fact that under experimental conditions, a large number of fully fertile interspecific hybrids are obtained and show that the offspring of different species are often capable of normal reproduction. This situation is true of all classes of terrestrial vertebrates.

In a large number of cases, however, it proves to be impossible to obtain interspecific hybrids, or the hybrids obtained possess lowered viability and fertility. The inability of different species to cross is explained by diverse causes: the structure of the reproductive organs, which excludes the possibility of pairing; the low vitality of sperm cells in the female's reproductive tract; differences in the rate and nature of embryonic development; the inability of spermatozoa to penetrate and fertilize the egg; the resorption of paternal chromatin; and, finally, marked disturbances in development, which are the consequences of a hereditary incompatibility of the species crossed.

The enormous significance of the factor of hereditary incompatibility is shown by experiments on the transplantation of embryos to "recipient mothers." Thus, during the transplantation of a sheep embryo into a goat, the embryo dies in the early stages of development in the majority of cases. In individual cases, the embryo completes its development, but there is a definite diminution in growth during development (Lopyrin *et al.*, 1951).

In some Lepidoptera, hybrid males develop normally, but the females die in the pupal stage. If, however, hybrid pupal females are given a transfusion of hemolymph from the father or the paternal species, they continue to develop normally and female Lepidoptera emerge (Meyer, 1955). This in itself shows that normal development of the hybrid embryo depends directly on the biochemical features of the internal environment of the organism. Vojtiskova (1959–1960) especially has studied this question. The data she obtained deserve a detailed exposition. Parallelism between tissue incompatibility and the results of experiments on distant hybridization led the author to the supposition that phenomena connected with the immunological reactivity of the crossed animals lay at the basis of disturbances of normal fertilization and of the decreased viability of hybrids. This supposition was verified experimentally. The author's experiments on skin transplantation in different birds showed that injection of marrow cells from the donor into the embryos of future recipients facilitates the mutual transplantation of tissues later on. In the process of development, the recipient adapts to the antigens of the donor and ceases to produce antibodies against them. Results analogous in principle were obtained in experiments on crossing guinea fowl with roosters. The blood of guinea fowl was introduced into an experimental group of roosters, and the blood of roosters into the guinea fowl. A study of the agglutination titer showed that the suffusion of foreign blood led to a sharp curtailment in the production of antibodies against the foreign erythrocytes. When roosters and guinea fowl from the experimental groups were crossed, about 50% of the hybrids possessed normal viability, although they were found to be sterile. In the control group, all the hatched chickens (less than 6% of the number of fertilized eggs) died within 5 days. The experimental results confirmed the presence of a connection between the immunological reactivity of animals and their capability to hybridize, although the mechanism of this connection remains unelucidated.

Numerous experiments showed that the capability for normal fertile crossing corresponds to the capability for mutual acceptance of tissues during transplantation (Gotronei and Perri, 1946).

The most general cause determining the possibility of successful crossing is the interrelationship of the nucleus and the cytoplasm (Zeller, 1956; Perri, 1954; Chilingaryan and Pavlov, 1961; and others). This problem was studied experimentally in especially great detail by Weiss (1960) in amphibians. The chromosomes of *Bufo bufo* contain a diploid genome.* Accordingly, when females of this species are crossed with males of *B. viridis*, the hybrids have a diploid maternal genome. They complete metamorphosis fully viable, but sterile. When the reciprocal crosses are made, gastrulation in the hybrids is disturbed, and the larvae die at early stages of development. The explanation of the experimental results is that in the first

* Recently, this has been corroborated by a study of the nuclear DNA content of different species of toads (Ullrich, 1965).

variant of crossing, the interaction of the diploid genome of *B. bufo* with the species-specific plasma ensures normal development of the embryo (the harmful action of the paternal genome is suppressed). Doubling of the nucleus of *B. viridis* was successfully elicited experimentally (by the action of high temperatures). When crossed with similar individuals, the diploid genome of *B. viridis* corresponds to their diploid genomes. The development of the hybrids proceeds well, but metamorphosis is not completed. This shows that for normal development of the embryo, the ratio of 2:2 of its own to the foreign genome in the plasma is still insufficient. Doubling of the *B. calamita* genome also elicited an improvement in the development of hybrids of the latter species with *B. viridis*. Disturbance in development elicited by the alien genome is compensated by the diploid genome of the maternal cell.

Under natural conditions, the basic obstacles to interspecific hybridization are often causes of not so much a physiological as an ecological order—different habitats, different breeding periods, sexual aversion, and so forth. Nevertheless, a large number of interspecific hybrids are known, many of which give fully fertile offspring and therefore could take part in the evolution of their engendered species if fertile crossing were the sole determinant of the possibility for the evolution of the given community of forms as a whole.

Natural interspecific hybrids are by no means a rarity, as is sometimes suggested. They are widely distributed in all classes of terrestrial vertebrates. Natural hybrids among the amphibians are known for many species of toads (genus *Bufo*). Where *Bufo americanus, B. terrestris, B. towleri, B. woodhousei,* and *B. vallipes* meet, they give natural hybrids. It has been shown experimentally that the hybrids produced are fertile (A. P. Blair, 1941). In areas of joint habitation of *B. woodhousei* and *B. vallipes,* they form mixed groups; 8% of the pairs consist of different species, but the number of hybrids does not exceed 1% (Thornton, 1955). Hybrid tadpoles are viable and grow and develop rapidly, but the adult individuals are infertile. In some cases, hybridization between morphologically distinctively differentiated species becomes a mass phenomenon. *Hyla cinerea* and *H. gratiosa* can serve as a good example. There is a great deal of overlap in the ranges of these species. Reproductive isolation is maintained by ecological mechanisms. In the state of Alabama, the construction of artificial reservoirs caused a sharp reduction in the numbers of *gratiosa* and an increase in the numbers of *cinerea*. The reproductive isolation was disturbed. In the hybrid population, all transitions from the typical parental forms were observed, including the traits by which the compared species were distinctly differentiated (hiatus!). Typical *gratiosa* and *cinerea* and individuals that, according to their traits, corresponded to 1st-generation hybrids were predominant, however. This finding suggests the possibility of reproduction within the hybrids. This possibility was confirmed by observations that showed that the hybrids possessed an increased mutual sexual attraction, probably determined by the specificity of their mating calls.

The hybridization of *B. americanus* × *B. fowleri* illustrates a somewhat different situation (Cory and Manion, 1955). Crossing between these species occurs often, but the diminished viability of the hybrids is clearly discernible. Typical *americanus* are confined within forests, *fowleri* in open places. Hybrids are encountered in mixed geographical areas, but their adaptability in no way achieves the perfection of the initial forms. The authors attribute the relatively small numbers of the hybrids to this imperfect adaptability. It is suggested that the species were previously isolated, and that the contiguity of their ranges was a consequence of the felling of the forests.

Natural hybrids are known among reptiles, e.g., *Vipera aspis* × *V. ammodites* (Schweizer, 1941). Some more recent investigations (Darevskii, 1967; Cooper, 1965; Mertens, 1950, 1956, 1963; Taylor and Medica, 1966; and others) permit us to think that the interspecific hybridization of reptiles in nature occurs more often than is determined by direct observations.

Among birds, as early as the end of the last century, hundreds of natural interspecific hybrids had been observed (for a survey, see Gray, 1958). Suchetet (1890) showed that in the orders of gallinaceae, passerines, lammelirostres, and diurnal birds of prey, there number no fewer than 20 hybrids. There are a very large number of natural hybrids among hummingbirds (Berlioz, 1929, 1937) and pigeons, for example, as was already mentioned, a large part of them being fertile. Dement'ev (1939) cites a large body of data on hybrids in lammelirostres. Two woodpeckers—*Colaptes auratus* and *C. cafer*—give hybrids where they meet (Amadon, 1950).

Analogous observations have been cited in regard to the honey-eaters of New Guinea—*Melidectes leucostephes* and *M. belfordi* (Mayr and Gilliard, 1952). Just as in amphibians, hybridization is sometimes observed in birds as a regular phenomenon. Sibley (1954) shows that in Mexico, as a result of the felling of forests, the ecological barriers between *Pipilo ocai* and *P. erythrophtalmus* were disturbed, and this disturbance led to the massive appearance of hybrids (24 stages of hybridization were noted). Hybridization between these species has already been going on for 300 years.

It would be possible to cite a great many more such data, showing that among birds, natural interspecific hybrids are by no means a rare phenomenon. It is not always possible to establish the degree of their fertility, but the material cited above on the fertility of hybrids obtained in captivity leads one to think that a small number of natural bird hybrids are fertile.

Of the mammals, perhaps the best known is the kidus, a hybrid of the sable and marten (*Martes zibellina* × *M. martes*). It has been shown (Starkov, 1947) that when the kidus is crossed with the original forms, it is completely fertile. The fertility of the kidus females has a practical application in fur-breeding: to obtain the so-called "secondary hybrids" (kidus–sable). The offspring from the kidus females and the Barguzin sable are valued as sable of very high quality. In the literature, there are many indications of other interspecific hybrids in the marten family, e.g., *M. americana* × *M. caurina* (P. Wright, 1953). In captivity, hybrids of the light polecat with the African (*Putorius eversmanni* × *P. furo*) have been obtained. The hybridization of the polecat and the mink has been noted. Hunters know well the hybrid between the common hare and the white hare (the so-called "tumak," *Lepus europaeus* × *L. timidus,* often encountered in regions of common habitation for both species). Among insectivores, there is the hybrid of *Erinaceus rumanicus* × *E. europaeus* (Herter, 1935). Among mouselike rodents, interspecific hybrids are noted relatively rarely, but individual cases have been determined—e.g., the hybrid of *Microtus californicus* × *M. montanus* (Hatfield, 1935). Hybrids of the suslik—*Citellus pygmaeus* × *C. major, C. major* × *C. fulvus* (Bazhanov, 1945), *C. armatus* × *C. beldingi oregonus* (R. E. Hall, 1943), and other species (Denisov, 1961)—are known, but the degree of their fertility has not been studied. A hybrid of the fallow deer (*Dama dama*) and red deer has been described. In all probability, these hybrids are fertile, since it has been indicated in the literature that hybrid females with young have been encountered (Hagendorf, 1926). A better studied natural hybridization is that of the spotted deer, *Cervus nippon,* with the "izyubr" (a large eastern Siberian deer), *C. elaphus xantopigus* (Mirolyubov, 1949; Menard, 1930). In captivity, hybrids of these species are obtained easily, and they are fully fertile. The full fertility of natural hybrids, known to the Chinese by the name *chin-da-guiza,* has also been shown. The izyubr takes over the harem of the spotted deer, which leads to the appearance of a large number of hybrid individuals. Just how common these hybrids are in the Ussuriysk territory is evident from the fact that R. K. Maak and N. M. Przheval'skii considered them an independent species.

Analysis of the research on hybridization leads to two important conclusions: First, it clearly shows that the degree of genetic incompatibility of different species varies from complete isolation to almost complete fertility. Second, it makes it clear that *genetic incompatibility develops under the conditions of independent development of closely related forms*: it is the result of their development under distinctive environmental conditions, the result of their adaptation to the specific conditions of existence. When the ranges of similar forms become contiguous, full reproductive isolation is developed on the basis of this genetic incompatibility, since the production of inviable offspring is disadvantageous for both species, and selection therefore works against their origin.* If, however, the decisive stage of development of genetic incompatibility had already been passed through, selection would be powerless. If the offspring from the crossing of two closely related forms are not distinguishable from their parents in degree of viability, this leads to the integration of the gene pool on a new basis. As was indicated earlier, this integration takes place during the crossing of populations and subspecies, but not of species. Thus, *hybridization research does not and cannot answer this question: Why do separate forms acquire features of species rank, and what are the mechanisms of this process?* Analysis of the very interesting research on comparative karyology leads to conclusions that are similar in principle.

It is natural that the specificity of the role of the chromosomes in the transfer of hereditary information led some investigators to the idea that the karyological features of separate forms may be used as decisive criteria in the determination of their taxonomic rank. The first work in this direction gave hopeful results. Differences in the number of chromosomes and in their structure were established in near species of susliks (Nadler, 1962), of the shrew (Halkka and Skaren, 1964), of pigeons (Sharma *et al.*, 1961), of water voles (Matthey, 1955), of chameleons (Matthey and van Brink, 1960), of thrushes (Udagawa, 1955), of squirrels (Nadler and Block, 1962), and so forth. Data were obtained to the effect that subspecies of one species are not essentially distinguishable according to karyological traits (Monroe, 1962). All these data (a fair number have now been obtained) led some authors to the conviction that it is possible to construct taxonomic conclusions on the basis of a study of the morphology of the cell's chromosomal apparatus. The interesting work by Vorontsov (1967 and others) is characteristic in this regard. Here are the conclusions of one of his investigations (Vorontsov *et al.*, 1967; p. 705): "The presence of differentiation between allopatric and geographically separate populations of *Phodopus sungorus sungorus* and *P. s. cambelli* testifies to their far-reaching divergence. It would be more correct to view these two forms, which are genetically isolated from one another, as allopatrically arising species *in statu nascendi*: *P. sungorus* and *P. campbelli,* belonging to the superspecies *Phodopus sungorus*." The karyological differences between the compared forms reduce to differences in the structure and sizes of the X chromosome. Vorontsov illustrates an

* The correctness of this position is confirmed by the following rule: spatially divided populations of two closely related species are very similar, and it is difficult to distinguish them; in places of joint habitation, the same species are clearly distinguishable (for numerous examples and a theoretical analysis of the problem, see W. L. Brown and Wilson, 1956).

analogous approach to the problem of species in other investigations as well (Vorontsov, 1958, 1960, 1967, and others).

As we have tried to show above, such conclusions go beyond the limits of methodological questions (is it or is it not possible to distinguish species on the basis of comparative karyological data?) and have very serious general biological significance for, in particular, an understanding of the role of the ecological factor in the evolutionary process. Even from a purely technical viewpoint, however, the problem touched on turns out to be complex. The difficulties are related, first of all, to an obvious chromosome polymorphism in a number of forms (Matthey, 1963, 1964; Pasternak, 1964; Rao and Venkatasubba, 1964; Nadler, 1964). It is possible, finally, to treat chromosome polymorphism as a manifestation of the beginning stages of sympatric speciation, but in a great many cases, this explanation clearly seems to be stretching the point. This has been understood for a fairly long time. Thus, Hamerton (1958), on the basis of a study of questions on the cytotaxonomy of mammals, came to the conclusion that even differences in the number of chromosomes (to say nothing of differences in the structure of individual chromosomes) are insufficient to subdivide morphologically homogeneous populations into taxonomically important units. This conclusion seems important, since cytological differences are detected even among forms that are absolutely indistinguishable morphologically (Valentine and Löve, 1958, and others). That many species have an identical number of chromosomes and sometimes are generally not distinguishable at the karyological level is well known (Sandness, 1955; Schmidtke, 1956; Tappen, 1960; and others). Significantly more important are some results of contemporary investigations, which show that important karyological differences can be observed in forms that are not always deserving of even the lowest taxonomic rank. Marked cytological differences were detected among subspecies of *Bufo boreas* (Sanders and Cross, 1964). Chromosomal races (distinctions in chromosome numbers) are described in Orthoptera (M. J. D. White *et al.*, 1964). Dyban and Udalova (1967) established that the karyotypes of two lines of rats are distinguished by the structure of two pairs of chromosomes (including one pair of autosomes distinguishable by a greater stability of structure). The cytological mechanisms and genetic consequences of these differences lie outside the sphere of our interests. It is exceedingly important, however, to emphasize that the special experiments of the authors showed that not the slightest trace of reproductive isolation between these lines was detected. It seems clear that karyological differences in themselves cannot serve as the basis for making decisions about either the degree of intraspecific divergence or, even less, the beginning stages of speciation. This is especially true in those cases in which the matter at hand is the differences in structure of the X chromosome that are detected in different lines of rats when the data of different investigators are compared (Pogosyants *et al.*, 1962; Fitzgerald, 1961; Yosida, 1955; and others). Moreover, in different individuals of the same line, both subtelocentric and acrocentric chromosomes are detected (Hungerford and Nowell, 1963). Differences in the variability of the structure of autosomes and sex chromosomes seems understandable, since the latter contain mainly heterochromatin, which is characterized by relatively little genetic activity. It is possible, in particular, that large and small Y chromosomes contain identical quantities of genetic material (Nadler, 1962). Data exist that testify to the possibility of convergence at the karyological level (Matthey, 1961).

The data cited show that karyological methods can be applied only as auxiliary aids. Data obtained with their help do not remove the necessity for broad biological analysis of the process of speciation. Formal criteria are found to be useless in defining the concept of "species," even when the latest research techniques, which permit the analysis of elementary biological structures, are applied.

Based on purely logical considerations, it could be possible that morphophysiological differentiation of a species occurs both before and after its genetic isolation (Vorontsov, 1967). This is a profoundly fundamental question. If genetic isolation of a species is the result of the morphophysiological reorganization of an ancestral population of the original species, the process of speciation is regarded as strictly determined, as the result of progressive adaptations to distinctive environmental conditions. If, however, morphophysiological divergence precedes genetic isolation (the result of "chromosomal evolution"), the process of speciation appears as a random process. The sum of knowledge about the living world testifies in behalf of the first solution of the problem: evolution is an adaptive process.

A comparison of the karyological data with the results of observations on hybridization also confirms this conclusion: genetic isolation is the result of the adaptive reorganizations of populations, not a defense against the crossing of forms that have diverged further than some "species norm." The best evidence in behalf of such a solution to the problem is the general conclusion of the founder of comparative karyology, Matthey. He correctly asserts that "chromosomal evolution" and "morphological evolution" are independent processes (Matthey and van Brink, 1960). Investigations attacking the problem of species with the help of different methodological means lead to analogous general conclusions.

The special interest in karyological investigations is psychologically understandable. The hereditary information is written in the chromosomes. The hope is natural, therefore, that the key to the meaning of genetic differences among compared forms—and thus the key to the species problem as well—could be found in the structure of the chromosomes. Indeed, if it were found that genetic differences are precisely reflected in the differences, determined by modern means, in the structure of chromosomes, the criteria of species would be found. It has been found that there is no strict correspondence between taxonomic distance and chromosomal differences. There remains the hope, however, that the same problem can be solved by another means—analysis of the protein composition of an organism's tissues. This method proceeds from the assumption that proteins are the direct product of the realization of hereditary information (Peakall, 1964). The study of proteins can therefore be used to determine the degree of hereditary differences among forms and groups of animals. Thus, there arose "protein taxonomy" (Sibley, 1960) and "taxonomic biochemistry," the problems of which served as the subject of discussion at a special symposium (Geone, 1964).

There is no need for us to go into detail about the investigations along this line; we can merely emphasize the main points. The successes of biochemical taxonomy are indisputable. Application of the methods of chromatography, electrophoresis, and immunology have helped to approach the solution of some important questions of macroevolution and systematics; in particular, it has helped to show the isolated (at the level of orders) position of the lagomorphs (Moody *et al.*, 1949); to show the relationship of the chordates and echinoderms, of the molluscs and annelids, and of the cetaceans and artiodactyls (for a review, see Sebek, 1955); to show the relative closeness of sea birds (Alcidae) and penguins (Gysels and Rabaev, 1964); and to solve a number of questions on the systematics of birds (Sibley, 1960; Peakall, 1960; Sukhomlinov *et al.*, 1966), of primates (Goodman, 1961), and of some other groups of animals.

With the help of these methods, the taxonomic position of separate species in different groups was specified (W. J. Miller, 1964; Bertini and Rathe, 1962; Fox *et al.*, 1961; Goodman and Poulik, 1961; Schmidt *et al.*, 1962; Picard *et al.*, 1963; Maldonado and Ortiz, 1966; Kaminski and Balbierz, 1965; Cei and Erspamer, 1966; Manwell and Kerst, 1966; Nadler and Hughes, 1966; and others). Data obtained with the help of biochemical and immunological methods served as the basis for broad generalizations concerning both specific questions and general rules of evolution (Blagoveshchenskii, 1945; Florkin, 1947; Goldovskii, 1957; Belozerskii, 1961; Fedorov, 1966; Florkin, 1966; and others).

These successes, however, have been accompanied by difficulties. The main one is that in those cases in which the biochemical differences concern traits having a direct adaptive value, biochemical convergence is a usual phenomenon (Mayr, 1965); their variability is determined by the general laws of variability of organisms. This fact leads to the situation that phylogenetically close but ecologically different forms can differ more among themselves according to biochemical traits than do forms that belong to more distant taxa, but are leading similar forms of life. This becomes completely clear when the matter concerns, for example, such biochemical traits as the chemical affinity of fats, the oxygen affinity of hemoglobin, the vitamin content of tissues, the activity of enzymes, and so forth. It seems superfluous to cite corresponding examples, which are well known. It is important to note only that this matter of variability in traits also concerns variability at the molecular level. Thus, it was shown that the most important part of the hemoglobin molecule, in a physiological sense (the α chain), is distinguished by great stability; the functionally less essential part of the molecule is variable (Ingram, 1962).

We will analyze in great detail those works that were implemented with the help of methods used to establish differences among closely related forms at the molecular level that have a direct relationships to the species problem. Bertini and Rathe (1962), using the method of paper electrophoresis, studied the mobility of blood plasma proteins in a large number of amphibian species. It was established that in a number of cases, the species were well distinguished by the number of protein fractions, their mobility, and their density. The results of these investigations agree well with the data of other authors; there is no need, therefore, to describe them in detail. It seems more important to us that within the limits of the genus *Bufo* (*B. marinus, B. arenarum, B. paracnemis*), it is practically impossible to establish differences among the species. At the same time, two subspecies of *B. granulosus* (*major* and *fernandezae*) are distinguished by the mobility of their protein fractions. The general conclusion of the authors is interesting (Bertini and Bathe, 1962; p. 184): "Within *granulosus,* speciation has resulted in pronounced differentiation of the blood proteins, while in the *marinus* group, little differentiation has taken place." Differences among species are maximal within the genus *Pleurodema,* but as the authors emphasize, *P. nebulosa* and *P. bibrani* are very similar electrophoretically, although they are by no means closely related species. The general conclusions of this work corroborate investigations implemented with the help of more precise methods. During the study of the stereochemical features of esterases in a large number of animal species, Bamann *et al.* (1962) established that some species are clearly distinguished from one another, while they did not succeed in detecting differences among others (man and chimpanzee, goat and chamois, and others). On the other hand, stereochemical specificity of esterases was established in some intraspecific forms (breeds of dogs, sheep, and swine; subspecies of tigers; and others). It was not always possible to establish differences between man and orangutan with the help of immunoelectrophoresis (Picard *et al.,* 1963). A serological study of fowl (Sasaki and Suzuki, 1962) gave unclear results. Differences were not detected between the wild *Gallus bankiva* fowl and the Japanese domestic fowl; differences between *bankiva* and leghorn were easily established. The possibility of using the immunological method to determine the taxonomic rank of closely related forms could not be unconditionally demonstrated in a comparison of aurochs, bison, and big-horned cattle. Finally, interesting data showing that in interspecific hybrids it is possible to find the traits of both parents with the help of immunological and electrophoretic methods (Beckman *et al.,* 1963; Crenshaw, 1965; and others) somewhat lose their significance, since analogous results were obtained in the hybridization of subspecies (Dessauer *et al.,* 1962). Sokolovskaya (1936) used serological methods (precipitation reaction) for a study of the phylogenetic relationships of 13 species of lammelirostres. The specific conclusions basically coincided with the ideas of systematists. The conclusion about the correlation of serological and hybridological data deserves attention:

> The fact that in the majority of cases serological data coincide with hybridological data deserves special attention. Cases are known in which hybrids were obtained between species that were found to be closely related serologically. Such pairs are: goose–swan, Nile goose–peganka (*Tadorna tadorna* L.), Nile goose–musk duck, gray goose–Chinese goose (found everywhere), musk duck–mallard duck, gray goose–white-fronted bernicle goose, and others, to say nothing of the closer duck species.
>
> On the other hand, there are species whose sera react badly together and do not give hybrids (at least, there are no documented observations of hybridization). Such pairs are: mallard–gray goose, ruddy shelduck (*Tadorna ferruginea*)–pintail duck (*Anas acuta*), swan–mallard, and so forth. Sazaki, who has worked with big-horned cattle, also indicates the coincidence of serological and hybridological indices.

This close correspondence is by no means universal, however. With the help of modern immunological methods (microcomplement fixation), the features of homologous proteins in various groups of amphibians were studied (Salthe and Kaplan, 1966). The authors succeeded in obtaining data of great significance for systematics and phylogeny (the marked antigenic closeness of *Caudata* and *Anura* in comparison with the reptiles; the relationship of *Amphiuma* with the Plethodontidae, the Racophoridae, and the Ranidae) that permitted them to make one general conclusion of direct interest to us. It was found that the average rate of reorganization of proteins remains constant in the phylogenetic series investigated. This material shows that the rate of reorganization of proteins does not depend on the degree of morphophysiological divergence; the "coefficient of dissimilarity" of birds and reptiles in comparison with amphibians was found to be identical. The concurrence of the authors' general conclusion with Matthey's conclusion is striking: "protein evolution" and "morphophysiological evolution" by no means always coincide.

Analysis of the data cited leads to the conclusion that "protein taxonomy," like comparative karyology, cannot be used as the decisive criterion in the determination of taxonomic rank of closely related forms (species, subspecies?). The known parallelism of the general conclusions of "protein taxonomy" with the results of hybridization and karyological investigations attracted our special attention, however. Experiments were conducted in our laboratory that permitted us to verify the conclusions in the type of investigations examined in thoroughly detailed material. Zhukov (1966, 1967*a,b*) studied the immunological interactions of different species of voles. In the first series of work, the complex of erythrocytic antigens with heteroimmune sera obtained from rabbits was investigated. Antisera against *Microtus oeconomus, Arvicola terrestris, Ondatra zibethica,* and *Clethrionomys frater* were obtained. It was established that there is a greater immunological closeness between any species of vole and another than between any species of vole and any species of mice. The ratio between species is determined by Mainardi's formula of immunological distance (Mainardi, 1961). The data obtained deserve attention. The species of one genus represent a fairly compact group immunologically, well differentiated from closely related genera. The immunological distance between different genera, however, does not correspond to their morphological differentiation and phylogenetic interrelationship. This is clearly evident in the diagram presented (Fig. 9). It was found that the immunological distance between representatives of the extremely close genera, *Microtus* and *Arvicola,* and of two sharply differing genera, *Microtus* and *Ondatra,* is practically identical. At the superspecies level, the degree of immunological remoteness does not correspond with morphological differences. During the comparison of species within the genus, the author investigated absorbed antisera, a method that permitted him to differentiate the compared forms more clearly. It was established that the close species (*M. oeconomus, M. arvalis,* and *M. agrestis*) differed less among themselves, immunologically, than from a representative of the independent subgenus *Stenocranius*. In this case, the correspondence of immunological and morphological data is clearly expressed.

Antigenic differentiation within a species was studied in our laboratory by Syuzyumova (1967) by the method of transplantation. In the first series of experiments, laboratory colonies of two subspecies of the *oeconomus* vole (*M. oeconomus oeconomus* and *M. oeconomus chachlovi*) were used. The second series of experiments was conducted on common voles (*M. arvalis*), taken directly from the field in areas situated not more than 40–50 km apart. The immunological relationship of the wild animals was determined

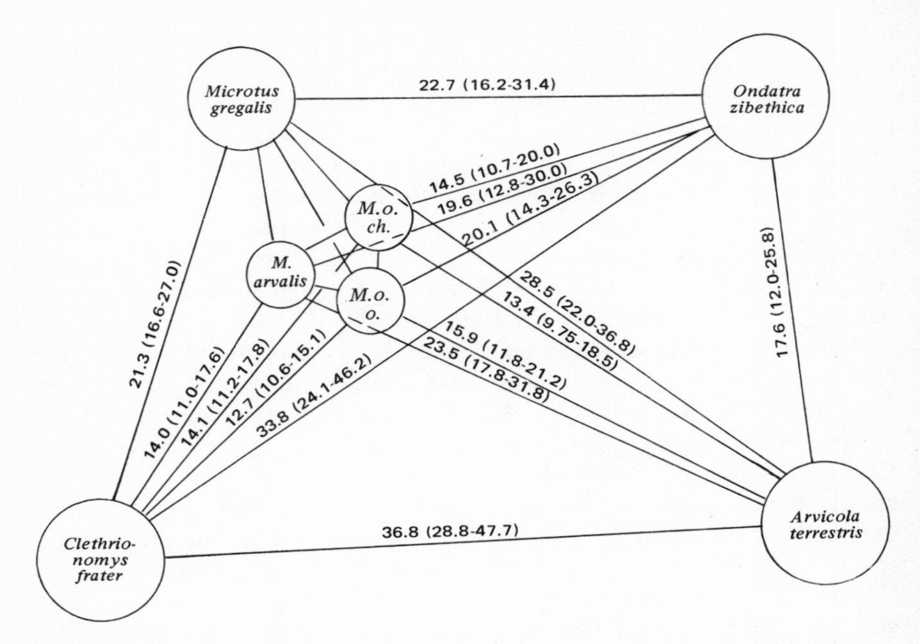

Fig. 9. Immunological distances among the voles of four genera (*Microtus, Ondatra, Clethrionomys*, and *Arvicola*) of the subfamily Microtinae.

by an antigenic complex of histocompatibility. Reaction to transplantation was used to determine differences among voles in the histocompatibility genes and made it possible, with a definite degree of probability, to distinguish the "strongest" locus among them (according to the length of time for rejection and the humoral indices of immunity). The clearly expressed incompatibility reaction of the tissues during the heterotransplantation between *oeconomus* and the narrow-skulled vole is characterized by a very short survival period for the skin patch (4–5 days) and the appearance of antibodies in the recipient that were strictly specific to the erythrocytes of animals of the donor species. It may be assumed that the whole complex of transplantation antigens participated in the reaction, reflecting their species-specific features as a whole. The antigenic relationship among subspecies of *oeconomus* approximates that of interspecific relationships. The terms of survival of the skin patches are limited to 6–7 days. The specificity of the humoral reaction of the southern subspecies to the erythrocytic antigens of the northern is maintained.

In the voles of one subspecies, transplants last 8–9 days. The genetic diversity within the colonies of both subspecies is approximately equivalent. The voles of one litter are more homogeneous. The viability of skin patches was maintained for 10 to 20 days. Transplantation within a subspecies was accompanied by a weaker humoral reaction. It was noted basically only as a reaction to the donor erythrocytes. Statistically significant differences among voles of neighboring populations were obtained. The formation of antibodies was noted only in a small number of wild animals, however. In the latter, a clear specificity of antibodies was not observed. Sera reacted with the erythrocytes of some voles from both populations.

The noted manifestations of a reaction to transplantation allow us to surmise the presence in voles of an incompatibility due to a strong antigenic complex, similar to *N-2* in mice. Some intraspecific features of antigenic relationships also began to show up, possibly conditioned by the structural features of this strong locus of histocompatibility. It seems to us that they reflect a degree of intraspecific differentiation. Differences in a series of structures of the locus are strong indices of incompatibility at the level of subspecies. These differences are smoothed over among neighboring populations, and appear in a community of factors that determine the presence of a broader group of uniform antigens in both populations. The relationships within the populations and within the colonies, apparently, are limited only by individual differences among antigens, responding to the cellular reaction in the rejection of transplants. The antigens elicited by the humoral reaction are possibly identical.

The results of these experiments are very significant. All degrees of antigenic differentiation are observed within species, from hardly noticeable to significant, approaching specific differences. If one takes into consideration that Syuzyumova used as controls the reaction of tissue incompatibility in two species belonging to different subgenera, her data can be viewed as additional evidence for the possibility of "specific" antigenic differentiation at the intraspecific level. The forms compared (*M. o. oeconomus* and *M. o. chachlovi*) were comprehensively studied in our laboratory. Morphological distinctions among them were insignificant (Pyastolova, 1967), nor were small disturbances of normal crossing detected, as was confirmed by special experiments (Ovchinnikova, 1967). The geographic isolation of these forms occurred long ago, however. The duration of their independent existence is measured in tens of thousands of years (Shvarts, 1963*c*). This duration, apparently, explains the origin of clear antigenic differences among them. In correspondence with the data from the work of Salthe and Kaplan (1966) that were cited, these data compel us to assume that "differentiation at the protein level," in those cases in which it does not have direct adaptive value, is, to a significant extent, a function of the duration of the independent development of the forms compared (neutral changes of the cytogenetic apparatus arise at a relatively constant rate). The rate of morphological reorganizations depends on selection pressure and the conditions under which selection operates, and it may actually vary within extensive limits. It is clear, therefore, that the rate of morphological and immunological divergence cannot coincide.

The same question was studied in our laboratory by Mikhalev (1966), by means of the investigation of electrophoretic mobility of protein fractions from closely related forms of rodents. He studied the same pair of subspecies (*M. o. oeconomus* and *M. o. chachlovi*), as well as two very sharply differentiated subspecies of the narrow-skulled vole (*M. g. gregalis* and *M. g. major*). Even though morphological differences between the second pair of subspecies are considerably greater than between subspecies of the *oeconomus* voles, the electrophoretic differences were clear in both cases, although expressed in different features of the electrophoretic pattern.

We have looked at three of the most important and most rapidly developing directions at the present time in the study of the species problem (hybridization research, comparative karyology, and protein taxonomy), and are convinced that this problem cannot be solved on the basis of formal criteria.

The old definition of the "species" concept, elaborated by systematists and zoogeographers on a broad biological basis, objectively reflects the correlation of phenomena in nature. A species is distinguished from any intraspecific forms of biological independence: it does not lose its morphophysiological distinctiveness during any changes in the external environment (hiatus); species-specificity is manifested in any individual of the species (subspecies are defined only on the basis of a series of individuals); species possess an independent range. To apprehend the process of speciation means to apprehend those rules that lead to the genetic and morphophysiological isolation of closely related forms. In different groups, the independence of closely related forms arises against a background of different development of the genetic isolation (karyology and "protein indices" reflect the degree of these differences). Hence, it follows that the species problem can be solved only as a complex biological problem, viewing the process of speciation as a stage in the progressive mastery of life on our planet.

In speaking about the three criteria of species, one usually forgets about the existence of a fourth, which is seemingly more diffuse, but in essence is more important and universal. *Every animal species is ecologically specific and is adapted to certain environmental conditions.* It is impossible to name even one exception to this rule, since even the utmost ecological plasticity is one of the forms of specialization. The adaptability of a specialized species is always higher than the adaptability of any specialized intraspecific form. It seems to us that there is not a single exception to this rule either. It is a law. In speaking of specialization, one should not treat this concept in the narrow sense. The question is not only (and not so much) about narrow specialization as about adaptations of very broad significance, up to adaptations to the conditions of one or several physical–geographic regions. In this sense, every species (even such as the wolf or the "pasyuk"—the Norway rat) is specialized, and its adaptive specialization is higher than the adaptive specialization of intraspecific forms. An ecological approach to the species problem becomes unavoidable.

We especially studied this question in the example of subarctic mammals (Shvarts, 1963*c*). The theriofauna of the tundra is made up of a small number of indigenous species (Siberian lemming, hoofed lemming, Middendorf vole, barren grounds caribou, musk-ox, polar fox, and a few others), and a number of specialized forms of widely distributed species. One is easily convinced that the adaptations of subarctic species are immeasurably more profound and diverse than the adaptations of specialized populations of species of wide distribution. To show that they are, it is sufficient to compare the lemming with any widely distributed species of voles (including even such specialized forms as *M. g. major*), the polar fox with the subarctic subspecies of foxes, the barren grounds caribou with the northern forms of the red deer or the elk, and so on. The material presented by Danilov (1966) in a monograph on the birds of the subarctic leads to fully analogous conclusions. The same phenomena are distinctly evident in a comparison of mountain or desert species with mountain or desert forms. The greater adaptability of species compared with specialized populations is manifested in a fuller utilization of the environment's resources for a smaller expenditure of energy, in a fuller mastery of the diverse biotopes, in greater numbers, and in other ways. It is natural that the superiority of species over specialized intraspecific forms that have not achieved the species level of differentiation is manifested especially clearly during an analysis of the means by which animals adapt to extreme environmental conditions. The special analysis that we attempted to make in Section 2 of Chapter V, however, shows that this phenomenon has universal significance. This conclusion is confirmed during a comparison of any species, in any environmental conditions, and only shortcomings in our knowledge make this rule not always obvious.

Speciation completes the microevolutionary process. A new species is distinguished from the original population of the ancestral species by a more complete adaptability to the specific environmental conditions. Naturally, this perfection of species adaptation is boundlessly diverse. The adaptations of the lemming to the conditions of existence in the far north consist of the perfection of physical thermoregulation, economization of the metabolic rate, an increased capacity to accumulate energy reserves, anatomical features of the intestines, the specificity of vitamin exchange (in particular, vitamins A and C), the capability to reproduce during the winter, the relative autonomy of seasonal and diurnal rhythms of life activity from changes in the external environment, a heightened capability to change physiological functions during a sharp change in the conditions of existence, and so on. The adaptive features of specialized Arctic, desert, and mountain species are just as diverse. The species' adaptations to certain types of feeding, locomotion, and defense against enemies are just as boundlessly diverse. It may be considered accidental that the species' adaptations (in the sense indicated) are more complete than that of populations (intraspecific), and that the species, which is distinguished from any intraspecific forms of genetic and morphophysiological independence, is also distinguished by the highest degree of adaptability. It seems to us that there exists a fundamental internal link between these phenomena. When the adaptive specialization of one of the geographic populations of an ancestral species achieves a high level, a new species arises, a new genetically isolated and independently evolving system.

Is it possible to detect something in common behind the boundlessly diverse species' adaptations that distinguish them from the adaptations of the most specialized intraspecific forms? Special investigations conducted in the author's laboratory over more than 15 years permit us to answer this question affirmatively.

The perfection of any adaptation is determined not only by its functional effectiveness, but also by its energy cost. Other conditions being equal, the adaptation that does not demand additional expenditures of energy is the more effective. The development of these ideas led us to the conviction that the perfection of any animal adaptations affects, with the inevitability of law, the tissue level of organization of the living organism (Shvarts, 1959). This, in turn, led us to certain general considerations regarding the process of speciation.

The microevolutionary process begins with the origin of ecologically irreversible reorganizations of populations and concludes with speciation. Is the process of speciation a qualitatively different process, in comparison with the first stages of microevolution?

We have already had the opportunity to substantiate our views in detail on this question in a number of special works, including a monograph (Shvarts, 1959), and will therefore limit ourselves to brief conclusions that have significance within the limits of our present theme and will add some new material to them.

During an experimental analysis of the theory of speciation, we proceeded from the following basic premises: Any change in the living conditions of animals is directly or indirectly connected with a change in the conditions for the maintenance of the energy balance (Kalabukhov, 1946;

von Wahlert, 1965; Bock, 1965). This change, of course, is reflected in the morphophysiological characteristics of the animals. Such a connection has long ago been established. Thus, for example, in the beginning of this century, it was shown that an intensification of metabolism leads to an increase in the dimensions of the heart. Rules of such a nature are expressed so distinctly that in the modern literature, they are raised to the rank of "laws of evolution," which limit and direct the evolutionary process (Rensch, 1961). In our research, we approached the investigation of these laws with strict quantitative criteria (Shvarts, 1954, 1959; Shvarts *et al.*, 1964*a*; and others). It was found that the quantitative approach to the investigation of what seemed to be a clear problem led to completely unexpected results. It was established that all conditions eliciting an intensification in metabolism (in the simplest cases, an increase in motor activity) actually lead to corresponding morphofunctional improvements (an increase in the dimensions of the heart and kidneys, an increase in the concentration of hemoglobin in the blood, and so forth). These improvements, however, were found to be sharply expressed only in the case in which different intraspecific forms were compared. In specialized species, it was in many cases generally not possible to detect morphofunctional improvements, and in all cases, they were expressed much less sharply than in intraspecific forms. Ecological investigations thus come close to the problems of theoretical taxonomy.

If we were dealing with single observations, perhaps we could attribute them to chance or to insufficiently precise judgments about the level of energy expenditure in the forms compared. During a 15-year period of investigations conducted in our laboratory along the lines indicated, however, hundreds of animal species of all classes of terrestrial vertebrates in very different environmental conditions (from the tundra to deserts and from sea level to high mountains) were studied according to a complex of very diverse indices (relative weight of the heart, the kidneys, the brain, and the liver; length of the intestines; hematological indices; vitamin content in tissues; intensity of gas exchange; and others). Despite this intensive study, we literally cannot name even one exception to the generalization formulated above: a change in the form and conditions of the life of animals elicits significantly more sharply expressed morphofunctional improvements within the limits of a species than in different species. It has become clear that we have encountered a rule of general significance.

For additional verification of our conclusions, investigation of the morphophysiological characteristics of animals of the far north, on one hand, and of some mountain species, on the other, were undertaken. Since, the material obtained on the first question is included in a special monograph (Shvarts, 1963*c*), we will limit ourselves to a citation of its general results.

The investigations of Dobrinskii (1962) have shown that in the far north, the relative heart weight of a large number of bird species has proved to be significantly greater than in related southern forms of comparable dimensions (Tables 7 and 8). The most extreme hypertrophy of the heart, however, is manifested in species that inhabit a basic range lying outside the limits of the subarctic and that master the subarctic least completely (e.g., the mallard and the treskunok teal (Garganey), *Anas querquedula,* of the ducks). In typical polar bird species, the index of the heart only insignificantly exceeds the corresponding indices of southern forms (see Table 8). This seemingly paradoxical result, which has been corroborated by the author in a very large body of material, shows that typical polar species maintain their energy balance under the conditions of the far north without sharply expressed morphofunctional adaptations.

The investigations on mountain species were conducted on voles of the genus *Clethrionomys,* which have different degrees of adaptations to dwelling at great heights. The data obtained are presented in Table 9. Analysis of these data allows us to draw the following conclusions:

In widely distributed species (*C. rutilus, C. glareolus*), mountain dwelling is connected with a significant increase in the heart index. Differences among the populations compared are statistically significant. During an investigation of the reddish-gray vole (*C. rufocanus*), a species that is often encountered in the upper mountain zones, somewhat different data were obtained. The dimensions of the heart are significantly less than in the Asiatic red-backed vole and European red-backed vole. Even taking into account the large dimensions of the reddish-gray vole, it would have been expected that in the mountains, its heart index would be greater. It is evident that already in this species, which one cannot in all truth call a mountain species, some kinds of mechanisms are manifested that allow it to

TABLE 7. Heart Index (‰) of Birds Typical of the Subarctic and of Related Species of More Southerly Distribution

	Yamal–Nenetskii region			Temperate latitudes	
Species	*n*	*M*	Species	*n*	*M*
Nyroca marila	6	10.0	*N. ferina*	11	9.4
			B. buteo	3	6.75
Buteo lagopus	1	7.9	*B. ferox*	2	6.9
			Alauda arvensis	28	14.95±0.5
Chionophilos alpestris	44	15.4±0.47	*Saxilauda yeltoniensis*	32	14.3 ±0.85
Anthus cervina	25	17.5±0.64	*A. campestris*	37	17.9 ±0.4
Acanthis flammea	43	17.2±0.3	*Linaria flavirostris*	28	17.2 ±0.8

TABLE 8. Heart Index (‰) of Subarctic Birds and of Subarctic Populations of Widely Distributed Species

Species	Subarctic species		Species	Subarctic populations	
	n	*M*		*n*	*M*
Nyroca marila	6	10.0	*Anas platyrhyncha*	5	14.0
Anser albifrons	4	10.2	*Cygnus cygnus*	2	11.3
Clangula hiemalis	12	11.0	*Mergellus albellus*	5	12.9
Oidemia nigra	25	10.9±0.27	*Anas acuta*	32	14.6±0.4
			Bucephala clangula	2	14.0
Lagopus lagopus Koreni	41	12.4±0.3	*Lyrurus tetrix*	4	14.8

maintain normal life activity in the mountains with relatively small heart dimensions. This phenomenon is evidenced with utmost clarity in the typical mountain species—the Tyan' Shan' vole (*C. frater*). The heart index in this species was found to be lower than in all mountain populations of the Asiatic red-backed and European red-backed voles of comparable size that were investigated, even though the Tyan' Shan' vole lives many hundreds of meters higher than the other two species. Moreover, Table 9 shows that the heart index of the Tyan' Shan' vole obtained at a height of more than 2000 m is lower than in plains populations of the Asiatic red-backed and European red-backed voles from the Central Urals. The indicated dif-

TABLE 9. Relative Heart Weight of Different Species of the Genus *Clethrionomys* Under Different Environmental Conditions[a]

Species	Collection site	Height above sea level (m)	Relative heart weight (‰)
C. rutilus	Central Urals	foothills	6.5±0.12
		600–800	7.9±0.24
C. glareolus	Central Urals	foothills	6.9±0.08
		600–800	7.4±0.17
C. glareolus	Southern Urals	100–150	5.5±0.13
		500–600	5.9±0.11
		800	6.1±0.23
C. rufocanus	Southern Urals	800	4.9±0.26
C. flater	Zailiiskoe Ala-Tau	2300	5.7±0.21
Alticola argentatus	Zailiiskoe Ala-Tau	2500–3000	4.9±0.14

[a] For comparison, data on the mountain silver vole are also cited.

ferences are statistically significant. Thus, the material on four species of one genus clearly shows that specialization to life in the mountains is connected with a relative reduction of the heart index.

A very low index is also characteristic of the highly specialized mountain species belonging to a genus close to *Clethrionomys*—the mountain silver vole (*Alticola argentatus*). The heart dimensions of this species do not exceed the dimensions of plains populations of widely distributed vole species.

Investigations along this line were conducted in our laboratory by Birlov (1967). He compared a series of internal indices of two subterranean forms, the common "slepushonka" (*Ellobius talpinus*) and the Prometheus vole (*Prometheomys schaposhnikovi*) under high mountain conditions. It turned out that the mountain population of a widely distributed species in comparison with a specialized mountain species (the Prometheus vole is endemic to the Caucasus) is distinguished by a larger heart (6.1±0.36‰ vs. 4.3±0.27‰; t = 12.0), kidneys (5.2±0.36‰ vs. 4.7±0.33‰; t = 3.85), and a significantly greater content of hemoglobin (16.5±0.57 mg/100 ml vs. 11.7±0.66 mg/100 ml; t = 21.81). We obtained analogous results in a comparison of two rodent species: from the southern plains (*Ochotona pusilla*) of the steppes of the Chelyabinsk region and from the northern mountains (*O. alpina*) of the polar Urals. The specialized species compared did not differ in their heart dimensions: in *O. alpina,* the relative heart weight was found to be 5.32±0.09‰; in *O. pusilla,* 5.58±0.38‰ (animals of the same age and dimensions were compared). The species compared did not differ in their kidney dimensions either (7.85±0.18‰ and 8.1±0.45‰). Broader investigations conducted by V. N. Bol'shakov in our laboratory confirmed this conclusion (Table 10). It is especially interesting that the high mountain species of rodents (red and big-eared) are characterized by small heart indices.

In this connection, it is interesting to note that the northern populations of hares (*Lepus timidus*) are distinguished by very large heart dimensions. The relative mean heart weight we obtained for hares in the tundra forest of Yamal was found to be 10.4±0.46‰. This substantially exceeds the corresponding index of more southern populations of this species, in particular, if one takes into consideration the extremely large dimensions of the polar hares (the weight of summer specimens goes up to 5 kg).

The hypothesis we have developed is also confirmed by an analysis of the material obtained by other authors. The research of Morrison and coworkers (Morrison and Elsner, 1962), conducted in the mountains of South America, showed that in mountain populations of the house mouse, the relative weight of the heart, on the average, exceeded the corresponding index of plains populations of the same species by 53%. The heart index of

TABLE 10. Internal Indices of the Rodent Fauna of the USSR

Species collection site	Brief ecological characterization of the species at the collection site	Body weight (g)	Heart index (‰)	Kidney index (‰)	Relative length of the intestines (%)	Relative length of the cecum (%)	Liver index (‰)	Remarks
O. hyperborea Polar Urals Yakutiya (lower Lena River)	Inhabitant of rock deposits in the Gol'tsy Mtns. (mountainous tundra) in the lower ranges	121.0	5.32±0.09	7.8±0.18	998±11	23±2	52.0±1.6	Material from N. S. Gashev
O. alpina Altai Mtns., Kolyushty ranges	Inhabitant of rock deposits below and within Gol'tsy Mtn. zone (up to 2000 m)	214.1	4.8	5.8	1041	19	57.3	
O. pusilla Transurals	Inhabitant of the steppe plains zone of the Transurals	64.9	5.5±0.38	8.1±0.45	1010	25	43.2	Material from V. N. Pavlinin
O. pricei Northern Pribalkhash	Inhabitant of rocky outcrops populating mounds from the base to a height of 1000 m	117.0	5.2	7.7	1021	23	68.7	
O. rufescens Kopet-Dag Mtns., Balkan Mtns.	Inhabitant of the foothills of the Balkans, confined to the rocky talus	215.0	3.57	6.2	—	—	47.0	Material from M. Sapargel'-diev
O. rutila Zailiiskii Ala-Tau	Inhabitant of the mountains at elevations of 1700–4000 m	183.4	3.4±0.2	6.3±0.3	880±26	21±2	32.1±1.6	
O. roylei Central Pamir	Inhabitant of high mountains at an elevation of 2500–4000 m	149.7	3.9	6.7	980	20	56.3	

typical mountain species of rodents (*Phillotis darwini, P. osilae, P. pictus,* and several species of *Acodon*), dwelling at an elevation of 4500 m, however, exceeds the heart index of closely related species living in the plains by only 20%. Complex investigations by the authors cited showed that the increase in the hemoglobin of the blood characteristic for man, domestic, and laboratory animals when climbing a mountain is not detected in typical mountain species. We are encountering here basically the same phenomenon: the nature of the adaptations of specialized species to a specific habitat does not correspond to intraspecific adaptations.

The facts cited corroborate the hypothesis we expressed earlier: the adaptations of specialized species and the adaptations of individual populations of widely distributed species proceed along fundamentally different paths.

In the adaptations of species to certain conditions of existence, a leading role is played by profound biochemical changes that make superfluous the expression of changes in anatomical features, so characteristic for intraspecific forms. Adaptations of an anatomical and physiological order have, in this case, subordinate significance (of course, we are speaking only of anatomical adaptations directly connected with the maintenance of metabolism at a certain level). On the other hand, these adaptations play a leading role in the process of adaptations of separate individuals, populations, and subspecies. In other words, we see the basis of differences between closely related animal species in the biochemical differences that determine their interrelationships with the external environment. The internal characteristics of animals that we have studied are indicators of these differences (Shvarts, 1959). The specific mechanisms of adaptations of closely related species at the tissue level remain unstudied, in the majority of cases. Some of the latest research permits us to approach an understanding of the biochemical processes of phenomena of interest to us.

One of the paths of tissue adaptation in species specialized for living in the mountains may consist of an increase in the content of myoglobin in the tissues (Reynafarie and Morrison, 1962). The myoglobin content of the tissues of typical mountain forms (various *Phillotis, Acodon, Hesperomys, Chinchillula, Conepatus rex,* the alpaca, vicuna, and others) is very high (in the vicuna, up to 8 mg/g), which has a clear adaptive significance under conditions of lowered partial pressure of oxygen. An increase in the myoglobin content of the tissues also occurs in the process of acclimatization of plains species to the mountains. The specificity of the mountain species is that their tissues maintain a high myoglobin content even when they are reared on the plains. Perhaps it is precisely for this reason that in not one of the mountain species investigated did a transfer even to very great heights and back elicit changes in the rate of the heartbeat or in the rate of breathing. Biochemical research showed, also, that the tissues of mountain species are distinguished by high activity of cytochrome-C-reductase. An increase in the enzyme activity of the tissues has special significance under conditions of work in individual organs.

Unfortunately, we did not conduct a study of the morphophysiological characteristics of animals from a third type of extreme conditions—deserts. We did have the opportunity, however, to analyze the very interesting material of M. A. Amanova in manuscript form, with which the author kindly acquainted us.

Amanova studied four species of sparrows living in Karakumy: house, European tree, saxaul, and desert (*Passer domesticus, P. montanus, P. ammondendri,* and *P. simplex*). It was found that in the desert populations of widely distributed species (European tree and house), the heart was significantly smaller than in the populations of the same species from central or high latitudes. This finding is understandable: a high temperature leads to a lowering of metabolism and a corresponding lowering of the heart index. In desert and saxaul sparrows, however, the heart index proved to be approximately equal to the heart index of tree and house sparrows from the centers of their ranges. Amanova's data on the water content in the tissues of these species help to understand these observations. From a wealth of material, the author cited only the following data: The water content in brain tissue in the summertime is the same in *P. domesticus*: 77.4±0.11%; *P. montanus*: 77.8±0.15%; *P. ammondendri*: 72.7±0.31%; and *P. simplex*: 67.7±0.02%. In the desert sparrows, the water content in the tissues (in the hottest period!) is less than in desert populations of widely distributed species. This means that a specialized species acquired the capability to maintain normal life activity during very substantial dehydration of the organism (tissue adaptation!). Of course, this adaptation created the prerequisites for a more active form of life (this is corroborated by direct observations) and a corresponding increase in the heart dimensions.

Summing up this section, we can assert that in truth, a vast body of facts attests to the different paths of adaptation of species and intraspecific groups. Speciation is connected with tissue reorganizations. There are still few direct observations confirming this viewpoint, since comparative biochemistry concerns itself with the species level—and, even more so, the intraspecific level—only rarely. Despite their relatively small numbers, data directly confirming the system of views we have developed are highly significant.

A series of concordant observations shows that in specialized mountain species, in contrast to mountain settlements of widely distributed species, climbing in the mountains is not connected with an increase in the oxygen capacity of the blood (Barbashova and Ginetsinskii, 1942; Tsalkin, 1945;

and others). In mountain species of camels (llama and vicuna), climbing even to very great heights (up to 5000 m) is not connected with an increase in the oxygen capacity of the blood. An oxygen dissociation curve trending to the left is characteristic for these species. The blood is saturated with oxygen at relatively low tension (F. G. Hall *et al.*, 1936). These conclusions agree with the data of Kalabukhov (1954) on rodents. He writes: ". . . in the susliks, living since olden times in the high mountain regions of the Caucasus, adaptation to a lowering of atmospheric pressure occurs in a different way, and not simply by an increase in the oxygen capacity of the blood, as in mountain forest mice."

It is natural (as we have emphasized repeatedly) that phenomena that attract one's attention during a comparison of different species can also be detected during a comparison of sharply differentiated intraspecific forms. Domestic animals are a good model for the study of this question. Raushenbakh (1958, 1959, 1966, and others), having studied a vast amount of material over a long period of time on the process of adaptation of plains forms to dwelling in the mountains, arrives at the conclusion that "in aboriginal sheep and horses of mountain populations, the structure of reactions is sharply distinguished from the structure of reactions characteristic of both foothill and plains populations. In mountain animals, when climbing to great heights, adaptation to the lowered oxygen content of air breathed is realized basically not through regulation mechanisms maintaining the level of oxidative processes, but through the regulation of metabolism itself. Under low oxygen conditions, the oxygen demands of their tissues are lowered (Raushenbakh, 1966, p. 38). The author emphasizes that in specialized mountain races, "a clearly expressed tissue adaptation takes place" (p. 41). Raushenbakh's research clearly shows that in the process of perfecting the specialization of animals, morphophysiological adaptations are replaced by tissue adaptations.

No other specific functional improvements in response to changes in environmental conditions of another type are detected in specialized species. The zebu (*Bos indicus*) may be cited as an example. When the environmental temperature is sharply increased, not only does the body temperature of the zebu increase significantly more slowly than does that of the big-horned cattle (*B. taurus*), but also, at the same time, the zebu is not observed to breathe heavily either.

These observations are closely akin to data showing that the working hypertrophy of the heart is expressed more markedly, the less economically the organism works and the less it is adapted to work (Beickert, 1954). Applying Beickert's terminology, the results of our research can be expressed as follows: an organism of specialized species is more adapted to work (thermoregulation and other such phenomena—this is physiological work) than an organism of specialized intraspecific forms and, corresponding to this, an increase in the dimensions of the organs and other morphofunctional adaptations is less markedly expressed in the organism of specialized species.

The observations of Bertalanffy and some other authors also show that the ability of specialized species to adapt to work is expressed in characteristics of their tissues. During a study of oxygen consumption by mice, they established that the muscle tissue of small species consumes oxygen more energetically than the tissues of large species, but they did not succeed in establishing a similar dependence within species (Bertalanffy and Estwick, 1953; Krebs, 1950). Still more significant are the data (Bertalanffy and Pirozynski, 1953) showing that changes in the basal metabolism level that occur during a change in the body dimensions of the animal do not entail a change in tissue respiration. The authors arrive at the conclusion, which is an important one for us, that within a species, a drop in basal metabolism with an increase in weight is not determined by intracellular factors, but depends on the regulating influence of the organism as a whole. In different species, differences in basal metabolism are connected with differences in the intensity of tissue respiration. Aleksandrov (1952) and Ushakov (1955, 1964, and others) arrived at essentially similar conclusions: the tissues of thermophilic species possess elevated thermostability of proteins; the thermostability of tissues of different subspecies of one species (from very different climatic regions) remains practically unchanged. On the basis of a vast series of cytophysiological investigations, Ushakov arrives at the conclusion that an adaptive change of protein structure of different tissues (cellular adaptation) takes place in species, at the same time that the adaptation of different forms of a species "to new microclimatic conditions is achieved not at the price of obligatory change in all the cells of the organism, but by a 'cheaper' means—the maintenance of relative constancy of the physiological properties of a number of tissues."

Recently, the theoretical interpretation of the data obtained by Ushakov was subjected to a comprehensive analysis by Aleksandrov (1965). He showed that the observed differences in the thermostability of proteins of different species do not, in the majority of cases, have independent adaptive significance, but indicate differences in conformational lability of their protein molecules. The profound nature of intraspecific differences is thereby emphasized. Aleksandrov (1965) writes:

> The normal correlation between the flexibility and elasticity of macromolecules can be disturbed not only by a change in the temperature conditions of life, but also by displacements of other environmental factors. For adaptation to a change of hydrostatic pressure or to a new makeup of the environment (salinity, pH, and so forth), supramolecular homeostatic mechanisms may prove to be

insufficient. The necessity arises, then, for a change in the stability of the protein macromolecules in some aspect or other, so that their primary structure can be rearranged. Considering the universality of heat as a factor in disrupting intramolecular bonds, one might expect that a similar change will affect not only the stability of macromolecules to environmental factors provoking them, but also their stability to heat denaturation. Thus, according to the proposed hypothesis, in the process of the origin of a new species adapted to different temperature conditions of existence, because of hereditarily conditioned changes in the primary structure of the proteins, there is a reestablishment of the disturbed correlation between the lability and stability of macromolecules. This change can be diagnosed by a change in the stability of proteins to heat denaturation.

In sum, the data cited show that fundamental differences are evidenced between species (including the phylogenetically closest) and any intraspecific forms. It follows that the process of speciation should also be qualitatively distinguished from the process of intraspecific differentiation. We depict this process as follows:

During the movement of animals to a new habitat that demands intensification of a certain function, acclimatization of the animals occurs. Specific reactions by the animals of the given species to a change in environmental conditions lie at the basis of the acclimatization process. In parallel with this process, there occurs natural selection for individuals with a more perfected morphophysiological reaction, and the population acquires hereditarily fixed morphophysiological characteristics. This is not, however, the most perfect path to mastery of a new environment that demands intensification of the organism's metabolism. It is disadvantageous in energy terms, since an increase in the dimensions of an organ or an intensification of its function demands an increased expenditure of energy for the maintenance of its own vital activity, to say nothing of the possible disruption of the coordination of functions among different organ systems. The result of natural selection, therefore, is determined by the specific mechanisms of the animals' adaptive reactions. There occurs the selection of individuals capable of maintaining their energy balance without sharply expressed morphofunctional adaptations. This process, apparently, is facilitated by the fact that morphofunctional changes in an organ are often accompanied by biochemical changes as well (an increase in the dimensions of the heart, for example, is always accompanied by an increase in the content of myoglobin in the heart muscle). As a result, tissue adaptations are substituted for morphofunctional adaptations. This is inevitably accompanied by a change in the chemical affinities of the internal environment of the organism. This change, as is known, is one of the basic causes for the inability of different species to cross (tissue incompatibility). Highly specialized groups of populations of a defined species therefore become reproductively isolated. A new species arises.

Experimental reproduction of this process presents significant difficulties, for a number of reasons (experiments in this direction are now being done in our laboratory), but a comparison of the facts cited above, and also of a considerable body of facts we cited earlier (Shvarts, 1959), compel one to think that they play a very important role in many cases of speciation.

Thus, we arrive at the conclusion that the microevolutionary process begins with ecologically irreversible reorganizations of populations and concludes in their reproductive isolation (genetic isolation).

The energy cost for the development of adaptations is placed at the basis of the viewpoint on the process of speciation developed here. It is important, therefore, to try to establish the role played by the conservation of metabolism in the maintenance of the numbers of a species. Actually, even a small reduction in energy demands increases the chances of a species and the populations composing it in the struggle for life, or increased expenditures of energy can easily be compensated. In the final analysis, a correct approach to the problem of speciation depends on the solution of this ecological task, in the modern sense of the term. An evolutionary approach to the solution of ecological problems no longer allows us to be limited to ascertainment of a defined type of adaptation, but demands that we evaluate it in terms of energy requirements. The energy cost approach to ecological research has only very recently become incorporated into the practice of ecological work, in connection with the analysis of general problems of the energetics of ecological systems. Investigation of the energetics of populations has no less significance in work conducted in the area of evolutionary ecology. The principal problem in this case can be formulated thus: to what degree is the energy balance strained in nature (in different species, in different environmental conditions); what is the selective advantage of a small reduction in metabolism, with the maintenance of an optimal level of activity by the animal; and under what conditions are the less economical morphofunctional adaptations replaced in the process of selection by the energetically more advantageous tissue adaptations?

General field observations create the impression that under ordinary conditions, animals are adequately assured of the necessary food resources for maintenance of energy exchange, and a small economy of energy therefore has no substantial significance. Special investigations, however, show that this impression is mistaken. Testifying that it is, first of all, are investigations showing that when the general food reserves are reduced, they are inevitably dispersed. The energy spent in search and procurement of food exceeds the potential energy of the food, and animals die from hunger when there is apparently plenty of food.

There are not yet very many investigations of this type, but they compel a new approach to the evaluation of the interrelationships of an animal with its environment and make it evident that the conservation of energy, which seems insignificant, can have decisive significance in the process of a species' consolidation in a new habitat. In this regard, some research of a general character is even more significant. One of the most brilliant of these studies is the research of Kalabukhov (1951) on the ecology and distribution of forest and yellow-throated mice. The most general conclusion of this work is that the limits of the distribution of yellow-throated mice are determined by the limits of distribution of trees with large seeds (the oak, hazelnut, and so forth). The large size of the yellow-throated mouse, of course, demands a large amount of food, and if the food is dispersed (small seeds), the mice simply do not have enough hours in the day to defray the energy expended in life activities. It is not difficult to imagine what significance the economizing type of metabolism acquires under these conditions. Here, it is especially important to emphasize that the correlation of the energy demands of the animal and its expenditure of energy in the procurement of food constitutes a characteristic of the species that is as important as the limits of its distribution. It is not difficult to see that this rule is hardly less than universal and, in a number of cases, determines the paths of macroevolutionary reorganizations. It is sufficient to remember that all the insectivorous vertebrates are small. There are no large insectivorous birds. Insectivorous mammals are very small representatives of the class. Insects are a dispersed type of food; procuring them requires a great expenditure of energy and time. That is why there can be no large insectivorous species. Exceptions to this generalization only emphasize the rule. Relatively large insectivorous animals either eat colonial insects (anteaters, African anteaters, honey buzzards, and so forth) or are distinguished in comparison with related forms by a clearly increased level of metabolism (of our fauna, for example, the hedgehog).

Observations of a more general nature also attest to the exceptionally important significance of a reduction of energy expenditure in the process of the adaptation of animals to specific environmental conditions. In the extreme north (short breeding period), high fecundity acquires exceptional significance. It is understandable, therefore, that in northern populations of many widely distributed species, fecundity is exceptionally high. Thus, for example, in the water vole, the Middendorf vole, the Asiatic red-backed vole, and the narrow-skulled voles, the average number of embryos in the female exceeds 9 (Shvarts, 1959, 1963*b*, and others). This is approximately 1½ times higher than the average fecundity of the same species in the temperate zone. In typical arctic species, however, fecundity is not so high: in the Norwegian lemming, it is less than 7 (Frank, 1962); in *Lemmus obensis*, less than 8 (Dubrovskii, 1940; Tsetsevinskii, 1940; Dunaeva, 1948; Kopein, 1958; and others); in the hoofed lemming, less than 7 (Dunaeva, 1948, and others).

The paradoxical fact of a clear reduction in the average fecundity of typical arctic rodents, of course, is explained on the basis of the hypothesis we have developed. High fecundity increases the demand for energy and puts the energy balance under greater strain. It is advantageous for a species to make sacrifices as regards the number of young born, and to decrease the strain on the energy balance and increase viability. The animals most fully adapted to northern conditions proceed along precisely this path. The aim of reducing energy expenditure determines the path of adaptation to the conditions of entire physicogeographical regions. That it does attests with considerable eloquence to the absolutely exceptional significance of the energy assessment of adaptations. This assessment (which is conducted by natural selection) determines, in the final analysis, the paths of the evolutionary reorganizations of animals and, principally, the process of speciation.

Bergmann's rule, in essence, is also witness to the importance of this assessment. Even though this famous rule has more than a few exceptions (in that sense, it is indeed a rule!) and even though its general treatment is subject to a well-known transformation (Terent'ev, 1946, and others), the body of facts that support it remains, in truth, vast: during the advance of animals of very different groups to the north, it has become more important. This rule is in fact universal. It follows that even a trifling savings of energy has such great selective significance that it determines the character of geographical variability in the majority of homeotherms.

In conclusion, it should be pointed out that there are direct observations showing that the widespread notion that there is an insignificant strain on the energy balance of animals in nature is mistaken. It has been established by many authors that under a positive energy balance, animals quickly accumulate energy reserves. It is especially important to note here that when animals are maintained under the conditions of an optimal regime of food and thermoregulation for a total of only 24 hours, the glycogen content in their livers increases rapidly to a level that under natural conditions is observed only in exceptionally rare cases. There also exist more exact methods that make it possible to show that the energy balance of animals in nature is almost always under strain.

These methods are based on the different capability of males and females to accumulate glycogen. This differing capability is reflected in the weight of their livers. Special investigations conducted in our laboratory showed that in the females of all vertebrates during the breeding period, the relative weight of the liver is significantly greater than in males or nonreproductive females (Pyastolova *et al.*, 1966; Shvarts, 1960*b*, 1961*a*; and others). The reason is that during the breeding period, females create a supply of glycogen in the liver, even when the energy balance is negative and even to their own detriment. Only under optimal conditions of existence (this is corroborated by N. A. Ovchinnikova's experiments) are the relative weights of female and male livers identical. This situation is observed extremely rarely in nature, however. It is shown thereby that the occasion is very rare in nature when energy does not have great significance.

The capability of animals to maintain normal life activity with less expenditure of energy may turn out to be the decisive factor in the struggle for existence. This is important evidence on behalf of the views we have developed on the process of speciation, since tissue adaptations, from the energy viewpoint, are more advantageous than morphofunctional adaptations.

Another important confirmation of our hypothesis is the analysis of the process of race formation in domestic animals. As is known, all races of domestic animals produced from a common ancestor are fertile, without restriction, when crossed among themselves or with the ancestral species. The usual point of view is simply that man has not created even one new species of animal only because there has not yet been sufficient time for him to do so. This viewpoint cannot stand up to criticism. It is known that the rate of evolutionary reorganizations depends on selection pressure and isolation, to say nothing of other factors. The strength of artificial selection is completely incommensurate with natural selection pressure, and the isolation of the original cohort of reproductives is incomparably greater than under the usual natural conditions. This assertion might demand special substantiation, if it were not that the results of artificial selection themselves confirm its unconditional correctness. It is not difficult to see that during a negligible period of time (on the geological time scale, the period of existence of even ancient breeds of domestic animals is minuscule), man has created intraspecific forms with morphological differences exceeding the differences among any subspecies in the entire animal kingdom, including the most sharply differentiated. We are not touching here upon the interesting question of the essential distinctions between races and subspecies of wild animals, but the very fact of the enormous morphological differentiation of domestic animals is so clear that it does not require either evidence or illustrative examples. It has been asked why the morphological differentiation of domestic animals is not accompanied by the origin of hereditary incompatibility. This most important question is given a natural explanation by the theoretical positions we have developed.

Man was interested in raising the productivity of animals, without regard for the specific form in which this productivity was expressed. It is most convenient for analysis to use the simplest form of productivity—the capacity for work. Selecting horses distinguished by their fleetness of foot, man was interested only in the results of selection, and was indifferent to what kind of mechanisms were involved in achieving this fleetness. Since the simplest physiological path for raising the motor activity of animals is morphofunctional changes, it is natural that the animals that gave the best results were animals with the best expressed and most nearly perfected morphofunctional features. Man was not interested in the fact that this great effectiveness was achieved by a means that was in no way the most advantageous for the organism, and, to be sure, investigation of these means has become possible only in the most recent times. Selection, therefore, went along the lines of morphofunctional adaptations.

It may be objected that man became interested relatively long ago in the expenditure of fodder, which is determined not only by the productivity of the animals, but also by the means by which this productivity is achieved. The energy value of the adaptations should therefore be taken into consideration, even if it is not obvious. The selectionist, however, for completely understandable reasons, is interested in providing food for the more productive animals, and since these animals are individuals with expressed morphofunctional features, artificial selection at present actually ignores the physiological mechanisms that determine the productivity of the animals and operates only on the final results. On the other hand, domestic animals, in comparison with their wild ancestors, have been better provided with food, even from the very beginning of domestication (possible isolated, often even intentional, deviations from this rule do not change the general picture). It is not ruled out, therefore, that the energy aspect of adaptations in the evolution of domestic animals played a significantly lesser role than in that of their wild ancestors. As a result, artificial selection did not lead to the origin of forms specific at the tissue level, even though the degree of morphophysiological differences among many races of one species is clearly of generic rank. It is probably unnecessary to cite examples supporting this position, since they are so well known.

It seems to us that the considerations expressed remove serious difficulties that arise during a comparative analysis of the evolution of domestic and wild animals. As is known, these difficulties led prominent researchers to the very pessimistic conclusion that evolution and domestication are different processes (Klatt, 1948; Remane, 1948; Herre, 1959).

It is useful here to digress from our theme and touch on problems connected with the biological bases of the further progress of animal selection. It is all the more necessary to do this because evolutionary studies developed, to a significant degree, on the basis of a generalization of the experience of agricultural practice, and selection work still does not adequately utilize the achievements of evolutionary theory. All the accumulated modern zoological data show that specialized species possess a qualitatively higher adaptability to given conditions of the maintenance of the energy balance than do the most highly adapted intraspecific forms. Speciation turns out to be, in the vast majority of cases, the most effective means of adaptation. Any increase in the productivity of domestic animals, however, signifies a change in the conditions for the maintenance of the energy balance, independently of whether the increase of productivity consists of an increase in milk or meat production, or of an increase in the enduring of a harsh climate, or an increase in hard work. It follows that the possibility of breeding a new species of domestic animal (not the domestication of a new wild species, but the creation of a specialized domestic species with given specific properties) would signify fundamental progress in animal husbandry, the significance of which would be difficult to foresee, but impossible to overestimate. Proceeding from the hypothesis we have developed, the formation of new species should be based on a new type of selection—selection in which the results of an increase in productivity are compared with the mechanisms that determine them. Practically, this means that the animals to be subjected to selection will not be the most productive animals, but rather those in which an increase in productivity is not accompanied by substantial morphofunctional displacements. The new principle of selection we have advanced is so simple and its results can be so significant that it seems advisable to us to recommend it to practice. Of course, the specific criteria of selection will be different in each case, but discussion of this question would lead us too far from our subject.

Here, it is important to us to emphasize that many centuries of the practice of animal husbandry clearly show that even very sharp morphological differentiation of different intraspecific groups does not lead to speciation, if selection operates on a purely functional basis without taking into account the energy cost of the morphofunctional adaptations of animals. It is possible to take into account the energy cost of adaptations only on an ecological basis and by ecological methods of investigation. It is very important, therefore, to establish the actual conditions under which a tissue adaptation inevitably replaces a physiologically simple but energetically disadvantageous morphofunctional path of adaptation. Under what conditions is the process of speciation the inevitable path to the perfection of the adaptation of animals to actual environmental conditions? It seems to us that the solution of this task numbers among the basic problems of evolutionary ecology for the immediate future.

The process of speciation is not an obligatory process. It takes place only under fully defined conditions. The long periods of existence of a large number of species, measured in millions of years, sometimes in tens and even hundreds of millions of years, is evidence of this. Moreover, even many subspecies often exist morphologically unchanged for a long time. Simpson (1963), for example, cites basic data showing that some subspecies exist at least for the course of hundreds of thousands of years. It is becoming evident that speciation is elicited by a certain combination of external factors. It is not by chance, of course, that the youngest of the presently existing mammalian species—the tundra and taiga species—are species adapted to the geologically youngest biomes. A new physicogeographical region arose. Its mastery demanded the origin of a new type of adaptation. As a result, a literally massive number of new species and even genera have arisen, to say nothing of specialized intraspecific forms (Reinig, 1937; Rensch, 1961; Janossy, 1961; Zablotskii and Flerov, 1963; and others). The question posed by us at the beginning of the chapter remains valid: under what conditions does the process of speciation become inevitable?

First, it is probably a relatively rare case in which the tissue type of adaptation is the only possible type. Figuratively speaking, it is that case in which it is completely impossible to manage with half measures.

A good example, which we have already mentioned, is the crab-eating frog (p. 65): the adaptation of an amphibian to existence in water of high salinity is possible only at the tissue level. It can scarcely be doubted that the capability of northern and mountain species of amphibians to develop rapidly at low temperatures is also not limited to some morphophysiological adaptations, since they appear as early as the first stages of egg division. As was indicated (pp. 63–64), the most probable mechanism for the adaptation of northern frogs is a change in the reactivity of tissues to hormonal action. We therefore have the right to assert that northern populations of frogs proceed along the path of development of a species, even though some investigators do not vouchsafe them even subspecific rank. In speaking of different species, one can cite similar examples of equally as many species investigated at the tissue level. Ushakov's work on the thermostability of proteins (Ushakov, 1955, and others) speaks clearly to this point, as do the comparative investigations on the chemistry of enzymes, the mobility of tissue proteins, and comparative immunology (Taliev, 1935; Boyden and Noble, 1933; Grosset and Zontendyk, 1929; Avrekh and Kalabukhov, 1937; Stallcup, 1954; Leone and Wiens, 1956; and others).

Tissue differences among species can have both a very general nature (e.g., protein lability) and a special character, clearly connected with the specific conditions of existence for the species (e.g., an increased resistance of the tissues to dehydration in desert species; see p. 171).

It is not necessary for us to describe the specific results of the research indicated, since it all clearly shows that there is literally not one species (at least among higher vertebrates) in which the morphophysiological characteristics would not affect the tissue level as well. All forms that satisfy the demands that modern systematics makes on species (the most important of which is reproductive isolation) are already specific at the tissue level. In this respect, the differences between species and sharply expressed subspecies are fundamental. This leads us to the conviction that the process of speciation cannot be regarded as a simple accumulation of those features by which subspecies are characterized: in the process of speciation, a change in the direction of variability itself inevitably occurs, and it is connected with a change in the type of adaptation to the surrounding conditions of existence.

It is extremely important to emphasize that the beginning of the process of speciation occurs within the bosom of the old species, in the process of the formation of a subspecies that is distinguished from the remaining forms of the species by tissue features. There are no differences in the mechanism of the formation of the new form here: a species is a subspecies that is essentially distinguishable at the tissue level; the measure of essentiality is tissue incompatibility. Selection works with equal ease at levels of the organization of a living organism, including the molecular level. The molecular mimicry of helminths (Damian, 1964), for example, is testimony to this. Thus, in those conditions in which the process of the reorganization of the population at the tissue level turns out to be the most effective adaptation, selection works in the direction of speciation, but the actual mechanism of the process remains unchanged.

All that remains unclear is what conditions permit an animal to take the first step toward the mastery of a fundamentally new environment that demands fundamentally new types of adaptations. We suggest that one of the most important conditions is the well-known phenomenon called *preadaptation.* In the next chapter, we will have an opportunity to dwell in detail on the analysis of this interesting concept. Here, however, let us note only those of its aspects that have special significance for a solution of the problem of speciation. It has been fairly long since this concept entered the scientific literature, but it still elicits vigorous protest from a number of investigators. There arose a completely erroneous notion of preadaptation as an adaptation readied in advance for the future. Of course, such an interpretation of the concept of preadaptation sounds clearly teleological. There is, however, no basis at all for such an interpretation. It is hardly necessary to demonstrate that species cannot foresee the paths of their future development and cannot prepare for them. Preadaptations are widely distributed in nature, nevertheless, and, it seems to us, are one of the most important ecological mechanisms of the evolutionary process. Unfortunately, the concept of preadaptation was used primarily to explain some of the turning points in the phylogeny of animals that were clearly of a macroevolutionary order (the appearance of vertebrates on land, the origin of parasitism, and so forth; of the more recent literature, see especially Kosswig, 1962). The concept of adaptation was enlisted far more rarely as an explanation of species-level adaptations of animals. Experiments along this line, however, are very interesting and without question progressive. Dwight (1963), in particular, cites very interesting examples. He shows that the conditions of existence in the high Antarctic are so specific that its settlement by penguins (as an example, he cites the imperial penguin) before they had acquired their admittedly unique characteristics is theoretically inconceivable. The author makes the valid assumption that these characteristics first arose in penguins living under milder, but nevertheless rigorous, climatic conditions. This opened to them a path to the Antarctic. The same approach, in Dwight's opinion, is applicable during the analysis of some partial adaptations. The weddel seal (*Leptonychotes weddeli*), which lives under glacial fields that do not have natural air holes, gnaws a path to air with its sharp, elongated incisors. The incisors must already have been elongated, however, before the seal could have penetrated into the extreme south. It turned out that seals living in regions in which there are natural air holes also have the incisors that allow them to gnaw air holes.

The concept of preadaptation should be interpreted more broadly than it is at present. If it is, preadaptations are found to be significantly more widespread than it appears and can explain a great deal. We will avail ourselves of only one example from our practical experience.

Comparing the morphophysiological characteristics of a large number of duck species from the steppe forests and the transpolar region, we detected a very clearly expressed rule: in all noble ducks, the heart in northern species is significantly larger than in southern species. The pochard ducks from different geographical regions are virtually indistinguishable according to this most important trait (Table 11). An explanation for

TABLE 11. Geographical Variability in the Relative Heart Weight (‰) in Different Groups of the Lamellirostres[a]

Species	Forest–steppe zone (Transurals)	Subarctic (Yamal)
River ducks		
Anas platyrhyncha	8.3	14.0
A. penelope	5.8	12.0
Querguedula querguedula	11.3	19.0
Q. crecca	8.2	13.0
Pochard ducks		
Nyroca fuligula	11.2	12.0
N. ferina	9.4	11.0
Clangula hiemalis	11.25	11.0
Krokhal' (narrow-beaked waterfowl)		
Mergellus albellus	13.6	13.1

[a] According to the data of L. N. Dobrinskii and the author. Average sizes are given.

this seeming paradox is easily given, based on the concept of preadaptation. The pochard ducks in any geographical environment lead a form of life that demands the intensification of the activities of the cardiovascular system. In southern and middle latitudes, therefore, the pochard ducks always possess a larger heart than the river ducks. Residence in the north also demands an increase in heart size. In river ducks, the heart is actually enlarged; in the pochard ducks, it is not, since the size of their hearts proved to be preadapted to rigorous climatic conditions. Such an approach to the analysis of material shows that the phenomenon of preadaptation is widely distributed. It is necessary, therefore, to give this phenomenon a clear ecological interpretation. Preadaptation arises under conditions in which it promotes the well-being of the species—is *useful,* but not necessary. The perfection of this adaptation permits the species to penetrate into an environment in which the adaptation is a *necessary* condition of existence.

An example will help to clarify our viewpoint. In the northern tundra, only species possessing a complex of specific adaptations can exist: economy of metabolism, a superior capability to create reserves of nutritive substances in the organism, relative autonomy of the rhythm of life activity (both seasonal and diurnal) from fluctuations of external conditions, and so forth. In the northern tundra, these adaptations are *necessary* conditions for the existence of a species. Before penetrating into the northern tundra, the species should already have possessed a minimum of such adaptive characteristics. These characteristics could already have arisen (and, undoubtedly, did arise) in the southern tundra. In this locale, they are not necessary (a not inconsiderable number of species in the fauna of the southern tundra do not possess them), but they are highly useful.

In the southern tundra, therefore, selection will work in a direction promoting the penetration of animals into the northern regions. Analysis of this complex phenomenon is of great interest for an understanding of the mechanisms of the evolutionary process. In one and the same environment, the paths of adaptation of different species are different. To understand these differences means to understand the actual course of the evolutionary process. One of the most important tasks of evolutionary ecology is to learn to distinguish necessary and useful adaptations of animals in different environmental conditions.

Under the safeguard of preadaptations, there occurs the accumulation of tissue adaptations, which, in the final analysis, replace less perfected morphofunctional adaptations. We have given special attention to the role of tissue adaptations in the process of speciation for two reasons: their origin is the most universal difference between species, and tissue differences are the most general cause for the different reactions of animals to environmental changes. Nontissue adaptations, however, lead to the same result if they lead to fundamental change in the species' norm of reaction to a change of the conditions of existence. The specific reaction to external conditions not only makes the reorganization of the populations irreversible (the criterion of subspecies), but also provides the specificity of their future development. Speciation is the realization of this possibility. Hence, investigation of the specific reactions of different populations of a species to a change in external conditions reveals the path to experimental investigation of the process of speciation. Using this criterion, we will be able to study those conditions that lead to speciation before it has actually been realized. The simplest path for the solution of this task is the study of geographic variability in closely related forms.

If the traits of a given subspecies go in the direction characteristic for the species, there is reason to believe that we are dealing with the manifestation of typical intraspecific variability. In contrast, in those cases in which the traits of a given subspecies transgress the usual rules of geographic variability for the species, there is reason to assume that its development is going along the path of the reorganization of certain biological properties of the species.

For example, if the northern races within a species are larger than the southern, a subspecies representing an exception to this rule should be regarded as the rudiment of a new type of interrelationship with the environment, one that can serve as the basis for the origin of a new species. The same can be said in regard to other traits, such as coloring, body proportions, and so forth.

For the majority of mammal and bird species, for example, there is a characteristic increase in dimensions from south to north. In the tridactylous woodpecker, however, certain northern forms are smaller than southern. The largest subspecies of the willow ptarmigan, *Lagopus lagopus major,* is distributed in the south, in the forest–steppe zone. The largest subspecies of the squirrel—the teleut—settles the southern steppe pine forests (*Sciurus vulgaris exalbidus*). The smallest hare (*Lepus timidus gidhiganus*) lives under the most rigorous climatic conditions—in eastern Siberia, Yakut.

In forms that are close but already consolidated as species, differences in the rules of geographic variability are much more marked than is underscored by the differences in their reactions to change in the conditions of existence. Let us cite two examples: one from the class of mammals, the other from the class of reptiles.

The forest mice, *Apodemus sylvaticus* and *A. flavicollis,* are undoubtedly very closely related species. Individual populations of *flavicollis* are so close to *sylvaticus* that without the help of an analysis of geographic variability in the forms compared, it is difficult to establish their species independence. Such an analysis, undertaken by Argiropulo (1946), showed, for example, that the *Apodemus* from Armenia is *A. flavicollis*. On the basis of an isolated study of only one of these populations, however, it would be impossible to substantiate such a conclusion. Argiropulo noted that the compared species are clearly distinguished according to the character of their variability: geographic variability in *A. flavicollis* is expressed more noticeably than in *A. sylvaticus*. Moreover, as Herre (1964) recently reminded us, the different variability of these species leads to the situation that in different parts of their range, the comparative diagnosis is based on different traits (*A. flavicollis alpina* differs in coloring from the wood mice, but according to the length of their tails, they are "*superflavicollis*").

Two closely related species of lizards, *Lacerta agilis* and *L. viridis*, are represented in the east by the subspecies *L. a. exigua* and *L. v. strieata*, which are so similar that their species independence seemed more than doubtful (Cyren, 1924). Their geographic variability, however, differs markedly. As a result, in the western limit of their range, *L. agilis* and *L. viridis* appear to be two sharply demarcated species. Differences in geographic variability can lead to the situation that subspecies of one species can be distinguished more markedly than two different species. Herre (1964) shows that the differences between *Triturus cristatus* and *T. c. dobrogocus* are significantly greater than between *T. c. carnifex* and *T. marmoratus*. These observations led the author to the very important conclusion that it is necessary to study the nature of the variability when diagnosing species. In recent times, many cases of different geographic variability in closely related forms have been described (see, for example, A. H. Miller, 1956). This makes it unnecessary to cite analogous examples.

Finally, the different morphological reaction of different subspecies to a change in the conditions of existence can be shown experimentally. When *Microtus gregalis major* is bred in captivity for a number of generations, its coloring remains virtually unchanged. The coloring of *M. gregalis gregalis*, however, is changed; moreover, it is changed in such a way that the distinctions between the subspecies become even sharper. The determination of the animals' coloring was conducted with the help of colorimetry, which permitted us to establish the statistical reliability of the observed changes (Shvarts *et al.*, 1960).

A disturbance of the rules of geographic variability typical for a given species can, in our opinion, be regarded as an indication of the origin of fundamentally new interrelationships of the populations of the species with the environment.

The point of view we have developed may encounter a natural objection. We know how diverse are the distinctions among species. Can all this diversity really be reduced to tissue differences? This objection demands careful analysis.

We will examine the process of speciation as a process of the most perfect adaptation to specific environmental conditions, as an ecological process. Such a statement of the question has a fairly weighty basis, since the process of evolution—this process of adaptation to the radiation of Life—is accompanied by morphophysiological progress. It is not happenstance that with each succeeding geological epoch, the paleontologist encounters organisms that are ever more nearly perfected (from the morphophysiological viewpoint), an increasingly saturated biocenosis, and the progressive expansion of Life on the lands and waters of our planet.* This is very general and indisputable evidence of the adaptive nature of evolution. Concrete evidence of this position is inexhaustible, in the literal sense of the word. Every species of animal is characterized biologically, in the broad sense of the word, and is distinguished by its relationship to its habitat and by its position in the biocenosis. There is not one species that can be regarded as an exception to this rule. It is not a rule, but a law.

Speciation is preceded by a lengthy period of development for the population (or group of populations) in distinctive environmental conditions. This position does not ensue solely from theoretical considerations. It is strengthened by a veritable arsenal of facts from modern intraspecific systematics: the most sharply differentiated subspecies are those forms that are specifically adapted to distinctive environmental conditions. When the adaptations of such populations achieve a high degree of perfection, they also affect, with the inevitability of law, the tissue level of the animals' organization. They do so because, *independently of the actual form of adaptations of the species being engendered,* the individuals that are predominantly obtained are those that maintain their existence with a relatively more economical metabolism. Under any environmental conditions, animals are compelled to balance their intake and expenditure of energy. Independently of the actual adaptive characteristics of the populations examined, therefore, an adaptation that arises through more economical biochemical and tissue adaptations (but not morphofunctional adaptations) is progressive, and the higher the level of adaptability of the populations, the greater the role it plays in their lives.

We attach special significance to the tissue differences among species, not because they are in all cases the most important manifestations of these distinctions, but because they are universal: *in distinction from any intraspecific forms, any species (including the closest) is specific at the tissue*

* That this is so is attested to not only by general observations, but also by exact calculations (see, for example, L. C. Cole, 1966).

level, since, sooner or later, the newly arisen species encounters the necessity to economize its energy expenditures, and this leads, with the inevitability of law, to the origin of tissue adaptations. We attach special significance to this observation, since it helps to avoid misunderstandings during an evaluation of the hypothesis we have developed.

A change of the animals' characteristics at the biochemical level elicits tissue incompatibility. The data cited in the first part of this chapter provide sufficient justification for this assertion. It is completely natural, however, that tissue incompatibility arising, not as a special defense against interspecific hybridization, but as a result of the adaptive evolution of the populations, can be expressed in various degrees and form. It can be manifested in full reproductive isolation, but it can also be manifested in a reduction of the viability of hybrids or hybrid populations (it becomes impossible to integrate the gene pool on a new basis). At the genetic, cytological, and biochemical (protein) levels, these are distinctly different phenomena. That is why it is impossible, on the basis of hybridization experiments and "protein taxonomy," to distinguish sharply expressed subspecies from species. At the ecological and evolutionary levels, these phenomena are of the same order, since they ensure the independence of closely related forms, testifying to their species independence. The criteria for species, therefore, can be founded only on a broad biological basis.

Full reproductive isolation is a secondary phenomenon, provoked by close species (already species!) inhabiting the same place. The species is the elementary system. The integration of the species' gene pool is such that its "enrichment" by genetic material from another species leads to a reduction in the viability of hybrids or to a reduction in the viability and adaptability of hybrid populations. It is precisely this phenomenon that gives us the right to assert that differences between a species and any intraspecific forms are fundamental. It is possible, of course, to select examples that provoke doubts of the fundamental differences between species and subspecies (forms exist the taxonomic rank of which is difficult to determine; hence the terms "half-species" [polu-vid], "near-species" [pochti-vid], and so forth. Doubt, however, can always be provoked in regard to any biological phenomenon. There are always intermediate cases. Moreover, one must be amazed at how few "half-species" and "near-species" there are in nature. In any well-studied group, their number never reaches even 1%. In addition, the overwhelming number of doubtful species are doubtful only because they have been little studied. Hence, it follows that subspecies and species are objectively and clearly demarcated natural phenomena: species are the independent units of the evolutionary process; subspecies are forms of the manifestation of species. The existence of a small number of "half-species" compromises this assertion as little as the presence of intersexes compromises the assertion that there are only two sexes in animals.

It is theoretically conceivable that the causes for the origin of an exclusive system of integration of genotypes (a species) are so diverse that the very attempt to find something general in them is foredoomed to failure. Our investigations show that this is not so. They lead to the conclusion that when the genetic differences between closely related forms determine a different level of metabolic processes in cells and tissues for an equal degree of energy expenditure for the organism as a whole, this leads to their genetic isolation. The species splits into two daughter species. This process is accompanied by a change in the interrelationships of the animals with the external environment, with all the ensuing consequences (hiatus, distinctiveness of geographic variability, the origin of full reproductive isolation, and so forth). It is difficult to say why precisely that stage of the population's differentiation that applies to the tissue level leads to isolation. We may assume as a working hypothesis that a change in the "biochemistry" of cells also applies to the sex cells, and that this change leads to a disturbance of the balance between the nucleus and cytoplasm of the zygote, which is the main cause of disturbances in the development of hybrids (see p. 149). The beginning stages of this process, although they do not lead to reproductive isolation, do lower the viability and adaptability of hybrid populations (see p. 150). These very interesting questions deserve the undivided attention of geneticists. The phenomenology of the process seems clear to us. When a progressive adaptation of the differentiated population leads to a change in the energetics of the organism, this change entails the origin of a new species. The external manifestation of this process is a sharp reduction in the role of morphofunctional adaptations in the maintenance of the organism's energy balance.

Tissue specificity is easily detected with modern methods of research, on which "protein taxonomy" is based. It is not possible with the help of these methods, however, to distinguish between species and intraspecific forms, since the "substitution" of tissue adaptations for energetically disadvantageous morphofunctional adaptations is realized on the basis of the reorganization of different biochemical functions. It is natural that the changes entailed in immunological, electrophoretic, chromatographic, and other such indices are expressed to different degrees and in different forms. Thus, the first part of this chapter, treating as it does questions very far from ecology (it may be perceived as an extraneous inclusion in this book), acquires the forces of important evidence of the ecological essence of the process of speciation.

The accumulation of new material and the analysis of new data from

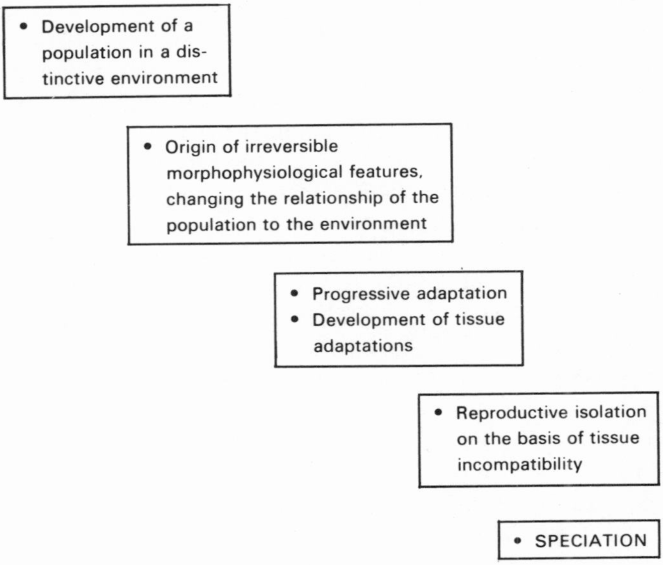

Fig. 10. A scheme for the process of speciation.

the literature only convinced us that "species are not species because they do not cross, but they do not cross because they are species" (Shvarts, 1959). This phrase, though quite imperfect in a literary sense, expresses our viewpoint well.* In the process of intraspecific differentiation, individual populations acquire those features that, as a corollary effect, lead to reproductive isolation. We have cited a great deal of evidence to support this position. Here, we think it important to cite yet one more, which has a very general character. If the basis of the process of speciation were the origin of reproductive barriers, it would be natural that the degree of genetic isolation among species of different organisms would be approximately equal, since species would arise after their isolation. In reality, this is far from so. Among mammals, species hybrids are relatively rare; among fish, they are a common phenomenon. The survey of data on hybridization cited above spares us the necessity of providing further examples. We will only recall that among species of plants, genetic isolation is significantly weaker than among animals. From the positions developed, these differences are very clear. In higher animals, as a characteristic of their physiology (strict constancy of the internal environment of the

* Beaudry arrived at a similar conclusion almost simultaneously with us. He shows that reproductive isolation is a second-order attribute of species (Beaudry, 1960).

organism), tissue incompatibility is sharper than in fish, and even more so than in plants. Changes in biochemical characteristics of the organism, which in mammals lead to genetic isolation, therefore do not create reproductive barriers in fish. If species of fish and mammals are approached with identical morphophysiological criteria, therefore, reproductive isolation of species of mammals should be immeasurably greater than that of fish. And precisely this takes place. The views developed here find support in the very general differences among classes. This question is analyzed specifically in a separate article (Shvarts *et al.*, 1966*a*).

On the basis of this exposition, we arrive at the scheme for the process of speciation presented in Fig. 10.

CHAPTER VII

The Ecological Essence of Macroevolution

Of what significance is evolutionary ecology for the analysis of problems of macroevolution, and to what conclusions for the study of the correlation of micro- and macroevolution do the results already achieved by evolutionary ecology lead?

From the very beginning of the discussion of this question, it should be noted that problems of macroevolution are investigated in basic systematics, and this has placed a characteristic imprint on the direction of these investigations. The main question is on the problem of the phylogenetic correlation of taxa of different rank. Its modern state was recently analyzed by Dement'ev (1965) and need not be discussed here. It is important to emphasize, however, that during an analysis of the question of the phylogenetic correlation of different taxa, undeservedly little attention is given to the no less, and in our opinion more, important question of the reality and biological essence of the taxa of macrosystematics.

In its concrete form, the question indicated takes this form: Under what conditions does the "conception" of new genera occur? What properties should a species possess in order to become a potential ancestor? As in many other cases, the answer can be purely formal: a genus is a group of species united by the greatest mutual resemblance. Under such a statement of the question, a "genus" is regarded as a purely conditional, artificial union. Many authors, taking their stand wholly on the position of phylogenetic systematics, adhere to such a viewpoint (Mayr *et al.*, 1956; Simpson, 1961; Huxley, 1940; and others). Nevertheless, this viewpoint seems mistaken to us, since it closes the path to the solution of the central question of evolutionary studies—the correlation of macro- and microevolution.

In this, we are proceeding from the following considerations: A species is an objective reality. Indeed, innumerable facts and many deep theoretical investigations so affirm. If the living world surrounding us did not consist of species limited to sexual reproduction within a relatively narrow circle of harmoniously developed living systems, evolution would probably have stopped at the level of bacteria, since universal panmixia based on the chance meeting of individuals would lead to the massive dying out of incompatible genotypes pairing by chance. To doubt the objective reality of species, therefore, means to doubt the objective reality of evolution. On the other hand, higher categories, too, are objective realities, not conventional categories. It may be doubted that classes are objective stages of evolution, objectively characterizing general types of mastery of the arena of life. The same can be said in regard to many orders and even families. It turns out that the lower taxonomic unit—the species—is a reality, and so too are the higher units, but the intermediate ones are conventional units. Is it not more probable to suppose that we simply have not yet found an objective criterion for evaluating the beginning links of the macroevolutionary process?

The working hypothesis we suggested for the solution of this question can be summarized as follows: In those cases in which a new species can give rise to a new type of adaptive radiation, we have the right to ask about the conception of a new genus. It is relatively easy to ascertain this process *ex post facto* (on closer examination, the majority of modern genera materialize by a certain type of adaptive radiation; recall such genera as *Parus, Sitta, Ardea, Falco, Ellobius, Lemmus, Felis,* and others), but apparently it is possible to detect it *in statu nascendi* only with the help of new methods of investigation.

A comprehensive study of a species permits the evaluation of the degree of not only its morphophysiological but also its evolutionary specificity, and permits us to foresee, with a certain degree of reliability, its possible role in the further evolution of the group. Species, the morphophysiological characteristics of which determine the high degree of specificity of relationships to the habitat, are, in the full and literal sense of the word, ancestors.*

What has been said is no more than a summary of the working hypothesis, but we are convinced that investigations in this direction are absolutely necessary for the creation of a theory of evolution in which *microevolution* and *macroevolution* will be no more than correlated chapters. On the other hand, in those investigations that gave rise to the possibility of

* In the original, rodonachal'niki; literally, founders of genera or families—Ed.

approaching a large series of species of the same order with a single measure, the specificity of lower taxa (not only genera, but also families) in higher vertebrates (especially notably—birds) appeared with full distinctness. In particular, we (Shvarts *et al.*, 1966*a*) detected this during a study of the correlation of morphophysiological characteristics of different species with an intensity of metabolism. Certain tendencies in the latest theoretical works inspire us with still greater confidence in the correctness of the chosen path. For example, W. L. Brown (1958) advanced the interesting concept of "adaptive extension." According to Brown, adaptations of this type are characteristic of so-called "potent species," which have a chance of giving rise to a new unit higher than the species. Moreover, in a summary work by Mayr (1963), it is shown that if a species arising anew opens a new "adaptive zone," it radiates until the zone is filled with its descendants. It seems to us that this conclusion ensues: it is necessary to elaborate the methods that would permit us to reflect this objective process in objective taxonomic categories.

These and some other investigations convince us that not only species, but also higher taxonomic categories, are biological realities, not conventional taxonomic units. In our eyes, this warrants every effort directed to the study of the evolutionary mechanisms for the establishment of new genera. Certain tendencies of modern taxonomic investigations permit us to hope that success in this direction may be expected in the coming years.

It is perfectly natural that the question of the reality of taxa above the species level may be finally decided only if a theory is elaborated that shows the general tendencies in which this reality is manifested. An attempt at the analysis of this question led to the conclusion, which is important for us, that the tendency consists in the fact that the character of phylogenetic reactions to a change in form or conditions of life in representatives of different genera, families, and so forth, is different. An ecological criterion, consequently, proves to be basic during the elaboration of a theory of macroevolution.

During an analysis of this question, it is difficult to avoid the temptation of being carried away by its purely phenomenological aspect, since a simple enumeration of the ecological and morphophysiological characteristics of different genera, families, and orders—which are often highly independent and precise and which affect every trait of the organism—is in itself testimony to the ecological nature of the macroevolutionary process. The origin of new taxa is connected with the capability of animals to master new environments of life and to conquer new ecological niches. The phenomenology of macroevolution, however, has repeatedly been the subject of special investigations (it is sufficient to recall the fundamental book of Rensch, 1959). The matter is significantly worse as regards analysis of the mechanisms in the process of the origin of taxa higher than the species, and worse yet as regards the philosophical analysis of the problem. If, from the geneticists' viewpoint, macro- and microevolutionary processes are often regarded as processes of a different nature (recall Dobzhansky's aphorism: "Macroevolution is unpredictable, not repeatable, and irreversible; microevolution is predictable, repeatable, and reversible"; see also p. 212), the majority of systematists and morphologists correctly regard macroevolution as a "continuation" of microevolution. Psychologically, this is understandable. The geneticist cannot reproduce the phylogenetic process, cannot reiterate it or foresee it, but he can, beforehand, plan the reconstruction of the genetic structure of a model population and foresee the result obtained, which is fully comparable with the reorganization of natural populations on a microevolutionary scale. The systematist and theoretician find themselves in a different position—that of an evolutionist compelled to construct his theory on the basis of the compared morphophysiological characteristics of forms of different taxonomic ranks. During this process, it is observed that there is no possibility of demarcating sharply expressed subspecies from species on the basis of morphophysiological criteria, and often the differences between subspecies are greater than between close species. It is well known how the modern systematists cope with the difficulties arising in connection with this, and we tried to give their general analysis in the preceding chapter. The systematist, however, encounters similar difficulties even when distinguishing genera. A large group of obviously close species can be regarded as a single genus, but it can be subdivided into several genera, since, among any group of close species, it is always possible to find subgroups of closer ones. In this connection, it is sufficient to recall the group of susliks, grey voles, runners, lizards, and so forth.

It would be possible to cite as many examples of this type as you please. Hence the conviction that microevolution and macroevolution are a single process. This viewpoint is not universal. The essentially saltational theory of Goldschmidt (1940, 1961) is far-famed. According to it, higher taxa arise suddenly, as a result of systemic macromutations. Other investigators also hold similar positions, in a less clear form. Thus, for example, Devillers (1965) asserts that large morphological differences among taxa can be elicited by insignificant quantitative changes in morphogenetic mechanisms. He supported his views with experimental data, showing that a reduction of the fibula can be elicited by a simple diminution of the quantity of mesenchyme in the bud of the extremity. He succeeded in achieving the formation of a fibula in chickens by putting an additional quantity of mesenchyme in the bud of the extremity. By putting successively smaller quantities of mesenchyme in the bud of the extremity of the frog, *Xenopus laevis*, he obtained a reduction of different digits. Devillers supposes that such changes of a phylogenetic scale as unpaired olfactory capsules and weak development of the frontal lobes of the brain in cyclostomata could be elicited by the reduced morphogenetic activity of the frontal end of the chord. A reduction of the morphogenetic activity of the nerve eminence of the lateral line, which in bony fish induces the development of the integumentary bones of the skull, could lead to the loss of these bones in cartilaginous fish.

Very similar viewpoints are maintained by several other modern authors, in particular Hecht (1965), who writes that significant morphological changes can be conditioned by small differences in morphogenetic mechanisms that depend on a few genes. This remark of Hecht's is especially indicative, since in his general article (Hecht, 1965), he asserts that in distinction from saltationists, the majority of biologists arrive at the conclusion that there are no sharp differences between the mechanisms of micro- and macroevolution. It is thought, however, that mutations that cause substantial disturbances of morphogenetic mechanisms are not distinguished in any substantial way from Goldschmidt's macromutations, and acknowledgment of the special role of such mutations in macroevolution signifies, in practice, acknowledgment of the special mechanisms of the origin of taxa of higher rank. The examples cited by Devillers and analogous examples, however, contribute almost nothing to an understanding of the essence of the problem. It is not enough to establish that disturbance of morphogenetic processes elicits changes of morphology on a macroevolutionary scale. It is necessary to show that animals with such changes not only are fully viable, but also can stand up to competition with normal individuals. This is not corroborated by factual data, but is theoretically achieved with the help of the hypothesis about systemic mutations, i.e., the hypothesis that arose from Goldschmidt's viewpoint. Factual corroboration of our evaluation of the role of morphogenetic changes in the evolutionary process can be found in an interesting article by S. C. King (1965), which shows that in an experiment on *Drosophila,* changes in the expression of diagnostic traits of order rank, even, were achieved within a few generations—whereas no one would take it into his head to assert that in nature animals could change so much in the course of a few years that they could be attributed to a new order. We would not want to enter into a specific polemic here on the question of the mechanisms of the macroevolutionary process. It is only important to show that even today, this very important theoretical problem arouses fundamental divergences in the views of experimentors, who, it would seem, are standing on one and the same theoretical platform. This is the best evidence that the problem as a whole has not been worked out.

Another important position, on which it is necessary to dwell before going on to an ecological analysis of the essence of the problem, is that according to the majority of authors, the rank of groups that have formed can be evaluated only in retrospect. In other words, the course of the macroevolutionary process is not predictable (Hecht, 1965; King, 1965; Bock and von Wahlert, 1965). This conclusion is extremely important. The process may be unpredictable in principle, but the lack of predictability may also ensue from the fact that it has not been well studied. The authors cited, whose views are typical of a large group of modern investigators, suggest that the unpredictability of the evolutionary process ensues from the fact that the transition to a new level of organization (the origin of a new taxon) demands a radical restructuring of the genome, which depends on the restructuring of a combination of many factors, the concrete manifestation of which is unpredictable. The unpredictability of the course of evolution ensues, also, in the opinion of the majority of authors, from the asynchrony of phylogenetic reorganizations. Different adaptive features of a new taxon do not appear simultaneously, and, as a rule, one cannot single out a determinant trait at the threshold level of adaptation (Bock's expression). This makes a clear demarcation of taxa impossible. The rank of the group that has formed also depends on the breadth of the new adaptive zone. All this makes it possible only in retrospect to evaluate the rank of the group that has formed.

The notion is created that the unpredictability of the course of macroevolution ensues not only from our ignorance, but also from the essence of the problem. This pessimistic conclusion seems mistaken to us. It may entail yet more serious theoretical and, in the longer perspective, practical mistakes.

Can it be said that if the environmental conditions under which a certain group of organisms develops and its general features are known, it is thereby possible, in principle, to predict the course of its development as well? It seems to me that the data of modern paleontology support an affirmative answer to this question. These data testify that the transition to a new level of organization in the course of phylogeny of individual groups occurred repeatedly. The discussion of the poly- or monophyletic origins of the mammals, and some other large taxa as well, which has arisen in recent times and shows no signs of dying down, ensues directly from these data.

Schaeffer (1965) recently generalized the results of these very interesting investigations. We will use his work for a brief exposition of these investigations and add some conclusions of other investigators. In the evolution of the cartilaginous fishes, the transition to modern elasmobranchs was accomplished from the hybodont level independently in the lamnoid and squaloid sharks and skates. In all these forms, amphistyly was replaced by hyostyly, bodies of vertebrae appeared, and the skeleton of paired fins was reorganized. In the evolution of the actinopterygian fishes, the transition from the palaeoniscoid level to the subholostean and then to the holostean and to the level of modern bony fishes was also accomplished in parallel in a number of closely related branches. This transition was attended by a radical restructuring of the jawbone apparatus and by progressive changes of the locomotor apparatus. In all probability, repeated experiments took place among the rhipidistian crossopterygian fishes to go to the "amphibian" level of organization. The reflection of such experiments can be seen in the structural features of different orders of the carbonaceous Stegocephali. The transition from the theriodontid to mammalian level of organization was accomplished in parallel (reorganization of the jawbone articulation and the middle ear). Parallelism also played a significant role in the formation of suborders of modern rodents—porcupines and, especially, the Myomorpha, each of which is characterized by an individual design for the disposition of the masculatory muscles.

Perhaps all that remains to be added to the data generalized by Schaeffer is that numerous works on the phylogeny of different mammalian groups (Thenius and Hofer, 1960; Westphal, 1962; Remane, 1964; and others) make it appear probable that the most important event in the history of life on earth occurred repeatedly: different groups of reptiles independently overcame that boundary that divides the two most important classes of vertebrates (reptiles and mammals). Fully analogous data have accumulated at the present time on other questions of the specific phylogeny of vertebrates (Romer, 1965, and others). It is not superfluous to recall that the viewpoint of Huene (1956) has not been refuted once and for all. According to it, the phylogenetic differences between caudate amphibians and all other groups of this class exceed the differences between the classes (the division of the vertebrates into the Eutetrapoda and the Urodelidia).

These data show that the possibilities of evolution—which are, in principle, unlimited—are limited in each case by the initial morphophysiological characteristics of the animals and by their habitats, the environments to which they are adapted or have begun to be adapted. If both the one and the other are known, the general course of evolution can thereby be foretold. The thesis of the unpredictability of macroevolutionary reorganizations, therefore, is based on another aspect of the process. There is the notion (which is not always formulated sufficiently clearly) that since the macroevolutionary process is a lengthy process, its course can be interrupted or changed by chance events that are not amenable to calculation. This is actually correct. It is impossible, however, to doubt that in those cases in which a vectorized process of development has begun, its course is limited within certain bounds. These bounds make the macroevolutionary process just as predictable (the quantitative differences are not substantive) as the process of microevolution. The progressive adaptation of any amphibian species to permanent residence on land should have led, with the inevitability of natural law, to the formation of powerfully developed organs for breathing the oxygen in air, to the appearance of special adaptations for the development of embryos outside of water reservoirs, to the perfection of locomotory organs, to changes in the sensory organs, capable of working in an air environment, and so forth.* In other words, amphibians should have acquired the features of reptiles, regardless of which specific group of amphibians provided the beginning of the new class. Chance events could lead to the situation that the process of reptilization of amphibians that had begun did not reach conclusion, but there are no chance events that could lead to the arrival on land of amphibians that did not possess those reptilian features that freed them from their ties to the water. In order to master a new arena of life—on land—amphibians should have become reptiles. A different path of development was closed to them by the preceding history of their development, which determined their main morphophysiological characteristics.

Analogous reasoning is correct in regard to any case affecting the mastery of a new environment of life by individual groups of animals, be it adaptation to a parasitic form of life or to the mastery of a new biome, a transition to a new type of nutrient, mastery of a new type of movement, or other case. It is not by chance that convergence is one of the most striking manifestations of the most general laws of evolution. And it is not by chance that in the history of science, we so often encounter different variants of the theory of orthogenesis. Meanwhile, the history of life on earth does not give any basis for resorting to the concept of orthogenesis in its idealistic sense. The direction (predictability) of the evolutionary process is determined by the reactions of organisms of a certain morphophysiological type to certain environmental conditions.

The fundamental importance of a question that already has a direct relationship to evolutionary ecology arises, however. The questions arise: Can one define that boundary that determines the whole future course of the evolution of a given taxon? At which taxonomic level is the future course of evolution determined? How far ahead is evolution anticipated?

As was already mentioned, the macroevolutionary process is connected

* The principal distinctions between reptiles and amphibians are not exhausted with the list of examples enumerated. Certain special adaptations to an air environment can have no less significance. The type of nitrogen exchange should be attributed to the number of such adaptations. On land, a ureotelic type of exchange is necessary. In reptiles, therefore, uric acid predominates among the final products of nitrogen exchange (Packard, 1966). This example emphasizes the complexity of changes preceding the conquest of a new arena of life.

with the acquisition by animals (or plants) of adaptations of broad significance. This concept is not equivalent to Severtsov's concept of aromorphosis,* since the latter concept, although it indicates adaptations of the broadest significance, is connected with raising the morphofunctional level of the animals' organization, whereas adaptations of broad significance are not obligatorily connected with morphophysiological progress (the best example is parasitism). Can we evaluate the significance of individual adaptations at the moment of their origin, or is it possible to do this only retrospectively? As we will try to show below, this is by no means a purely academic question. In order to answer it, it is useful to emphasize that any modern taxon can be characterized not only morphophysiologically, but also ecologically. Even higher taxa represent no exceptions.

We will use classes of a subtype of vertebrates for an example. Fishes are animals that have conquered the environment of water, but are incapable of mastering land. Amphibians are animals that are capable of mastering significant territories, but are incapable of reproducing out of water. Reptiles are animals that have fully mastered land, but are incapable of conquering northern regions of the earth (more than half of all earth's land is in practice closed to reptiles). Birds have mastered the environment of the air and overcome the gravitational barrier. Mammals have fully mastered land.

The question arises: can we select that boundary that determines the transition of animals from one ecological group (in the sense indicated) to another, or is it, in principle, impossible to determine this boundary, since the transition from one ecological group to another at once demands complex reorganizations?

The origin of saltationist theories speaks to the fact that a large number of investigators adhere decisively to the second point of view. Others hold more moderate views, but consider that the complexity of adaptations arises gradually. Many of the difficulties of such theories are removed by the modern treatment of the theory of preadaptations. It is presented in perhaps its clearest form in the works of Osche (Osche, 1962; see also Chapter VI). The basic positions of his views can be summarized as follows:

In order to master a new environment, the organism should possess some characteristics that will permit him to take the first steps in this direction. These characteristics are necessary prerequisites for subsequent adaptations; they are preadaptations. Any morphological structures, apart from the main ones, also possess accessory functions. Some of these accessory functions can create the conditions for penetration into new

* A change in organization that increases an organism's energy or life activity—Ed.

habitats. For illustration, Osche analyzes the possible paths for the emergence of rhipidistian fishes (Rhipidistia) onto land and the origin of amphibians. The modern *Periophthalmus,* and *Boleophthalmus,* which dwell in the coral reefs and silt-containing littoral zones and at times lead an amphibian form of life, can serve as a distinctive model for this process. The possibility of these fishes coming onto land is determined by the development of extremities adapted to movement on hard ground, a rich blood supply to the epidermis (skin breathing), and binocular vision. All these characteristics arise as adaptations to conditions of existence in the waters of the coral reefs, but they create the possibility for the fish to come out onto land. The degree to which these characteristics of *Periophthalmus* and *Boleophthalmus* are expressed is found to be under the control of the thyroid gland, and is strengthened by the influence of thyroxine (experimental data of Herms, 1952). Osche suggests, therefore, that the hyperfunction of the thyroid could have been a "conclusive mutation" during the emergence of fish onto land. Osche also considers possible the origin of "preadaptive modifications" (in marine fish, distillation of water elicits a hyperfunction of the thyroid), which later become hereditarily fixed. The preadaptation hypothesis also proves to be useful during the analysis of the possible origin of parasitism. Osche (1962) illustrates this with the example of the nematode. Preadaptations to parasitism could have arisen in the saprobiontic nematodes, since their development occurs in the decomposing organic substrate with high temperatures and low oxygen pressure, and in the presence of decomposing proteins produced by bacterial enzymes. This hypothesis is supported by the repeated origin of parasitic nematodes from nematodes leading the life of saprobionts. Parasitic nematodes are encountered often among the earth forms (*Phasmidia*), but very rarely among the water forms (*Aphasmidia*). This difference can be explained naturally by the fact that in the water medium, conditions for the saprobiontic life form do not exist, and consequently, the conditions for the development of preadaptations to parasitism do not exist either.

We intentionally set Osche's views and system of evidence forth in great detail, since they express the possible role of preadaptations in macroevolution to the fullest degree. The facts synthesized by the preadaptive theory clearly show that evolution in a determined direction is always based on morphological and ecological prerequisites, which, in their turn, arise only under fully defined circumstances. It follows that any morphological and ecological characteristics of individual groups of animals can be appraised from the point of view of their possible evolutionary significance. This corroborates the fundamental importance of resolving the question we posed at the beginning of this chapter: at which taxonomic level are the perspectives of evolution determined, and what are the actual mechanisms of the beginning stages of macroevolutionary reorganizations?

Any adaptation capable of increasing the numbers of a species in a given habitat or of permitting it to expand its already conquered ecological niche will inevitably be picked up by selection. If one approaches the analysis of the problem from this viewpoint, it develops that different populations within a species (to say nothing of different species) do not really so rarely possess those characteristics that create the prerequisites for broadening progressive evolution. It is important, however, that in the majority of cases, these characteristics are not morphological, but physiological. We have already had the opportunity to touch on this question. Recall only that the capability of crab-eating frogs to withstand high salinity in water actually opens for amphibians the path to the sea—or at least to littoral waters. Since the number of ecological niches within the limits of this part of the arena of life is actually boundless, one may suppose that on the basis of the characteristic acquired by the crab-eating frog, there becomes possible a broad adaptive radiation of *Anura,* which may well lead to the origin of new genera or even families. As was indicated, analogous reasoning is correct in regard to subarctic populations of frogs, the speed of development of which permits them to penetrate into the tundra. And in this environment, the conditions favor adaptive radiation, and consequently the origin of new taxa.

How can one represent the beginning stage of macroevolution? We will use an example well known to us from our personal experience of the subarctic fauna. Many years of investigation showed (Shvarts, 1963*c*) that the paths of adaptations of different animal groups to life in the far north are very similar. Moreover, a feature characteristic of the subarctic—an increased capability to create reserves of nutrient substances in the organism—is detected in representatives of all classes of land vertebrates. Frogs are not distinguished from the highest mammals in this regard! Another almost universal feature of all vertebrates of the north is an increase in the rate of development and the relative autonomy of the rhythm of reproduction from seasonal changes of the external environment (rodents reproduce in winter; frogs undergo development at low temperatures; many birds begin to reproduce in the north at almost the same time of year as in the temperate climatic zone). The causes of all these coincidences in the adaptation of different animal groups to living in the Arctic are understood. The very same rule is reflected in these coincidences: the course of evolution is directed.

The majority of the aboriginal mammals of the Arctic are represented by specific genera or higher taxa (*Ovibos, Rangifer, Alopex, Lemmus,*

Dicrostonyx). There are also many endemic Arctic genera among birds. We have before us a suitable example for analysis of the actual course of the beginning stage of macroevolution.

One of the most typical Arctic species of mammals is the lemming (*Lemmus sibiricus*). This is the only small rodent that has mastered the tundra to the fullest degree, from both a geographical and an ecological point of view (circumpolar range; capability of inhabiting biotopes of any type, including the parts of the low-lying moist tundra least favorable for rodents). The characteristics that opened a path to the tundra for it are: highly efficient thermoregulation at low environmental temperatures (the thermoneutral zone is more than 10°C lower than in close forms of more southern origin) and the capability to reproduce in the winter, under snow. However important the characteristics of the lemming that raise its resistance to cold are, they are less essential in the evolutionary scheme than the capability to reproduce in winter. Under the conditions of a short polar summer, the small rodent is not capable of realizing a geometric progression in reproduction in the course of one season, and is not capable of maintaining its numbers over a long period. The lemming, from the beginning, must have acquired the capability of reproducing in the winter and penetrating into the deep tundra (utilizing favorable microclimatic features of individual biotopes during this process), and then acquired those morphophysiological characteristics that made it possible to utilize diverse habitats (including the least favorable).

Of the species of broad distribution that do not possess special adaptations for life in high latitudes, the most numerous is the *oeconomus* vole. We will attempt, using the method of a Gedanken experiment, to represent the path by which *oeconomus* could have acquired the property of the lemming—to begin to reproduce under the snow. As was mentioned, the *oeconomus* in the north begin to reproduce at the same time of year as in the south (in the forest steppe), but in a completely different phenological situation. The young of the first litter are born, probably, at a very unsuitable time—just when the snow is beginning to melt. A study of the population age structure of subarctic populations of *oeconomus* shows that this unfortuitous circumstance makes itself known: the majority of young in the first generation die. In all populations followed, the young of the first generation yielded in numbers not only to the young of the second litter, but also to their parents. Considering the exceptionally high fecundity of northern *oeconomus*, this finding can be regarded as indisputable evidence of the disproportionately high mortality of the young of the first litter (Shvarts, 1959; Pyastolova, 1964). To illustrate this very interesting phenomenon, we present some tables reflecting the age structure of northern and southern populations of *oeconomus* and of a typical subarctic dweller—the lemming (Tables 12 and 13). We will try to understand what should have occurred in order for *oeconomus* to have begun to reproduce in the manner of the lemming.

The early reproduction of northern *oeconomus* indicates that the rhythm of its reproduction is determined primarily by an "internal clock," not by changes in the external environment. The biological clock of *oeconomus* is established according to the demands of southern populations, and arouses the northern ones at a very inappropriate time. It is possible that some part of the population is oriented to a change in their external conditions and not to the clock. Such individuals begin to reproduce later: their first litter coincides in time with the second litter of the greater part of the animals. In V. N. Boikov's collection, we found a fairly large number of female *oeconomus*, added in the beginning of June, that had not yet begun to reproduce. Such a deviation from the norm under tundra conditions obviously has gloomy prospects. No matter how small the number of surviving individuals of the first generation, they play a positive role, even in the middle of the year, in the dynamics of the population's numbers, since they mature rapidly, and as early as the end of June to the beginning of July produce a second generation of young animals. In the years with a very early spring, when the first generation survives in large numbers, the first generation may become a decisive force in the dynamics of the population's numbers. It is possible that the massive reproduction of *oeconomus* in the far north (we observed it in 1956 in Yamal) is determined

TABLE 12. Dynamics of the Population Age Structure of the *oeconomus* Voles[a]

	Khadyta River				Polar Urals				Lower Sobya River		Lower Poluy River		Lake Sasykul'	
	1958		1959		1960		1961		1960		1962		1963	
	July 15–Aug. 12		July 10–Aug. 12		Aug. 14–Aug. 26		July 23–Sept. 2		July 21–Aug. 29		Sept. 7		May 16–June 27	
	n	%	*n*	%	*n*	%	*n*	%	*n*	%	*n*	%	*n*	%
Overwintering	45	35.4	9	26.5	2	7.1	4	6.4	13	28.8	—	—	63	52.5
Generation I	19	14.9	7	20.6	5	17.9	8	12.8	8	17.8	2	6.0	33	27.5
Generation II	27	21.3	9	26.6	6	21.4	22	35.6	19	42.3	7	21.2	24	20.0
Generation III	36	28.4	7	20.6	8	28.6	14	22.6	5	11.1	7	21.2	—	—
Generation IV	—	—	2	5.8	7	25.0	14	22.6	—	—	17	51.6	—	—
TOTAL:	127		34		28		62		45		33		120	

[a] Lake Sasykul' is forest steppe; the rest of the populations are subarctic.

TABLE 13. Dynamics of the Population Age Structure of *Lemmus obensis* in 1956–1957[a]

		Generation (litter)									
		Over-wintering		Winter		Spring		Summer I		Summer II	
		n	%	*n*	%	*n*	%	*n*	%	*n*	%
1956	July–August (II half)	—	—	6	25.0	1	4.0	17	71.0	—	—
	September	—	—	—	—	1	3.0	8	18.0	32	79.0
	October	—	—	—	—	—	—	2	10.0	17	90.0
1957	May–July	5	25.0	6	30.0	9	45.0	—	—	—	—
	September (other region)	—	—	—	—	2	5.0	4	11.0	32	84.0

[a] According to the data of K. I. Kopein.

by precisely this cause. We do not have actual data supporting this supposition, but it seems very probable to us. It is indisputable, in any case, that selection cannot support an increase in the degree of correlation between the rhythm of reproduction and signals from the external environment, since this would inevitably curtail the period of reproduction. It becomes clear that selection should work in another direction.

It is hard to doubt that the course of the biological clock is somewhat different in different animals. Under the conditions of the subarctic, those *oeconomus* with somewhat fast clocks gain a clear advantage. Such individuals begin to reproduce just slightly earlier than the rest, but this circumstance may become decisive: the young of the first litter have enough time before the thawing of the snow to attain an independent form of life, and this saves them, apparently, from inevitable death. Such cases actually occur. Pyastolova (1964) cites data showing that some individuals begin to reproduce as early as April. The young of such early litters avoid the fatal action of the melting waters. Certain considerations issuing from Waddington's theory of the assimilation of acquired traits cause one to think that in the conditions described, selection should work slowly (the direction of selection does not coincide with the nature of actions of the external environment; see Chapter I). That is why subarctic vole populations acquire the hereditarily determined property of high fecundity much more rapidly than changes in their original reproductive rhythm occur. Selection should occur and undoubtedly will occur sooner or later, however, so that *oeconomus* will acquire the type of reproduction that is now characteristic for forms more completely adapted to life in the north (*Lemmus, Dicrostonyx, M. gregalis major, M. middendorffi*).

Being a member of a group of species that has mastered the subarctic with the degree of completeness characteristic for the enumerated forms, *oeconomus* maintains its specificity. This is determined by the circumstance that its adaptations, which are similar to the adaptations of other rodents, will be perfected on a different general ecological basis. One can even predict what these differences will consist of. *Oeconomus* consumes the bark of trees to a greater degree than other species of *Microtus* (Formozov, 1948; Bannikov, 1953; Shvarts *et al.,* 1955; Karaseva, *et al.,* 1957; Snegirevskaya, 1961; Nazarova, 1958; and others). Under the conditions of typical tundra, therefore, its distribution in the biotopes will be specific, and its dietary regimen will be unique. In the final analysis, this also determines the unique zone of its adaptive radiation. A change in the phenology of *oeconomus'* reproduction can lead to macroevolutionary reorganizations. Having established itself under the conditions of the far north, *oeconomus* will have the opportunity to diffuse widely in a new environment, and the direction of selection will change. This elicits substantive changes in morphology and physiology, which will be settled by systematists. If the displacements prove sufficiently substantive, a new genus will arise. It should not be understood from what has been said that the preconditions for macroevolutionary displacements inevitably arise. Of course they do not. But it can be assumed that each step in macroevolution has such ecological preconditions as its basis. Their biological essence consists in the acquisition by species of those characteristics that open a path to a new habitat, that open new possibilities for adaptive radiation and, consequently, for the origin of new genera and taxa of higher rank. It becomes clear that the mechanisms of macroevolutionary reorganizations are the same mechanisms that govern the reorganizations of populations and speciation. We intentionally used an example in which the gates to macroevolution were opened by a very insignificant (from a morphophysiological point of view) displacement in the biology of the animal, in order to make this point even clearer. One and the same mechanism in different actual conditions can lead to microevolutionary reorganizations of populations, to speciation, or to displacements on a macroevolutionary scale. Macroevolution can be regarded as the result of adaptive radiation in a new environment. Since, however, the interrelationship of the organism with the environment is determined not only by the characteristics of the environment, but also by the characteristics of the organism itself, the path to macroevolutionary reorganizations is opened not only by those adaptations that make possible the mastery of a new habitat, but also by those morphophysiological changes that change the character of the organism's interrelationship with the environment—change it so substantially that the organism, remaining in the old environment, is already subject to new conditions of existence. Probably the best example is the snakes. Leglessness (which arose as an adaptation to feeding on capricious rodents) and the complex of anatomical changes connected with it established a new group (a suborder of reptiles) with distinctive relationships to the environment and led to a powerful adaptive radiation, as a result of which the snakes, which arose only at the end of the Cretaceous period, have produced a multitude of ecologically and morphophysiologically independent species. (At present, the number of species and genera of snakes and lizards is approximately equal.) We arrive at the conclusion that just as with microevolution, macroevolution is subject to ecological rules and proceeds on the basis of the very same mechanisms.

A certain body of facts makes this conclusion indisputable. Each taxon is specific, not only from a morphological but also from an ecological point of view. We have already turned our attention to this matter, and will add only a few examples, touching on the beginning stages of macroevolution. Let us use the subfamily Microtinae. The genus *Microtus* are voles of open spaces (even in the forest, they are numerous in sunlit parts), which eat predominantly low-calorie green food. The genus *Clethrionomys* are forest voles, which, to a great extent, utilize the higher-caloric bark and seeds of trees. *Alticola* are voles that are specifically adapted to life in the mountains. *Lemmus* are specialized arctic forms. *Ellobius* are highly specialized voles, leading a subterranean form of life. *Ondatra* are voles adapted to an aquatic form of life. Such a list could be continued, or a new one could be started, using any group of animals. This shows that genera become isolated in the process of adaptation by an ancestral species to a specific habitat and to a specific form of life.

On the other hand, the governance of the macroevolutionary process by ecological rules is evident in paleontological observations showing that the rate of macroevolutionary reorganizations is determined by changes in the conditions of the animals' existence. An interesting article by Kurten (1965) in the collection under the descriptive title *Ideas in Modern Biology* was devoted to this question. Referring to his previous work (Kurten, 1959), the author shows that the rate of evolution of morphological traits varies a thousandfold. The evolution of different groups occurs at different rates, but the decisive factor is, apparently, the conditions of development. This is evidenced by the following example: As a rule, animals with a rapid turnover of generations evolve *somewhat* more rapidly than animals with a longer life span.* In the Pleistocene, rodents evolved more rapidly, on average, than any other group of mammals. And the group of present-day elephants (in contrast to mastodonts) evolved significantly more rapidly than rodents. Still more interesting are the data that characterize the rate of

* That they do is natural, since the rate of evolution is measured in numbers of generations, not in years.

evolution of individual groups at different times.* The rate of evolution of mammals in the Pleistocene was ten times greater than at the end of the Tertiary period. This corresponds to changes in the environment that occurred in the lower Pleistocene. A sharp acceleration of evolutionary rates also took place in the Paleocene and in the transitional period between the Permian and the Triassic. Revolutionary changes in the conditions of existence of life on earth are also characteristic of these periods. The data cited by Kurten clearly show that the rate of macroevolution is determined by external conditions. In a period in which a greater number of forms were compelled to adapt to a new environment, an acceleration of the macroevolutionary development of a number of groups occurred. This shows that just as in the process of microevolution, the motive force of macroevolution is adaptation to actual conditions of existence. In those cases in which new adaptations opened a path to broad ecological conquests, an adaptive radiation began and new taxa arose. These evolutionary displacements are perceived as macroevolutionary. The dependence of the rhythm of evolution on changes in environmental conditions was recently analyzed by Yablokov-Khnzoryan (1963). He also advanced the interesting hypothesis that a relatively short period of aromorphosis is followed by long periods of idioadaptation.

It is necessary to add only one important observation to what has been said. In speaking of the initial morphophysiological characteristics of new taxonomic forms that open the gates to a new stage of macroevolution, one cannot attach to them the significance of aromorphoses, in the sense used by Severtsov (1939). Aromorphoses not only determine the possibility of broad adaptive radiation, but also (in this lies their essence) raise the general level of the animals' organization and open the path to morphophysiological progress. Far from every precondition of macroevolution is aromorphosis, although any aromorphosis has macroevolutionary significance. A good example is the South American ungulates (primarily *Notungulata*). Having penetrated into South America, the primitive ungulates underwent a very broad adaptive radiation. It is sufficient to say that many of them possessed traits of different orders, recalling the rodents, horses, rhinoceroses, elephants, and Chalicotheriidae. A number of forms were so distinctive that it was found there was not a name well known enough by which to describe them (Romer, 1939, and others). It is completely clear that with the penetration of ungulates into South America, a powerful outburst of macroevolution occurred. This was not evolution along the path of progress, however, and it ended with the extinction of a very rich and distinctive fauna.

* The botanists (Vasil'ev, 1964) arrive at analogous conclusions.

A similar phenomenon illustrates the powerful adaptive radiation and subsequent extinction of the American ground sloths. Already in the Pleistocene, they were represented by 14 genera, distributed from Alaska to Patagonia, that had mastered the deserts, cold steppes, mesophytic forests, and subarctic taiga (Martini *et al.*, 1961). Their capacity to utilize for food plant species that are inedible for modern herbivores indicates the specialization of individual forms (*Nothrotherium*). Environmental changes and the changes in the biocenotic situation connected with them, however, caused the complete extinction of the ground sloths, which had satisfactorily survived four glacial epochs. The last ground sloths (*Mylodon*) died out about 8000 years ago. The suggestion that they were annihilated by man is highly improbable, since at the time of their massive extinction, man had already penetrated into America, but was still very small in numbers.

The precondition of macroevolutionary reorganizations is that some initial group of animals find itself under conditions that promote adaptive radiation. As we attempted to show, *these conditions can be both external (changes of the environment, which elicit changes in the conditions of the animals' existence or promote their resettlement) and internal (changes of morphophysiological and ecological features, which permit them to master a new habitat)*. In the latter case, as we also tried to emphasize, the phenomenon of preadaptation plays a large role.

If the very possibility of macroevolutionary reorganizations is determined by the conditions for adaptive radiation, then the strongest influence on its actual course is exerted by interspecific relationships, by the conditions of the struggle for existence. In order to make this very important feature of macroevolution clear, we must deviate somewhat from our basic theme. Every adaptation is relative. This is illustrated especially clearly by the usual source of competition between insular and mainland forms. A truly vast number of facts are known that show that in those cases in which a mainland species finds itself on islands, it accomplishes a real revolution in the makeup of the island fauna. As a rule with almost no exceptions, the mainland forms prove to be the conquerors; the island endemics either die out or are driven back into suboptimal environmental conditions.

The complex of adaptations that ensured a species' flourishing in the absence of strong competition proves to be insufficient during a change of the biocenological situation, which inevitably arises during the introduction of mainland species to islands. The degree of adaptability is determined by conditions of the struggle for existence and is never absolute. This situation has very great significance for a correct understanding of the course of macroevolution in connection with the fact that adaptations to a certain form of life always affect the whole organization of the animal, not just those characteristics that externally characterize the animal's adaptation most directly. At first glance, it appears that the lemming's adaptations to conditions of life in the Arctic reduce to the perfection of physical thermoregulation. We know that this is not so, that the lemming proved to be capable of mastering the north thanks to a complex combination of morphophysiological characteristics, including even such "fine points" as biochemical characteristics of its cells and tissues.

Let us return now to the problem of macroevolution. In the example of *Notungulata,* we saw that under conditions of isolation, the ungulates produced adaptive types, which in the conditions of the holarctic are characteristic for different orders. Having developed under conditions of isolation, the marsupials "imitate" the adaptive radiation of the placentals, but at a different taxonomic level. This rule is illustrated even more clearly by the rodents of Australia and New Guinea. The investigations of Simpson (1961) showed that the adaptive radiation of the subfamilies Pseudomyinae and Hydromyinae (family Muridae) proved to be commensurate with the adaptive radiations of two orders (rodents and lagomorphs) of the holarctic! Pseudomyinae are represented by adaptive types of squirrels, muskrats, rabbits, mice, rats, voles, jerboas, and shrews. How are these interesting facts explained? Let us attempt to answer this question by analyzing a specific example.

In order to occupy the ecological niche of the adaptive type "hare" under the conditions of the impoverished biocenoses of islands, it is sufficient to acquire a complex of anatomical characteristics that permit the development of high speed in running short distances and satisfaction with a relatively low-calorie vegetative diet. That same adaptive type under the conditions of the mainland biocenoses (and consequently, in an immeasurably harder struggle for existence) should possess more highly perfected adaptations. It should possess the capability of utilizing the most indigestible food with very low caloric value (in connection with this, certain special adaptations are coprophagy); it should possess running speed commensurate with the running speed of its most highly developed predators, and young individuals should possess the capability of evading predators (recall the remarkable physiological and behavioral reactions of young hares); it should possess the capability of rapidly reestablishing its numbers after a sharp reduction; it should possess perfected sense organs and perfected behavioral reactions. Finally, in order not only to consolidate itself in a certain part of the arena of life, but also to struggle successfully to broaden its range, our generalized hare should possess great ecological plasticity and should be a biologically distinctive animal. Its distinctions from less specialized ancestors should include different aspects of morphology, physiology, and biochemistry. These profound distinctions are also reflected in the nomenclature: hares are justly assigned to an independent order. Analogous reasoning (but at another taxonomic level) is justified in regard to such adaptive types as squirrels, muskrats, mice, and so forth. Under the conditions of impoverished biocenoses (relatively easy struggle for existence), the formation of these types does not demand the cardinal restructuring of the organization of the animals. They can be formed on the basis of a general schematic structure of one subfamily or genus. Under the conditions of a harsher struggle for existence, the formation of such adaptive types demands a change of the schematic structure. Macroevolution is manifested in the origin of taxa of higher rank.

Thus, both the specific course and the scale of macroevolution, in the final analysis, are determined by ecological causes.

The differences between micro- and macroevolution are not contained within the mechanisms and rules of the reorganization of the initial forms, but within the result of these reorganizations. Here, the complete analogy with the process of speciation is manifested. A new species arises on the basis of the same rules to which intraspecific formation is subordinate, but under different conditions they lead to different results, in dependence on which the reorganization of certain traits and properties of the organism leads to the isolation of the new form. Under those conditions in which the formation of new specific features opens the path for conquering a new part of the arena of life, for a new adaptive radiation, the new species becomes a potential progenitor. We have the basis for broadening the conclusion at which we arrived earlier: *evolution is the process of progressive adaptation of organisms to the environment, which consists not only of fuller utilization of natural resources with the lowest expenditure of energy* (*speciation*), *but also of the progressive mastery of different environments of life by the organisms* (*macroevolution*).

There is, however, still one circumstance that leads many authors to counterpose microevolution and macroevolution. It is that in the process of macroevolution, not only are "deeper" features changed (although the measure of depth of these changes remains unknown), but also another "sort" of trait is affected. From the large number of works that could illustrate this point of view, we will use an interesting article by Evans (1965), based on a study of the Australian Cicadolloidea. Careful investigations led the author to the conclusion that species, genera, and tribes are characterized by small distinctions, but that genera are distinguished by a larger number of them, since the duration of their independent development is greater, than are species of one genus. The differences among taxa of higher rank (families) are concerned with more substantial distinctions. The impression is created, Evans concludes, that they "arise on the basis of an evolutionary process of a different type." Similar reasoning also leads other investigators to counterpose the mechanisms and motive forces of macro- and microevolution.

We will use Evans' work for illustration, since it, in essence, contains an answer to the rhetorical question posed by the author. He clearly shows that the beginning stages of macroevolution (which are nevertheless macroevolution) are not distinguished in any fundamental way from "ordinary" speciation (genera are distinguished by the same characteristics as species of one genus). Even though the author's general conclusion differs radically from that which we are striving to substantiate, his partial conclusion agrees with ours: "the path in macroevolution" is opened by the animals' acquisition of characteristics that are insignificant from a morphophysiological point of view. Their individual significance becomes evident only with an ecological approach to the problem.

A similar conclusion is not found on the surface of phenomena only because genera of the same family are usually compared. Among such genera, really close forms constitute an obvious minority. The impression is created that the first stages of macroevolution (the formation of new genera) are already fundamentally distinct from the microevolutionary process. Indeed, if within the subfamily Microtinae the genera *Microtus, Ellobius, Prometheomys, Clethrionomys,* and *Ondatra* are compared, the impression is created that their distinctions are based on changes in a different type of trait than those by which species within a genus are distinguished. In this case, the genera that are compared have differentiated and long ago diverged, each being specialized to different conditions and a different form of life. If we compare really close genera, however, this impression substantially vanishes. Differences between the genera *Microtus* and *Arvicola* are of the same order as differences between individual species of *Microtus*. Moreover, the sum of morphophysiological distinctions between *M. oeconomus* and *M. socialis,* in any case, is not less than between *M. oeconomus* and *A. terrestris,* but the point is that *Arvicola's* characteristics opened a path for it to a new environment, opened a path to adaptive radiation. Water voles acquired the capability of utilizing the richest aquatic plants as food. Settling near water, *oeconomus* utilizes, on the whole, riverside thickets of sedge. Both *Arvicola* and *Microtus* descended from root-toothed voles. There is the suggestion that in the beginning of the Pleistocene, two different species of the root-toothed *Mimomys* in the same place and at the same time produced two species of *Arvicola* (Heim de Balsac and Guislan, 1955). There is no unconditional evidence of the correctness of this conclusion, but it in itself testifies that the transition from one genus to another does not demand any special conditions, and in any case is not regarded as an exception, as a chance happening (macromutation is a chance event).

Evans (1965) correctly noted the great importance of the fact that in the process of macroevolution, changes embrace a new complex of traits. This occurs, however, not because they "arise on the basis of an evolutionary process of a different type," but because their origin is dictated by evolutionary expediency, which arises immediately after the first stage of macroevolution has occurred.

Let us use an imaginary example. Let us assume that *Arvicola's* adaptations to life in water will be perfected. It is evident that water voles, just like muskrats, acquire morphological characteristics that permit them to swim rapidly, stay under water a long time, and gnaw the underparts of plants and partially to eat plants underwater. Changes in their fur coats arise, the speed of molting is changed, the capability of regulating the entry of blood into different organs during dives arises, and so forth. All these characteristics really demand changes in traits by which, as a rule, species of one genus are not distinguished. But they arise at the level of families, not because they include some new mechanisms of evolution, but because it is expedient for them to appear only after the first stage of macroevolution has already occurred. It is also important to note that the progressive evolution of water voles does not lead to the appearance of a "new muskrat," since the muskrat was adapted for life in an aquatic environment, while maintaining characteristics of the primitive voles—rooted teeth.

General observations ostensibly testify that variability within a genus usually does not affect those traits by which animals of different families are distinguished. On closer examination, this assertion proves to be without foundation. We have already recalled the experimental work showing that variability of model populations can serve as a basis for changes in traits characteristic of different orders (S. C. King, 1965; see also p. 197). Study of domestic animals gives a great deal of material on this question (for a summary of the data, see Herre, 1959). In many cases, intrapopulational variability even affects traits used for determination of genera (Herre, 1964). It is known that the variability of domestic animals is greater than that of wild ones. It is also known, however, that under new environmental conditions, variability is markedly increased (for a theoretical analysis of the problem, see Kosswig, 1962). The species that are founders of genera find themselves exactly thus under new conditions. Their variability should therefore be high, which will facilitate their further course of evolution and subsequent divergence.

Analysis shows that there is no basis for counterposing micro- and macroevolution. Evolution is the single process of the progressive mastery by organisms of the arena of life.

Conclusions

I. Ecological analysis of different stages of the phylogenetic development of animals (intraspecific formations, speciation, macroevolution) shows that evolution is the single process of the progressive adaptation of organisms to the environment, which consists of the perfected utilization of vital resources with the least expenditure of energy and of the progressive expansion of life on the land and waters of the earth.

II. Ecological analysis of the basic phenomena of the evolutionary process permits us to express general ideas of the motive forces and mechanisms of evolution more concretely, and to substantiate ideas about the ecological mechanisms of the evolutionary process. The main task of evolutionary ecology is the comprehensive investigation of these mechanisms.

III. The beginning stage of the evolutionary process consists in the irreversible adaptive reorganization of populations.

1. *A population is the form of existence of the species, and it possesses all the necessary conditions for independent existence and development* during the course of an indefinitely long period of time; it is capable of reacting adaptively to changes in the external environment. The capability for indefinitely long independent existence is the basic property of the population; it determines the genetic and morphophysiological specificity of every population of the species. A group of adjacent populations makes up a geographic form, which is characterized by common morphophysiological and ecological features. A subspecies is a geographic form that possesses irreversible morphophysiological characteristics that determine the specificity of reactions to environmental changes. The specificity of a subspecies consists in the singularity of its evolutionary fate.

2. The capability of a population to maintain its numbers in a constant dynamic equilibrium, despite changes in the external environment, is determined by adaptive homeostatic reactions of separate individuals, by the dynamic ecological structure of the population, and by change in its genetic makeup. *Fluctuations in the quality of a population are just as characteristic an attribute of it as are fluctuations in its numbers.* Reversible changes of the genetic makeup of a population are a regular phenomenon; they are not part of the microevolutionary process, and should be regarded as one of the manifestations of population homeostasis—*as the homeostatic reorganization of the populations's genetic makeup.*

3. The morphophysiological singularity of a population is manifested in every trait and property of the animals. *Every population of the species is morphophysiologically specific.* The genetic nature of differences among populations differs (phenotypic differences, monogenically and polygenically determined differences), but in the majority of cases, differences among populations are determined by a complex of genetic mechanisms.

4. In distinction to homeostatic changes of the population genetic structure, *the irreversible reorganization of populations characterizes the beginning stages of the microevolutionary process.*

a. The irreversibility of intraspecific reorganizations is determined, in the final analysis, not by morphophysiological and genetic characteristics of the population (the degree of morphophysiological differences, the degree of genetic differentiation), but by their ecological characteristics, by the change in their norm of reaction to a change of environmental conditions. This change is the basis for the development of the singular morphogenetic reactions of the animals, which make a return to the initial form not only practically, but theoretically, impossible (the irreversibility of microevolutionary reorganizations). Subspecies are geographical forms in which the microevolutionary process materializes.

IV. An indispensable condition for the maintenance of a population's viability under changing environmental conditions is a high degree of its genetic heterogeneity.

1. The maintenance of the population's genetic heterogeneity is ensured by ecological mechanisms (a different form of life for different intrapopulational groups of animals, strict regularity in the formation of pairs, a different rate of sexual maturation for males and females, a different sex ratio in different age groups of animals, and others).

2. The general principles by which ecological mechanisms act to maintain the genetic heterogeneity of populations are: intrapopulational groups of animals, developing under different conditions of existence, are subjected to different forces of selection; their genetic structures inevitably become different; when genetically different groups of animals are combined into a single population, the common gene pool is enriched, and heterogeneity is continuously maintained at a high level. *There is little probability of the irreversible impoverishment of the gene pool of a population with a normal ecological structure.*

a. Impoverishment of a population's gene pool by chance (i.e., through occurrences not controlled by natural selection in its different forms) is possible only in so-called "evolutionary traps" [*lovushka*], during the isolation of a small groups of animals—founders of a new population—and under the conditions of an impoverished (simplified) biocenosis. This chance impoverishment leads to evolutionary blind alleys.

3. When numbers are sharply reduced (possible impoverishment of the gene pool, homozygosis of the population), ecological mechanisms for the maintenance of the population's genetic heterogeneity work with special effectiveness (the population is reestablished due to the reproduction of animals that underwent development under markedly different environmental conditions).

4. The character of the manifestation of such universal phenomena as the sexuality of the whole living world and the common mechanisms for the determination of sex in different groups of animals attest to the exceptional biological significance of ecological mechanisms in the maintenance of a population's genetic heterogeneity.

V. The basis of the microevolutionary process is the irreversible reorganization of the genetic structure of populations. *Side by side with natural selection, the most important motive force of evolution is the ecological mechanisms of the reorganization of a population's genetic structure.*

1. The general principle of the action of ecological mechanisms in the reorganization of a population's genetic structure is that *a change in a population's ecological structure entails, with the inevitability of law, a change in its genetic makeup.* This very important rule is a consequence of the different genetic structure of every intrapopulational group of animals.

a. Since the singularity of the genetic structure of intrapopulational groups of animals is determined by natural selection, the ecological mechanisms for the reorganization of the population's genetic structure may be regarded as a singular form of natural selection, indirectly expressed as the ecological structure of the population. The principles of action of different forms of natural selection differ markedly, however. Designating the ecological mechanisms of the reorganization of a population's genetic structure as an independent factor of evolution is therefore justified, both in essence and from a terminological point of view.

2. The ecological mechanisms of the evolutionary process are manifested in three important forms, based on a change in the population age structure (age selection), a change in numbers (nonselective elimination), and a change in the spatial structure of the population.

3. Age selection is understood to mean a marked change in the population age structure, which occurs during a change of environmental conditions. *As a result of age selection, the general genetic structure of the population is sharply changed, and a rapid mobilization of the population's reserves of genetic variability occurs.* This mobilization serves as a prerequisite for the further directional reorganizations of the population under the action of natural selection. The rate of evolutionary reorganizations occurring under the influence of age selection repeatedly surpasses the maximum rate of microevolution under the influence of natural selection. *In many cases, age selection works successfully under conditions in which the usual form of natural selection is powerless.* A change in environmental conditions that decreases the pressure of natural selection can lead to the especially effective action of age selection and to rapid evolutionary reorganizations. This phenomenon explains many paradoxical cases of rapid evolution.

4. Sharp changes in numbers are the most important factor in the reorganization of populations. In contrast to the almost universally accepted notion, *nonselective elimination, as a rule, exerts a strictly selective action on the ecological structure of a population and reorganizes it in definite directions,* which correspond to the general changes in environmental conditions.

   a. The directed action of nonselective elimination is determined by the marked ecological differences among different age groups of animals and animals of different micropopulations, on the one hand, and by the seasonality of the action of factors of elimination, on the other (nonselective elimination acts against a background of a certain age structure). This makes a change in population ecological structure inevitable during a sharp reduction in numbers. Accordingly, the genetic makeup of the population is changed. Nonselective elimination, like age selection, facilitates the rapid mobilization of the population's genetic reserves and is, as a rule, the strongest factor in its adaptive evolution.

a′. A sharp reduction in the population's numbers as a result of nonselective elimination can produce the beginnings of a random (nonadaptive) direction in evolution only in extremely impoverished biocenoses (a weakening of the struggle for existence). The reorganization of populations under these conditions serves as a convenient natural model of the evolutionary process, but the magistral path of evolution is determined by other mechanisms.

5. The dynamics of the spatial structure of a population (or of a group of combined populations) is one of the most important mechanisms for the adaptation of a species to changed environmental conditions. A complex spatial structure ensures the population against chance impoverishment of its gene pool and, at the same time, creates the conditions for the rapid adaptive reorganization of its genetic structure.

a. The temporary isolation of structured subdivisions of the population during a period of reduced numbers and the subsequent reunification of these subdivisions into a single unity is a universal phenomenon. Often, neighboring, independent populations are involved in this process. In such cases, the groups of combined populations evolve as a single unity.

b. During their reunification (usually during a period of a peak in the species' numbers), there is recovery from the random disturbance of the gene pool of temporarily isolated structured subdivisions of the population, subject to the laws of genetic automatic processes. This limits the action of genetic automatic processes and ensures the maintenance of the population's gene pool in a state of dynamic equilibrium.

c. Individual settlements of the species, developing under conditions of temporary isolation, are subjected to similar (but not fully identical) selection forces. During a change in environmental conditions, they undergo similar parallel changes determined by different genetic mechanisms. (The ecological mechanisms of the reorganization of populations ensure their change during a short period of isolated development.) The subsequent reunification of micropopulations possessing similar morphophysiological characteristics of a different genetic nature leads to a strengthening of the phenotypic manifestation of their characteristic features (in cases of additive gene action), or to the appearance of populations distinguished by new traits. This mechanism, in the final analysis, markedly increases the effectiveness of the action of natural selection. It can ensure the progressive development of a population during a sharp reduction of selection pressure or even after the cessation of its action. This makes understandable many mysterious cases of the development of individual traits of animals "beyond the limits of usefulness." This mechanism also explains many cases of *the intensification of variability in the direction of selection.* Selection itself creates the conditions for increasing the effectiveness of its action. The joint action of different forms of natural selection (which also include ecological mechanisms of the reorganization of populations) ensures rapid microevolutionary changes. It has been shown that changes on a microevolutionary scale can occur during the lifetime of a single generation of animals.

VI. Microevolution begins with the origin of irreversible reorganizations of the population's genetic structure and is completed with speciation. The essence of the process of speciation is the progressive adaptation of animals to the conditions of existence that ensures the fullest utilization of the resources of the environment with the least expenditure of energy.

1. In distinction from every intraspecific form, a species is biologically independent (reproductive isolation, hiatus, independent range). From the genetic point of view, a species is that system of integration of genotypes the "enrichment" of which by genetic material from another analogous system leads to a reduction in its viability. The evolutionary independence of the species is determined by its reproductive isolation.

   a. On the genetic and morphophysiological levels, different forms of reproductive isolation often appear as phenomena of a different order. Therefore, *the specificity of a species appears only at the evolutionary and ecological levels*; a formal criterion for the definition of a species, even one based on the acceptance of the newest methods (karyology, "protein taxonomy", immunology), is, in principle, useless. The problem of the species can be decided only on an ecological basis.

2. The most important characteristic of a population is ecological. Any species of animals, without exception, is characterized by a higher degree of adaptability to that complex of conditions that led to its origin than any specialized intraspecific form. There are no exceptions to this rule. This means that an animal species in the process of its development passed through the stage of a specialized subspecies, characterized by perfected morphophysiological adaptations. The subsequent stage of microevolution consists of a reduction in the energy cost of these adaptations, which in the final analysis leads to the origin

of distinctions at the tissue level, to tissue incompatibility, and to reproductive isolation.

3. *The motive forces of speciation are not distinguished, in principle, from the mechanisms of intraspecific differentiation.* These processes differ in their results, but not in their mechanisms of action.

4. In distinction from intraspecific reorganizations, *in the process of speciation, natural selection evaluates not only the morphofunctional perfection of the animals' adaptations, but also their energy cost.* And this process leads, in the final analysis, to the formation of a new species.

VII. Microevolution and macroevolution are a single process. Macroevolutionary reorganizations are accomplished on the basis of the same mechanisms and rules as is speciation: *species possessing characteristics that open a path to a new adaptive radiation are potential progenitors, initiators of the macroevolutionary process.*

1. At the beginning stages of phylogeny of individual groups, the type of evolutionary reorganizations (microevolution, speciation, macroevolution) is determined not by the degree of morphophysiological distinctions among the diverging forms, but by those ecological consequences that the distinctions elicit.

2. The past history of any taxon canalizes its future development within certain limits. In this process, the phenomenon of preadaptation plays a large role: adaptations that are useful (but not necessary) under given conditions of existence open a path to a new environment. Modern methods of ecological research permit us to evaluate not only the actual but also the potential significance of individual adaptations and the complex of adaptive features of individual forms.

VIII. Investigations of the ecological mechanisms of the reorganization of populations create the basis for the elaboration of a theory on the control of the evolutionary process.

1. *By means of a directional change in a population's ecological structure, a new population with a given genetic makeup can be created in a short period of time.*

   a. The specific means of influencing the population's ecological structure under different environmental conditions and in conjunction with specific functional aims may be different: nonselective elimination at different stages of the population life cycle, age selection, and a change of the population's spatial structure. A complex application of different means of influencing the population should be especially effective.

2. A combination of a complex of measures directed at a change in the age structure of the population (or of the micropopulation) with a change in its spatial structure and with a disturbance of the ecological barriers among neighboring populations should lead to the creation of a new population with fundamentally new biological properties.

a. *Through the utilization of ecological mechanisms in the reorganization of populations, it is theoretically possible to exercise direct conscious control over the evolutionary process in nature.* It is also possible to create new forms of animals that utilize environmental resources more efficiently and to increase the effectiveness of the energetic and geochemical work of the biogeocenoses. On this basis, it is possible to elaborate a theory on the creation of especially productive biogeocenoses.

3. *Artificial selection based not only on a functional but also on an energy evaluation of the adaptive reactions of animals can lead to the creation of new animal species with given properties.* Mastery of the process of speciation signifies simultaneous mastery of the process of macroevolution.

IX. At present, the main task of evolutionary ecology is the experimental study of the dependence between the ecological and genetic structure of natural populations, on the one hand, and their productivity and adaptability, on the other. These investigations will create the basis for elaborating methods for the directed change of the structure of populations of different species under different environmental conditions and, in the not too distant future, for the control of evolution.

References and Bibliography

In the Russian edition, the references were presented in two listings: (1) those in Russian, and (2) those in English and other languages employing the roman alphabet. Here, they are combined in one listing. In order to resolve any possible ambiguities concerning the language of the original publications, those references that appeared in the second listing are here designated by bullets (•). In accordance with accepted bibliographic practice, the author's name has here been transliterated "Shvarts." His own preferred spelling, however, was "Schwarz," which form generally appears in the German literature and occasionally elsewhere.

• Adamczyk, K., and Petrusewicz, K., 1966, Dynamics and intrapopulation differentiation of a free-living population of house mouse, *Ecol. Polska* **14**(36).

• Adlersparre, A., 1938, Einiges über Pigmentstoffwechsel und andere Farbenmodificationen bei Gefangenschaftsmilieu, *Ornithol. Monatsbericht* **46**(1).

• Aebli, H., 1966, Rassenunterschiede im Bezug auf Entwicklungsgeschwindigkeit und Geschlechtsdifferenzierung bei *Rana temporaria* in den Tälern des Kantons Glarus (Schweiz), *Rev. Suisse Zool.* **73**(1).

Aleksandrov, V. Ya., 1952, On the connection between heat resistance of protoplasm and the thermal conditions of existence, *Dokl. Akad. Nauk SSSR* **88**(1).

Aleksandrov, V. Ya., 1965, On the biological significance of the correspondence between the level of heat resistance of proteins and the thermal conditions of existence for the species, *Usp. Sovrem. Biol.* **60,** No. 1 (4).

• Allen, H. W., 1954, Propagation of *Horogenes molestae,* an asiatic parasite of the oriental fruit moth, on the potato tuberworm, *J. Econ. Entomol.* **47.**

• Allison, A. C., 1956, Aspects of polymorphism in man, *Cold Spring Harbor Symp. Quant. Biol.* **20.**

Alpatov, V. V., 1927, Biometric characteristics of Middle Russian and Ukrainian bees, *Russ. Zool. Zh.* **7**(4).

• Amadon, D., 1950, The species—then and now, *Auk* **67**(4).

• Anderson, P. K., 1966, The role of breeding structure in evolutionary processes of *Mus musculus* populations, *Mutat. Populat. Prague* (Czechosl. Acad. Sci.).

• Andrzeijewski, R., Petrusewicz, K., and Walkowa, W., 1963, Absorption of newcomers by a population of white mice, *Ecol. Polska* **11**(7).

Anorova, N. S., 1958, Influence of the age of hens on the development of their offspring, *Ptitsevodstvo,* No. 1.

Anorova, N. S., 1959, Age of the parents and development of offspring in birds, *Ornitologiya,* No. 2 (MGU).

Anorova, N. S., 1960, Influence of a bird's age on the formation of eggs, *Ornitologiya,* No. 3 (MGU).

Anorova, N. S., 1964, Age and fertility of birds, in: *Problems in Ornithology,* Lvov State University, Lvov.

Argiropulo, A. I., 1946, On the question of the individual and geographic variability of some species of the genus *Apodemus, Tr. Zool. Inst. Akad. Nauk SSR* **8**.

Arnol'di, K. V., 1957, On the theory of range in connection with the ecology and origin of species populations, *Zool Zh.* **36**(11).

Astaurov, B. L., 1963, Problem of the regulation of sex, in: *Science and Mankind,* Znanie, Moscow.

Astaurov, B. L., 1965, Two landmarks in the development of genetic concepts. On the republication of the article by S. S. Chetverikov, "On some features of the evolutionary process from the viewpoint of modern genetics" (1926), and the article by N. K. Kol'tsov, "Hereditary molecules" (1935), *Byull. Mosk. O. Ispyt. Prir. Otd. Biol.* **70**(4).

Averkina, R. F., 1960, Comparative study of the antigenic properties of heart tissues in different species of animals and man, *Zh. Obshch. Biol.* **21**(3).

Aver'yanov, I. Ya., Malyshev, P. P., and Budagov, S. M., 1952, On the sex ratio of Karakul lambs under different developmental conditions of their parents, *Karakulevod. Zverovod.,* No. 1.

Avrekh, V. V., and Kalabukhov, N. I., 1937, Consanguinity of mountain and plains forms of the meadow mouse (*Apodemus sylvaticus*) and other close species of mice, *Zool. Zh.* **16**(1).

• Bader, R. S., 1965, A partition of variance in dental traits of the house mouse, *J. Mammal.* **46**(3).

• Badr, F. M., and Spickett, S. G., 1965, Genetic variation in the biosynthesis of corticosteroids in *Mus musculus, Nature London* **205**(4976).

Balakhnin, I. A., 1964, Experiment on the use of precipitation reactions for the establishment of relationships in fish, *Vopr. Ikhtiol.* **4**(3).

• Bamann, E., Ullmann, E., and Gebler, H., 1960, Die stereochemische Spezialität der Leberesterase als Grundlage für die systematische Ordnung des Tierreiches am Beispiel der Simiae (Affen). I. Mitteilung über Fermentsysteme und ihre Specialität in Beziehungen zur systematischen Ordnung des Tier und Pflanzenreiches, *Hoppe-Seyler's Z. Physiol. Chem.*

• Bamann, E., Gebler, H., Schub, E., and Forster, H., 1962, Die stereochemische Spezifität der Esterase der Leber, der Lunge und der Niere—ein biologisches Merkmal zur Klärung entwicklungsgeschichtlicher Zusammenhänge, *Naturwissenschaften* **49**(12).

Bannikov, A. G., 1953, *Definer of Mammals of the Mongolian Peoples' Republic,* Izd. Akad. Nauk SSR, Moscow.

Barbashova, Z. I., and Ginetsinskii, A. G., 1942, Features of adaptation to height in Gissarskii sheep, *Izv. Akad. Nauk SSSR Ser. Biol.,* No. 5.

• Barber, H. N., 1954, Genetic polymorphism in the rabbit in Tasmania, *Nature London* **173**(1227).

Barcroft, J., 1937, *Basic Traits of the Architecture of Physiological Functions,* Biomedgiz, Moscow and Leningrad.

• Barnett, S. A., 1965, Mice at −3°C, *New Sci.* **27**(461).

• Barnett, S. A., and Coleman, E. M., 1960, Heterosis in F1 mice in a cold environment, *Genet. Res.* **1**(1).

• Barrows, C. H., Yeinst, M. J., Shock, N. W., and Chow, N. E., 1957, Age differences in cellular metabolism of various tissues of rats, *Fed. Proc. Fed. Amer. Soc. Exp. Biol.* **16**(1).

• Bartlett, A. C., Bell, A. E., and Anderson, V. L., 1966, Changes in quantitative traits of *Tribolium* under irradiation and selection, *Genetics* **54**(2).

Bashenina, N. V., 1957, Some features of the ecologicogeographical variability of the common vole, in: *Materials for the Conference on Questions of the Zoogeography of Land.*

• Battaglia, B., 1965, Advances and problems of ecological genetics in marine animals, in: *Genetics Today,* Vol. 2, Pergamon Press, Elmsford, New York.

Bazhanov, V. S., 1945, On the systematics of two species of susliks from the Kazakh SSSR, *Vestn. Kaz. Fil. Akad. Nauk SSSR,* No. 5/8.

Baziev, Zh. Kh., 1967, Present distribution and numbers of the Caspian ular in the Transurals, *Zool. Zh.* **46**(5).

• Beale, G. H., 1954, *The Genetics of Paramecium auralia,* Cambridge University Press, Cambridge, Massachusetts.

• Beaudry, R., 1960, The species concept: Its evolution and present status, *Rev. Can. Biol.* **19**(3).

• Beckman, L., Conterio, F., and Mainardi, D., 1963, Studio elettroforetico sulle proteine del Siero di ibridi di uccelli, *Atti Assoc. Genet. Ital.* **8**(137).

• Beickert, A., 1954, Zur Entstehung und Bewertung der Arbeitshypertrophie des Herzens, der Nebenniere un Hypophise (Tierexperimentelle Untersuchungen), *Arch. Kreislaufforsch.* **21**(1–4).

Beklemishev, V. N., 1951, On the classification of biocenological (symphysiological) ties, *Byull. Mosk. O. Ispyt. Prir. Otd. Biol.* **6**(5).

Beklemishev, V. N., 1960, Spatial and functional structure of populations, *Byull. Mosk. O. Ispyt. Prir. Otd. Biol.* **65**(2).

Belozerskii, A. N., 1961, Species specificity of nucleic acids, in: *International Biochemical Congress on Evolutionary Biochemistry, Symposium 3,* Izd. Akad. Nauk SSSR, Moscow.

• Benedict, F. G., 1932, *The Physiology of Large Reptiles, Carnegie Inst. Washington Publ.,* **No. 425.**

• Benedict, F. G., 1938, *Vital Energetics, A Study in Comparative Basal Metabolism, Carnegie Inst. Washington Publ.,* No. 503.

• Benett, I., 1960, A composition of selective methods and test of the preadaptation hypothesis, *Heredity* **15**(1).

• Beninde, J., 1940, Die Fremdblutkreuzung (sog. Blutauffrischung) beim deutschen Rotwild, *Z. Jagdkunde,* Sonderheft 3.

• Bentvelzen, P., 1963, Some interrelations between density and genetic structure of a *Drosophila* population, *Genetica* **34**(4).

Beregovoi, V. E., 1963, Rules of geographic variability and intraspecific systematics of birds (in the example of three species of the genus *Motacilla*), author's Candidate dissertation, Sverdlovsk.

Beregovoi, V. E., 1966, Variability of natural populations of the common spittlebug (*Philaenus spumarius* L., Cercopidae: Homoptera), *Genetika,* No. 11.

Beregovoi, V. E., 1967*a*, The problem of species and populations of polymorphic species, *Zh. Obshch. Biol.* **28**(1).

Beregovoi, V. E., 1967*b*, Variability of populations of the common spittlebug, in: *Materials for the Summary Session of the Laboratory of the Population Ecology of Vertebrates for 1966* (Inst. Ékol. Rast. i Zhiv., UFAN SSSR), Sverdlovsk.

Beregovoi, V. E., and Danilov, N. N., 1964, Intraspecific variability of birds and phenogeography, in: *Theses of the Conference on Intraspecific Variability of Terrestrial Vertebrates and Microevolution,* Sverdlovsk.

• Berlioz, J., 1929, La hibridation entre les Torguilidos Rewista, *Soc. Colomb. Cient. Nat.* **18**(102).

• Berlioz, J., 1937, Three new cases of presumed natural hybrids among Trochilidae, *Ibis* **14**(1).

• Berry, R. J., 1963, Epigenetic polymorphism in wild populations of *Mus musculus, Genet. Res.* **4**(3).

• Berry, R. J., 1964, The evolution of an island population of the house mouse, *Evolution* **18**(3).

• Bertalanffy, L., and Estwick, R., 1953, Tissue respiration of musculature in relation to body size, *Amer. J. Physiol.* **173**(1).

• Bertalanffy, L., and Pirozynski, W. J., 1953, Tissue respiration, growth and basal metabolism, *Biol. Bull.* **105**(2).

• Bertini, F., and Rathe, G., 1962, Electrophoretic analysis of the hemoglobin of various species of anurans, *Copeia,* No. 1.

• Birch, L. C., 1960, The genetic factor in population ecology, *Amer. Nat.,* No. 94.

• Birch, L. C., 1961, Natural selection between two species of tephritid fruit-fly of the genus *Dacus,* Evolution **15**(3).

• Birch, L. C., 1965, Evolutionary opportunity for insects and mammals in Australia, in: *The Genetics of Colonizing Species* (H. G. Baker and G. L. Stebbins, eds.), Academic Press, New York and London.

Birlov, R. I., 1967, On the process of adaptation to high mountain conditions by two species of shrews, in: *Materials for the Summary Session of the Laboratory of the Population Ecology of Vertebrates for 1966* (Inst. Ékol. Rast. i Zhiv., UFAN SSSR), Sverdlovsk.

Blagoveshchenskii, A. V., 1945, Biochemical factors of natural selection in plants, *Zh. Obshch. Biol.* **6**(4).

• Blair, A. P., 1941, Variation, isolating mechanisms and hybridisation in certain toads, *Genetics* **26**(4).

• Blair, W. F., 1958, Call difference as an isolation mechanism in Florida species of hylid frogs, *Q. J. Fl. Acad. Sci.* **21**(1).

Blyumental', T. I., and Dol'nik, V. R., 1962, Evolution of the energy indices of birds under field conditions, *Ornitologiya,* No. 4.

Bobrinksii, N. A., 1944, Basic information on systematics, in: *Definer of Mammals of the USSR,* Sov. Nauka, Moscow.

• Bock, W. J., 1965, The role of adaptive mechanisms in the origin of higher levels of organization, *Syst. Zool.* **14**(4).

• Bock, W. J., and von Wahlert G., 1965, Adaptation and the form–function complex, *Evolution* **19**(3).

• Bogert, C. M., Blair, W. F., Dunn, E. R., Hall, C. L., Hubbs, C. L., Mayr, E., and Simpson, G. G., 1943, Criteria for vertebrate subspecies, species and genera, *Ann. N.Y. Acad. Sci.* **44**(2).

Bol'shakov, V. N., 1962, Rules of individual and geographic variability in voles of the genus *Clethrionomys,* author's Candidate dissertation, Sverdlovsk.

Bol'shakov, V. N., 1972, *Means of Adaptation of Small Rodents to Mountainous Conditions,* Nauka, Moscow.

Bol'shakov, V. N., and Shvarts, S. S., 1962, Some rules of geographical variability of rodents in continuous parts of their range (in the example of voles of the genus

Clethrionomys), in: *Questions of the Intraspecific Variability of Mammals, Tr. Inst. Biol. UFAN SSSR,* No. 29.

Bol'shakov, V. N., and Pokrovskii, A. V., 1966, Features of the blood of mountain species and of mountain populations of widely distributed species of rodents (hemoglobin content of the blood of Tyan-Shan voles under natural and experimental conditions), in: *Experimental Study of the Intraspecific Variability of Vertebrates, Tr. Inst. Biol. UFAN SSSR,* No. 51.

• Bourns, T. K. R., 1967, Serological relationships among some North American thrushes, *Can. J. Zool.* **45**(1).

• Boyden, A. A., and Noble, G. K., 1933, The relationships of some common Amphibia as determined by serological study, *Amer. Mus. Nat. Hist.,* No. 606.

• Bragg, A. N., 1962, Predator–prey relationship in two species of spadefoot tadpoles with notes on some other features of their behaviour, *Wasmann J. Biol.* **20**(1).

• Bresse, E. L., and Mather, K., 1957, The organisation of polygenic activity within a chromosome in *Drosophila, Heredity* **11.**

Briges, O. I., 1953, Influence of the age of sires and dams on the pedigree and astrakhan quality of their offspring, *Karakulevod. Zverovod.,* No. 1.

• Bronson, F. H., and Eleftheriou, B. E., 1963, Influence of strange males on implantation in the deermouse, *Gen. Comp. Endocrinol.* **3**(5).

• Brown, A. W., 1958, Insecticide resistance in arthropods, *WHO Bull.,* Geneva.

• Brown, W. L., Jr., 1958, General adaptation and evolution, *Syst. Zool.* **7**(4).

• Brown, W. L., Jr., and Wilson, E. O., 1956, Character displacement, *Syst. Zool.* **5**(2).

• Buddenbrock, W., 1934, Über die kinetische und statische Leistung grosser und kleiner Tiere und ihre Bedeutung für den Gesamt-Stoffwechsel, *Naturwissenschaften* **22**(40).

Bulatova, N. M., 1962, Features of the blood of high mountain animals, *Tr. Inst. Morfol. Zhivotn. Akad. Nauk SSSR,* No. 41.

• Buzzati-Traverso, A. A., 1955, Evolutionary changes in components of fitness and other polygenic traits in *Drosophila melanogaster* populations, *Heredity* **9.**

• Camin, J. H., and Erlich, P. R., 1958, Natural selection in water snakes (*Natrix sipedon* L.) on islands in Lake Erie, *Evolution* **12**(4).

• Carson, H. L., 1961, Relative fitness of genetically open and closed experimental populations of *Drosophila robusta, Genetics* **46.**

• Cei, J. M., and Bertini, F., 1962, Proteinas sericas de *Bufo spinulosus*: Estudio electroforetico y variacion geografica, *Arch. Bioquim. Quim. Farm.* **10**(1).

• Cei, J. M., and Erspamer, V., 1966, Biochemical taxonomy of South American amphibians by means of skin amines and polypeptides, *Copeia,* No. 1.

Chel'tsov-Bebutov, A. M., 1965, Biological significance of the breeding grounds of the black grouse in the light of the theory of sexual selection, *Ornitologiya,* No. 7 (MGU).

Chernov, S. A., 1941, The species problem, *Tr. Zool. Inst. Akad. Nauk SSSR* **6.**

Chetverikov, S. S., 1965, On some features of the evolutionary process from the viewpoint of modern genetics, *Byull. Mosk. O. Ispyt. Prir. Otd. Biol.* **70**(4) [reprinted from *Zh. Eksp. Biol.* **2**(1) (1926)].

Chilingaryan, A. A., 1966, The biology of the development of hybrids from crosses between species and between races, in: *On the Ontogenesis of Hybrids, Erevan,* Akad. Nauk Arm. SSR.

Chilingaryan, A. A., and Pavlov, E. F., 1961, Quantitative changes in the DNA content of erythrocyte nuclei in interspecific hybrids of birds and reptiles, *Dokl. Akad. Nauk Arm. SSR* **32**(1).

• Chitteleboraugh, R., 1955, Puberty, physical maturity and relative growth of the female

humpback whale, *Megaptera hodosa* (Bonna terre), on the Western Australian coast, *Aust. J. Mar. Freshwater Res.* **6**(3).
• Christian, J. J., 1961, Phenomena associated with population density, *Proc. Nat. Acad. Sci. U.S.A.* **47**(4).
• Christian, J. J., 1963, The pathology of overpopulation, *Mil. Med.* **128**(7).
• Cohen, R., 1962, La variacion estacional esqueletica coma caracter diferencial morfofisiologico en Leptodactilus, *Rev. Soc. Argent. Biol.* **38**(7–8).
• Cole, C. J., 1963, Variation, distribution and taxonomic status of the lizard, *Sceloporus undulatus virgatus* Smith, *Copeia*, No. 2.
• Cole, L. C., 1966, Man's ecosystem, *Bioscience* **16**(4).
• Cook, S. F., and Hannon, J. P., 1954, Metabolic differences between three strains of *Peromyscus maniculatus*, *J. Mammal.* **35**(4).
• Cooper, I. S., 1965, Notes on fertilization, the incubation period and hybridisation in *Lacerta*, *J. Herpetol.* **3**(9).
• Corbet, G. B., 1961, Origin of the British insular races of small mammals and of the Lusitanian fauna, *Nature London* **191** (4793).
• Corbet, G. B., 1964, Regional variation in the bank vole *Clethrionomys glareolus* in the British Isles, *Proc. Zool. Soc. London* **143**(2).
• Cory, B. L., and Manion, J. J., 1955, Ecology and hybridisation in the genus *Bufo* in the Michigan–Indiana region, *Evolution* **9**(1).
• Cowdry, E. V., 1954, Summary and general conclusions. Parental age and character of the offspring, *Ann. N.Y. Acad. Sci.* **57**(4).
• Crenshaw, J. W., 1965, Serum protein variation in an interspecies hybrid swarm of turtles of the genus *Pseudemys*, *Evolution* **19**(1).
• Cross, W., 1964, Polytipische Stamme im System der Wirbeltiere?, *Zool. Anz.* **173**(1).
• Crow, J. F., 1957, Genetics of insect resistance to chemicals, *Amer. Rev. Entomol.* **2.**
• Crow, J. F., 1960, Genetics of insecticide resistance: General considerations, *Misc. Publ. Entomol. Soc. Amer.* **2**(1).
• Crow, J. F., and Kimura, M., 1965, Evolution in sexual and asexual populations, *Amer. Nat.* **99**(909).
• Crowcroft, P., 1961, Variability in the behaviour of wild house mice (*Mus musculus* L.) towards live traps, *Proc. Zool. Soc. London* **137**(4).
• Cunha, A. B., and Dobzhansky, T., 1954, A further study of chromosomal polymorphism in its relation to the environment, *Evolution* **8**(1).
• Cunha, A. B., Dobzhansky, T., Pavlovsky, O., and Spassky, B., 1959, Genetics of natural populations, III, *Evolution* **13**(2).
• Cyren, O., 1924, Klima und Eidechsenverbreitung, *Göteborgs K. Vetensk. Vitterhets-Samh. Handl.* **27**(5).
• Damian, R. T., 1964, Molecular mimicry: Antigen sharing by parasite and host and its consequences, *Amer. Nat.*, **98** (900).
• Danielli, J. F., 1953, On some physical and chemical aspects of evolution, *Evol. Symp. Soc. Exp. Biol.*, No. 7, Cambridge University Press, Cambridge, Massachusetts.
Danilov, N. N., 1966, *Paths of Adaptation of Terrestrial Vertebrates under the Conditions of Existence of the Subarctic*, Vol. 2, *Birds*, *Tr. Inst. Biol. UFAN SSSR*, No. 56.
• Danneel, R., and Schaumann, F., 1938, Zur Physiologie der Kälteschwärzung beim Russenkaninchen, *Biol. Zentralbl.* **58**(5–6).
Darevskii, I. S., 1964, Natural parthenogenesis in vertebrates, *Priroda*, No. 7.
Darevskii, I. S., 1967, Rock lizards of the Caucasus, author's doctoral dissertation, Leningrad.

Darevskii, I. S., and Krasil'nikov, E. N., 1965, Some features of the blood cells of triploid hybrids of the rock lizard (*Lacerta saxicola* Evers), *Dokl. Akad. Nauk SSSR* **164**(3).
Darevskii, I. S. [Darevsky, I. S.], and Kulikova, V. N. [Kulikova, W. N.], 1961, Natürliche Parthenogenese in der polymorphen Gruppe der kaukasischen Felseidechse (*Lacerta saxicola* Evers.), *Zool. Jahrb. Abt. Syst.* **89**(2).
• Davies, L. L., 1963, The antitropical factor in cetacean speciation, *Evolution* **17**(1).
• Davis, W. H., 1959, Disproportionate sex ratios in hibernating bats, *J. Mammal.* **40**(1).
• Dawson, W. D., 1965, Fertility and size inheritance in a *Peromyscus* species cross, *Evolution* **19**(1).
• Dean, P. B., and DeVos, A., 1965, The spread and present status of the European hare *Lepus europaeus hybridus* (Desmarest) in North America, *Can. Field Nat.* **79**(1).
• Dehnel, A., 1949, Studies on the genus *Sorex, Ann. Univ. Marie Curie-Sklodowska Sect. C* **4**.
Dement'ev, G. P., 1939, On the question of hybrids of the lammelirostres birds under natural conditions, *Byull. Mosk. O. Ispyt. Prir.* **48**(1).
Dement'ev, G. P., 1946, Concepts of the integrity of the organism and some tasks of systematics, *Zool. Zh.* **25**(6).
Dement'ev, G. P., 1951, *Gyrfalcons*, Izd. Mosk. O. Ispyt. Prir., Moscow.
Dement'ev, G. P., 1965, Systematics of birds, in: *Modern Problems of Ornithology*, Ilim, Frunze.
Dement'ev, G. P., and Larionov, V. F., 1944, Investigations on the coloration of vertebrates, *Zool. Zh.* **23**(5).
Denisov, V. P., 1961, Relations of the little and spotted susliks at the juncture of their ranges, *Zool. Zh.* **40**(7).
Denisov, V. P., and Denisov, I. A., 1966, Seasonal changes in the protein composition of the blood serum of some hibernating animals, *Zh. Obshch. Biol.* **27**(3).
• Dessauer, H. C., Fox, W., and Pough, F. H., 1962, Starch-gel electrophoresis of transferrins, esterases and other plasma proteins of hybrids between two subspecies of whiptail lizard (genus *Cnemidophorus*), *Copeia*, No. 4.
• Devillers, C., 1965, The role of morphogenesis in the origin of higher levels of organization, *Syst. Zool.* **14**(4).
• Dice, L. R., 1936, Age variation in *Peromyscus maniculatus gracilis, J. Mammal.* **17**.
• Dice, L. R., and Liebe, M., 1937, Partial infertility between two members of the *Peromyscus truei* group of mice, *Contrib. Lab. Vertebr. Biol. Univ. Mich.* **5**(1–4).
Dmitrieva, T. V., 1949, Changes in the coloration of house mice as a possible criterion for forecasting their numbers, *Tr. Voronezh. Gos. Univ.* **18**, Zool. Vyp.
Dobrinskaya, L. A., 1966, On distinctions in the variability of interesting traits of fish and terrestrial vertebrates, in: *Questions of the Intraspecific Variability of Terrestrial Vertebrates and Microevolution, Theses of the Reports of the Conference on Intraspecific Variability and Microevolution*, Sverdlovsk.
Dobrinskii, L. N., 1962, Organometrics of birds of subarctic western Siberia, author's Candidate dissertation, Sverdlovsk.
Dobrinskii, L. N., 1966, Geographical variability of two close species of *Sterna*, in: *Intraspecific Variability of Terrestrial Vertebrates and Microevolution, Tr. Vses. Soveshchaniya*, Sverdlovsk.
• Dobzhansky, T., 1941, *Genetics and the Origin of Species*, New York.
• Dobzhansky, T., 1954, Evolution as a creative process, *Caryologia Suppl.*
• Dobzhansky, T., 1955, A review of some fundamental concepts and problems of population genetics, *Cold Spring Harbor Symp.* **20**.

• Dobzhansky, T., 1958, Genetics of natural populations, XXVII, *Evolution* **12**(2).

• Dobzhansky, T., and Levene, H., 1955, Genetics of natural populations, XXIV, *Genetics* **40.**

Drachevskii, K., 1961, Reconstruction of the producer fauna of Kirgiz, *Okhota i Okhot. Kh-vo.*, No. 3.

Druri, I. V., 1949, Wild northern deer of the Soviet Arctic and subarctic, *Tr. Arkt. Inst.* **200.**

Dubinin, N. P., 1966*a*, *Evolution of Populations and Radiation*, Atomizdat, Moscow.

Dubinin, N. P., 1966*b*, The bases of the genetics of populations, in: *Urgent Problems of Modern Genetics* (MGU), Moscow.

Dubinin, N. P., and Glembotskii, Ya. L., 1967, *Genetics of Populations and Selection*, Nauka, Moscow.

Dubinin, N. P., and Tinyakov, G. G., 1947, Climate and the distribution of inversions in the range of a species (*Drosophila funebris* F.), *Dokl. Akad. Nauk SSSR* **56**(9).

Dubrovskii, A. N., 1940, Fur-bearing animals of the Yamal National District, *Tr. Nauchno-Issled. Inst. Polyarnogo Zemledeliya, Zhivotnovodstva i Promyslovogo Khozyaistva*, No. 13.

Dunaeva, G. N., 1948, Comparative survey of the ecology of tundra voles from the Yamal Peninsula, *Tr. Inst. Geogr. Akad. Nauk SSSR* **41.**

• Dwight, D., 1963, Comparative anatomy and the evolution of vertebrates, in: *Genetics, Paleontology and Evolution* (G. L. Jepsen, G. G. Simpson, and E. Mayr, eds.), Atheneum, New York.

Dyban, A. P., and Udalova, L. D., 1967, Features of the morphology of the X chromosome and third pair of autosomes in different lines of the rat (*Rattus norvegicus*), *Genetika*, No. 2.

• Edwards, A. W. F., 1960, Natural selection and the sex ratio, *Nature London* **188**(4754).

• Eisentraut, M., 1929, Die Variation der balearischen Inseleidechse *Lazerta lilfordi* Günthr., *S. B. Ges. Naturf. Fr.*, No. 1–3, Berlin.

• Eisentraut, M., 1965, Rassenbildung bei Säugetieren und Vögeln auf der Insel Fernando Po, *Zool. Anz.* **174**(1).

Ėngel'gardt, V. A., 1960, The specificity of biological metabolism, *Vopr. Filos.*, No. 7.

• Espinasse, P., 1964, *Genetical Semantics and Evolutionary Theory*, Form Strat. Sci. Dordrecht & Reidel Publ. Co.

• Etkin, W., 1964, *Metamorphosis. Physiology of the Amphibia*, Academic Press, New York and London.

• Evans, J. W., 1965, The future of natural history, *Aust. J. Sci.* **28**(3).

Evropeitseva, N. V., 1962, The development, numbers, and question of the significance of dwarf males in ecologically different representatives of a genus, in: *Problems in Ecology*, Vol. 5, Kiev State University, Kiev.

• Falconer, D. S., 1965, Maternal effects and selection response, *Genetics Today*, Vol. 3, Pergamon Press, Elmsford, New York.

Fedorov, V. D., 1966, Biochemical evolution from the position of microbiology, *Tr. Mosk. O. Ispyt. Prir. Otd. Biol.*, No. 24.

• Fehringer, O., 1962, Rangordnung bei Vögeln, *Naturwiss. Rundsch.* **15**(1).

• Ferguson, D. E., 1963, In less than 20 years Mississippi delta wildlife developing resistance to pesticides, *Agric. Chem.* **18**(9).

• Fisher, R. A., 1930, *The Genetical Theory of Natural Selection*, Oxford.

• Fitzgerald, P., 1961, Cytological identification of sex in somatic cells of the rat (*Rattus norvegicus*), *Exp. Cell Res.* **25**(191).

• Fleming, C. A., 1962, History of the New Zealand land bird fauna, *Notornis* **9.**

Florkin, M., 1947, *Biochemical Evolution*, IL, Moscow.

• Florkin, M., 1966, *Aspects Moleculaires de l'Adaptation et de la Phylogenie,* Masson & Cie., Paris.

Folitarek, S. S., 1948, Predatory birds as a factor of natural selection in natural populations of small rodents, *Zh. Obshch. Biol.* **8**(1).

• Ford, E. B., 1963, Early stages in allopatric speciation, in: *Genetics, Paleontology and Evolution* (G. L. Jepsen, G. G. Simpson, and E. Mayr, eds.), Atheneum, New York.

• Ford, E. B., 1964, *Ecological Genetics,* Methuen, London.

• Ford, E. B., 1965, *Genetic Polymorphism,* Faber, London.

Formozov, A. N., 1948, Small rodents and insectivores of the Sharvin region of the Kostroma Province in the period 1930–1940, in: *Fauna and Ecology of Rodents, Tr. Mosk. O. Ispyt. Prir.,* No. 3.

• Foster, S. B., 1964, Evolution of mammals on islands, *Nature London* **202**(4929).

• Fouguette, M. J., 1960, Isolation mechanisms in three sympatric tree-frogs in the Canal Zone, *Evolution* **14**(4).

• Fox, W., Dessauer, H. C., and Maumus, L. T., 1961, Electrophoretic studies of blood proteins of two species of toads and their natural hybrid, *Comp. Biochem. Physiol.* **3**(1).

• Frank, F., 1962, Zur Biologie des Berglemmings, *Lemmus lemmus* L., *Z. Morphol. Oekol. Tiere* **51**(1).

• Franz, H., 1929, Morphologische und physiologische Studien an Carabus L. und den nächstverwandten Gattungen, *Z. Wiss. Zool.* **135.**

• Frederickson and Birnbaum, 1956, Competitive fighting between mice with different hereditary backgrounds, *Biol. Abstr.* **30**(27984).

• Frick, H., 1961, Allometrische Untersuchungen an inneren Organen von Säugetieren als Beitrag zur neuen Systematik, *Säugetierenkunde* **26**(3).

• Frier, G., 1959, Some aspects of evolution in Lake Nyasa, *Evolution* **13**(4).

• Fujino, K., 1960, Immunogenetic and marking approaches to identifying subpopulations of the North Pacific whales, *Sci. Rep. Whales Res. Inst.,* No. 15.

Galushin, V. M., 1966, Makeup and dynamics of settlements of predatory birds of the European center of the USSR, author's Candidate dissertation, Moscow.

• Gee, E. P., 1953, Wild buffaloes and tame, *Bombay Nat. Hist. Soc.* **51**(3).

Geodakyan, V. A., 1965, Role of the sexes in the transmission and reorganization of genetic information, *Probl. Peredachi Inf.* **1**(1).

• Geone, C. A., 1964, *Taxonomic Biochemistry and Serology,* Ronald Press Co., New York.

Geptner, V. G., Nasimovich, A. A., and Bannikov, A. G., 1961, *Mammals of the Soviet Union,* Vyshaya Shkola, Moscow.

Gerasimova, M. A., 1955, Commercial properties of the pelts of hares acclimatized to Siberia, *Tr. VNIO,* No. 15.

Gershenzon, S. M., 1945*a*, Distribution of black hamsters in the Ukrainian SSR, *Dokl. Akad. Nauk SSSR* **17**(8).

Gershenzon, S. M., 1945*b*, Seasonal changes in the frequency of black hamsters, *Dokl. Akad. Nauk SSSR* **18**(9).

Gershenzon, S. M., 1946, Role of natural selection in the distribution and dynamics of mechanisms in the hamster (*Cricetus cricetus* L.), *Zh. Obshch. Biol.* **7**(2).

Gershenzon, S. M., 1965, Investigations of mutability in the ichneumon fly, *Mormontella vitripennis* Wek, *Genetika,* No. 2.

• Gillham, N. W., 1956, Further thoughts on subspecies and trinomials, *Syst. Zool.* **5.**

Gilyarov, M. S., 1954*a*, Some theoretical positions of modern ecology, *Third Ecological Conference,* Part 1.

Gilyarov, M. S., 1954*b*, Species, populations, and biocenosis, *Zool. Zh.* **33**(4).

Gladkina, T. S., Meier, M. N., and Mokeeva, T. M., 1966, Features of the reproduction and development of three subspecies of the steppe spotted vole (*Lagurus lagurus*) and their hybridization, in: *Intraspecific Variability of Terrestrial Vertebrates and Microevolution, Tr. Vses. Soveshchaniya,* Sverdlovsk.

Gladkov, N. A., 1953, On species in zoology, *Zool. Zh.* **32**(5).

Goldovskii, A. M., 1957, On some rules of the evolutionary process, *Usp. Sovrem. Biol.* **43**(2).

- Goldschmidt, R., 1940, *The Material Basis of Evolution,* Yale University Press, New Haven, Connecticut.
- Goldschmidt, R., 1961, *Theoretische Genetik,* Berlin.
- Goodman, M., 1961, The role of immunochemical differences in the phyletic development of human behaviour, *Hum. Biol.* **33**(2).
- Goodman, M., and Poulik, E., 1961, Serum transferrins in the genus *Macaca*: Species distribution of nineteen phenotypes, *Nature London* **191**(4796).
- Gordon, M. S., Schmidt-Nielsen, and Kelly, H. M., 1961, Osmotic regulation in the crab-eating frog (*Rana cancrivora*), *J. Exp. Biol.* **38**(3).
- Gotronei, G., and Perri, T. I., 1946, Trapanti studiati in rapporto con la ibridazioni interspecifiche, *Boll. Zool.* **12.**
- Gould, S. J., 1966, Allometry and size in ontogeny and phylogeny, *Biol. Rep. Cambridge Philos. Soc.* **41**(4).
- Grace, L., 1951, An example of interspecific mating in toads, *Copeia,* No. 1.
- Grad, B., and Kral, V. A., 1957, Response of young and old mice to cold, *Fed. Proc. Fed. Amer. Soc. Exp. Biol.* **16**(1).
- Gray, P., 1954, *Mammalian Hybrids,* London.
- Gray, P., 1958, *Bird Hybrids,* London.
- Grosset, E., and Zontendyk, A., 1929, Immunological studies in reptiles and their relation to aspects of immunity in higher animals, *Publ. South African Inst.* **4.**
- Grünberg, H., 1961, Evidence for genetic drift in Indian rats (*Rattus rattus* L.), *Evolution* **15**(2).
- Gysels, H., 1963, New biochemical techniques applied to avian systematics, *Experientia* **19**(2).
- Gysels, H., and Rabaev, M., 1964, Taxonomic relationships of *Alca torda, Fratecula arctica* and *Uria aalge* as revealed by biochemical methods, *Ibis* **106**(4).
- Hacker, H. P., and Pearson, H. S., 1944, The growth, survival, wandering and variation of the long-tailed field mouse, *Apodemus sylvaticus, Biometrika* **33:**II.
- Hagendorf, F., 1926, Rot-Damwildkreuzung, *Wien. Allg. Forst-Jagdztg.,* No. 44.
- Haggerty, I., 1966, Erase scent trails—eliminate rats, *Pest Control* **34**(8).
- Hagmeier, E. M., 1958, Inapplicability of the subspecies concept to North American marten, *Syst. Zool.* **7.**
- Haldane, J. B. S., 1954, The measurement of natural selection, *Proceedings of the 9th International Congress on Genetics.*
- Haldane, J. B. S., 1956, The relation between density regulation and natural selection, *Proc. Roy. Soc. London* **145.**
- Haldane, J. B. S., 1957, Cost of natural selection, *J. Genet.* **55.**
- Halkka, O., and Skaren, U., 1964, Evolution chromosomique chez genre Sorex: Nouvelle information, *Experientia* **20**(6).
- Hall, B. P., 1963, The francolins, a study in speciation, *Bull. Br. Mus. Nat. Hist. Zool.* **10**(2).
- Hall, F. G., Dill, D. B., and Baron, E. S., 1936, Physiology of high altitudes, *J. Cell. Comp. Physiol.* **8.**

- Hall, R. E., 1943, Intergradation versus hybridisation in ground squirrels of the western United States, *Amer. Midl. Nat.* **29.**
- Hamerton, J. L., 1958, Problems in mammalian cytotaxonomy, *Proc. R. Soc. London* **169**(112).
- Harris, H., 1954, Biochemical individuality in man, *Biol. Hum. Aff.* **19**(2).
- Harrison, J. W., and Garret, F. C., 1925, The induction of melanism in the Lepidoptera and its subsequent inheritance, *Proc. R. Soc. London,* No. 99.
- Hatfield, D. M., 1935, A natural history study of *Microtus californicus, J. Mammal.* **16**(4).
- Hayne, D. W., and Thomson, D. Q., 1965, Methods for estimating microtine abundance, *Trans. N. Amer. Wildlife Conf.,* **30:**393–400.
- Hecht, M. K., 1965, The role of natural selection and evolutionary rates in the origin of higher levels of organization, *Syst. Zool.* **14**(4).
- Heim de Balsac, H., and Guislain, R., 1955, Evolution et speciation des campagnols de genre *Arvicola* en territoire français, *Mammalia* **19**(3).
- Herms, I. W., 1952, Die Realisation der Gene und die consecutive Adaptation, *Z. Wiss. Zool.* **146.**
- Herre, W., 1955, *Das Reh als Haustier,* Leipzig.
- Herre, W., 1959, Domestication und Stammesgeschichte, *Evol. Organism* **2.**
- Herre, W., 1963, Neues zur Umweltbeeinflussbarkeit des Säugetiergehirnes, *Naturwiss. Rundsch.* **16**(9).
- Herre, W., 1964, Zur Problematik der innerartlichen Ausformung bei Tieren, *Zool. Anz.* **172**(6).
- Herter, K., 1935, Ein Igelbastard (*E. roumanicus* × *europaeus*), *S. B. Ges. Naturf. Fr. Berlin,* No. 1–3.
- Herter, K., 1943, Die Beziehungen zwischen der Öcologie und dem Thermotaxis der Tiere, *Biol. Gen.* **17**(1–2).
- Hesse, R., 1924, Tiergeographie auf öcologischer Grundlage, Jena.
- Hesse-Doflein, 1943, Tierbau und Tierleben, Vol. 2, Jena.
- Hickman, K., and Harris, P., 1946, Tocopherol interrelationships, *Adv. Enzymol.* **6.**
- Hillman, R. E., 1964, Chromotographic evidence of intraspecific genetic differences in Eastern oyster, *Grassosterea virginica, Syst. Zool.* **13**(1).
- Howes, C. E., and Hutt, F. B., 1956, Genetic variation efficiency of thiamine utilization by the domestic fowl, *Poult. Sci.* **36**(6).
- Hsu, T. C., Rearden, H. H., and Luguette, G. F., 1963, Karyological studies of Felidae, *Amer. Nat.* **97**(895).
- Hubbs, C. L., and Raney, E. C., 1946, Endemic fish fauna of Lake Waleamaw, North Carolina, *Publ. Mus. Zool. Univ. Mich.* **65.**
- Hückinghaus, F., 1961, Die Bedeutung der Allometrie für die Systematik der Rodentia, *Z. Säugetierkunde* **26.**
- Hückinghaus, F., 1965*b*, Craniometrische Untersuchung an verwilderten Hauskaninchen von den Kerguelen, *Z. Wiss. Zool.* **171**(1–2).
- Huene, F., 1956, Paläontologie und Phylogenie der niederen Tetrapoden, *Jena.*
- Hungerford, D. A., and Nowell, P. C., 1963, Sex chromosome polymorphism and normal karyotype of three strains of laboratory rat, *J. Morphol.* **113**(2).
- Huxley, J., 1940, The New Systematics, London.

Imaidzumi, Iosinori, 1963, Endemic species of mammals of Japan and geographic isolation, *Sidzen Kagaku to khakuvutsukan* (cited in *Ref. Zh.,* No. 21, 1963).

- Immelmann, K., 1962, Vergleichende Beobachtungen über das Verhalten domestizierter

Zebrafinken in Europa und ihrer wilden Stammform in Australien, *Z. Tierz. Züchtungsbiol.* **77**(2).

• Ingles, L. C., and Wormand, J. B., 1951, The contiguity of the ranges of two subspecies of pocket gophers, *Evolution* **6**(2).

• Ingram, V. M., 1962, The evolution of a protein, *Fed. Proc. Fed. Amer. Soc. Exp. Biol.* **21**(6).

Ishchenko, V. G., 1966, Experiment on the use of allometric equalities for the study of morphological differentiation, *Tr. Inst. Biol. UFAN SSSR*, No. 51.

Ishchenko, V. G., 1967, Intrademe variability of allometric indices in the water vole, *Tr. Mosk. O. Ispyt. Prir.* **25.**

Ishchenko, V. G., and Dobrinskii, L. N., 1965, Relative growth of organs of two species of *Sterna*, in: *Contemporary Ornithology*, Nauka Kaz. SSR, Alma-Ata.

Ivanov, P., 1962, Urals boars in the USA, *Okhota Okhotn. Kh-vo*, No. 12.

• Iversen, E. S., 1965, Genetics of marine fish, *Sea Frontiers* **11**(2).

Ivlev, V. S., 1962, Physiological prerequisites of changes in the body size of animals in phylogenetic branches, *Fiziol. Zh. SSSR* **48**(11).

• Janossy, D., 1961, Die Entwicklung der Kleinsäugerfauna Europas in Pleistozän (Insectivora, Rodentia, Lagomorpha), *Z. Säugetierkunde* **26**(1).

• Jennings, H. S., 1911, Assortative mating, variability and inheritance of size, in the conjunction of *Paramecium, J. Exp. Zool.* **11.**

• Jocum, H. B., and Huestis, R. R., 1928, Histological differences in the thyroid glands from two subspecies of *Peromyscus, Anat. Rec.* **39.**

• Johnston, R. F., and Sealander, R. K., 1964, House sparrows: Rapid evolution of races in North America, *Science* **144**(3618).

• Junkins, B. L., 1963, Arsenic and its radioisotopes in the environs, in: *Radioecology*, Reinhardt Publishing Corp., New York.

Kalabukhov, N. I., 1935, Physiological features of mountain and plains subspecies of the meadow mouse (*Apodemus sylvaticus*), *Dokl. Akad. Nauk SSSR* **21**(1).

Kalabukhov, N. I., 1944*a*, On the dependence of the rate of breathing in mammals on temperature, *Dokl. Akad. Nauk SSR* **13**(9).

Kalabukhov, N. I., 1944*b*, Biological bases of measures in the struggle with mouselike rodents in enzootic places of rabbit fever, *Zool. Zh.* **23**(6).

Kalabukhov, N. I., 1946, Maintenance of an organism's energy balance as the basis of adaptation, *Zh. Obshch. Biol.* **7**(6).

Kalabukhov, N. I., 1950, *Animals' Ecological and Physiological Characteristics and Environmental Conditions*, Kharkov State University, Kharkov.

Kalabukhov, N. I., 1951, Methods of experimental investigations of the ecology of terrestrial vertebrates, Sov. Nauka, Moscow.

Kalabukhov, N. I., 1954, Ecological and physiological features of the geographical "forms of existence of species" and of close species of animals, *Byull. Mosk. O. Ispyt. Prir. Otd. Biol.* **9**(1).

Kalabukhov, N. I., and Ladygina, N. M., 1953, Origin of ecological and physiological features in mammals under the influence of the external environment, *Zool. Zh.* **32**(2).

Kalabukhov, N. I., and Raevskii, V. V., 1935, Study of mobile susliks (*Citellus pygmaeus*) in the steppe regions of the Northern Caucasus by the methods of banding, *Vopr. Êkol. Biotsenol.*, No. 2.

Kalabukhov, N. I., Kaliman, P. A., Mikheeva, E. S., Mumrin, V. I., Svistel'nikova, A. A., Mironov, N. P., Konnova, A. M., Volodina, O. A., and Pavlov, A. N., 1950, *Study of the Acceptability of Different Baits with Different Poisons by Little Susliks and the Effectiveness of Using This Method Against Susliks*, Rostov-on-the-Don.

- Kallmus, H., and Smith, C. A. B., 1960, Evolutionary origin of sexual differentiation and the sex-ratio, *Nature London* **186**(4730).
- Kaminski, M., and Balbierz, H., 1965, Serum proteins in Canidae: Species, race and individual differences, in: *Blood Groups of Animals.*

Kamshilov, M. M., 1960, On "systemic" and "cellular" adaptations, *Tr. Murm. Morsk. Biol. Inst.,* No. 2.

Karaseva, E. V., Narskaya, E. V., and Bernshtein, A. D., 1957, The *oeconomus* vole inhabiting the environs of Lake Nero in Yaroslavl Province, *Byull. Mosk. O. Ispyt. Prir. Otd. Biol.* **62.**

Karpinskaya, R. S., and Vizgin, V. P., 1966, On the biochemical approach to the problem of chemical evolution, *Nauchn. Dokl. Vyssh. Shk. Filos. Nauki,* No. 4.

Kashkarov, D. N., 1945, *Bases of the Ecology of Animals,* Uchpedgiz, Leningrad.

Keeler, C., Ridgway, S., Lipscomb, L., Fromm, E., 1968, The genetics of adrenal size and tameness in colorphase foxes, *J. Hered.* **59**(1).

Kein, A., 1958, *Species and Its Evolution,* IL, Moscow.

- Kelham, M., 1956, Report from Regents Park, *Zoo Life* **11**(2).
- Kennedy, J. P., 1964, Experimental hybridization of the green treefrog *Hyla cinerea* Schneider (Hylidae), *Zoologica U.S.A.* **49**(4).
- Kestner, O., and Plaut, R., 1924, Physiologie des Stoffwechsels, *Wintersteins Handb. Vergl. Physiol.* **11.**
- Kettlewell, H. B. D., 1956, Further selection experiments on industrial melanism in the Lepidoptera, *Evolution* **10**(1).

Khozatskii, L. I., and Églon, Ya. M., 1947, On one of the ways of burial and fossilization of the remains of vertebrates, *Priroda,* No. 1.

- King, I. A., and Eleftherion, B. E., 1960, Differential growth in the skulls of two subspecies of deermice, *Growth* **24**(2).
- King, I. A., Deshies, J. C., and Webster, R., 1963, Age of weaning in two subspecies of deer mice, *Science* **139**(3554).
- King, J., 1961, Swimming and reaction to electric shock in two subspecies of deermice (*Peromyscus maniculatus*) during development, *Anim. Behav.* **9**(3–4).
- King, S. C., 1965, Genetic implications in the origin of higher levels of organisation, *Syst. Zool.* **14**(4).

Kirikov, S. V., 1934, On the distribution of black hamsters, *Zool. Zh.* **13**(2).

Kirpichnikov, V. S., 1958, The degree of heterogeneity of populations of sazan and of hybrids of sazan with carp, *Dokl. Akad. Nauk SSSR* **122**(716).

Kirpichnikov, V. S., 1966*a,* The effectiveness of group and individual selection in fish-breeding, *Genetika,* No. 4.

Kirpichnikov, V. S., 1966*b,* Aims and methods in the selection of carp, *Izv. Vses. Nauchno-Issled. Inst. Ozern. Rechn. Rybn. Khoz.,* No. 62.

Kirpichnikov, V. S., 1967, Hybridization of the European carp with the Amurskii sazan and selection of species, *Dokl. Soiskanie Uchenoi Stepeni Dokt. Biol. Nauk.* (Zool. Inst. Akad. Nauk SSSR), Leningrad.

- Klatt, B., 1948, *Haustier und Mensch,* Hamburg.
- Kleiber, M., 1947, Body size and metabolic rate, *Physiol. Rev.* **27.**
- Klein, D. R., and Olson, S. T., 1960, Natural mortality patterns of deer in Southeast Alaska, *J. Wildl. Manage.* **24**(1).
- Kleinschmidt, O., 1926, *Die Formenkreislehre und das Weltwerden des Lebens,* Halle.
- Klomp, H., 1966, The dynamics of field population of the pine looper *Bupalus piniarius, Adv. Ecol. Res.* **3,** London–New York.
- Kluiyver, H. M., 1963, Über das Gleichgewicht in der Natur, *Angew. Ornithol.* **1**(3–4).

- Knight-Jones, E. W., and Moyse, J., 1961, Intraspecific competition in sedentary marine animals. Mechanism in biological competition, *Symp. Soc. Exp. Biol.*, No. 15, New York.

Kopein, K. I., 1958, Materials on the ecology of the Obskii lemming and the large narrow-skulled vole in Yamal, *Byull. Mosk. O. Ispyt. Prir. Otd. Biol. Ural. Otdel.*, No. 1, Sverdlovsk.

Kopein, K. I., 1964, Experiment on the study of natural selection under natural conditions, in: *Questions of the Intraspecific Variability of Terrestrial Vertebrates and Microevolution, Theses of the Reports of the Conference on Intraspecific Variability and Microevolution*, Sverdlovsk.

Kopein, K. I., 1967, Analysis of the age structure of populations of ermine, *Tr. Mosk. O. Ispyt. Prir. Otd. Biol. Ural. Otdel.* **25.**

- Korkman, N., 1957, Selection with regard to sex difference of body weight in mice, *Animal Breed. Abstr.* **26**(2).

Korzhuev, P. A., 1959, Hemoglobin as a factor of adaptation to lowered oxygen, *Usp. Sovrem. Biol.* **47**(3).

- Kosswig, C., 1962, Über präadaptive Mechanismen in der Evolution vom Gesichtspunkt der Genetik, *Zool. Anz.* **169**(1–2).
- Kramer, I., 1932, Die fortlaufenden Veränderungen der Amphibienleber im Hungerzustande, *Z. Mikrosk.-Anat. Forsch.* **28**(1–2).
- Kratochvil, J., 1965, Chronologische Grundlagen zur Kenntnis der Differenzierung und Herkunft der Formen der Gattung *Arvicola lacepede* (1799), *Biol. Rundsch.* **3**(5–6).
- Krebs, H. A., 1950, Body size and tissue respiration, *Acta. Biochem.* **4.**
- Krebs, H. A., 1964, Cyclic variation in skull–body regressions of lemmings, *Can. J. Zool.* **42**(4).
- Krech, D., Rosenzweig, M., Bennett, E., and Kruckel, B., 1954, Enzyme concentrations in the brain and adjustive behaviour patterns, *Science* **12**(3128).

Krivosheev, V. G., and Rossolimo, O. L., 1966, Intraspecific variability and systematics of the Siberian lemming (*Lemmus sibiricus* Kerr) of the Palearctic, *Byull. Mosk. O. Ispyt. Prir. Otd. Biol.* **71**(1).

Kubantsev, B. S., 1964*a*, Conditions of existence and sex in mammals, *Uch. Zap. Volgogr. Gos. pedagog. Inst.*, **No. 16.**

Kubantsev, B. S., 1964*b*, On the variability of the numerical ratio of the sexes during the breeding of mammals, depending on their systematic classification, *Uch. Zap. Volgogr. Gos. Pedagog. Inst.*, **No. 16.**

- Kurten, B., 1959, Rates of evolution in fossil mammals, *Cold Spring Harbor Symp. Quant. Biol.* **24.**
- Kurten, B., 1965, Evolution in geological time, *Ideas Mod. Biol. N.Y.*

Kusakina, A. A., 1965, Heat resistance of the hemoglobin and cholinesterase of the muscles and liver in representatives of three subspecies of the gray toad (*Bufo bufo*), *Sb. Rab. Inst. Tsitol. Akad. Nauk SSSR*, No. 8.

Kuz'minykh, 1968, *Skh. Biol.* **3**(4).

- Lack, D., 1965, Evolutionary ecology, *J. Appl. Ecol.* **2**(2).

Ladygina, N. V., 1964, On the comparative characteristics of the Kurganchik mouse and house mouse, in: *Problems in Genetics and Zoology*, Khar'kov State University, Khar'kov.

- LaMotte, M., 1959, Polymorphism of natural populations of *Cepaea nemoralis*, *Cold Spring Harbor Symp. Quant. Biol.* **24.**
- Landauer, W., 1946, Form and function in frizzled fowl, *Biol. Symp. London* **6.**

• Lantz, L. A., 1926, Essais d'hybridation entre different formes de lizards du sousgenre Podarcis, *Rev. Hist. Nat. Appl.* **1**(7).

Larina, N. I., 1955, On the question of speciation in rodents, *Nauch. ezhegodn. Saratovskogo Gos. Univ. za 1954 God.*

Larina, N. I., 1964, Geographical variability of some ecological and physiological traits of forest and yellow-throated mice, in: *Questions of the Intraspecific Variability of Terrestrial Vertebrates and Microevolution, Theses of the Reports of the Conference on Intraspecific Variability of Terrestrial Vertebrates and Microevolution,* Sverdlovsk.

Larina, N. I., and Denisova, I. A., 1966, Use of serological methods in research on the systematics of vertebrates, in: *Questions of the Innervation and Regulation of the Organism's Functions,* Saratov State University, Saratov.

Larina, N. I., and Golikova, V. L., 1959, Morphological and ecological characteristics of hybrid populations of *Apodemus* of the Caucasus and Transcaucasus, *Uch. Zap. Sarat. Gos. Univ. Vyp. Biol.-Pochv.* **64.**

Larina, N. I., and Kul'kova, A. V., 1959, On the question of the consanguinity of close species and of geographically distant populations of the same species, in: *Theses of the Reports of the Conference on Ecological Physiology,* Vol. 1, Leningrad.

Lavrova, M. Ya., and Karaseva, E. V., 1956, Activity of predatory birds and the settling of the common vole in agricultural lands south of Moskva Province, *Byull. Mosk. O. Ispyt. Prir. Otd. Biol.* **61**(3).

• Laws, R. M., 1956, Growth and sexual maturity in aquatic mammals, *Nature London* **178**(4526).

• Lederer, G., 1950, Ein Bastard von *Elaphe guttata* L. × *Elaphe quadrivittata quadruvittata* (Helbrook)–und dessen Rückkreuzung mit der mütterlichen Ausgangsart, *Zool. Gart. N. F.* **17**(1–5).

• Lehmann, E., 1966, Anpassung und "Lokalkolorit" bei den Soriciden zweier linksrheinischer Moore, *Säugetierkundl. Mitt.* **14**(2).

• Lenick, P., 1963, Action of adult tissue extracts and their fractions on the early development of the chick embryo, *Proceedings of the XVI International Congress on Zoology,* Vol. 3.

Lenin, V. I., 1934, *Philosophical Notebooks,* Pub. Tsk VKP(b).

• Leone, C., and Wiens, A. L., 1956, Comparative serology of carnivores, *J. Mammal.* **37**(1).

• Leraas, H. J., 1938, Variation in *Peromyscus maniculatus osgoodi* from the Uinta Mountains, *Contrib. Lab Vertebr. Genet. Univ. Mich.,* No. 6.

• Lerman, A., 1965, On rates of evolution of unit characters and character complexes, *Evolution* **19**(1).

• Lerner, I. M., 1954, *Genetic Homeostasis,* New York.

• Lerner, I. M., 1958, *The Genetic Basis of Selection,* New York.

• Lerner, I. M., 1965, Ecological genetics (synthesis), in: *Genetics Today,* Vol. 2, Pergamon Press, Elmsford, New York.

• Levine, L., and Krupa, P. L., 1966, Studies on sexual selection in mice. III. Effects of the gene for albinism, *Amer. Nat.* **100**(912).

• Levine, L., and Lascher, B., 1965, Studies on sexual selection in mice. Reproductive competition between black and brown males, *Amer. Nat.* **99**(905).

• Levy, A., 1965, Sex ratio in isogenic laboratory population of *Drosophila melanogaster, Amer. Nat.* **99**(908).

• Lewontin, R. C., 1955, The effects of population density and composition on viability in *D. melanogaster, Evolution* **9**(1).

• Lewontin, R. C., 1958, Studies on heterozygosity and homeostasis, II. Loss of heterosis in a constant environment, *Evolution* **12**(3).

• Lewontin, R. C., 1962, Interdeme selection controlling a polymorphism in the house mouse, *Amer. Nat.* **94**(1).
• Lewontin, R. C., 1965, Selection in and of populations, *Ideas Mod. Biol. N.Y.*
• Lewontin, R. C., and Birch, L. C., 1966, Hybridization as a source of variation for adaptation to new environments, *Evolution* **20**(3).
• Lewontin, R. C., and Dunn, L. C., 1960, The evolutionary dynamics of a polymorphism in the house mouse, *Genetics* **45**(5).
• Li, C. C., 1955, *Population Genetics*, University of Chicago Press, Chicago.
• Lidicker, W. Z., Jr., 1960, An analysis of intraspecific variation in the kangaroo rat *Dipodomys merriami, Univ. Calif. Publ. Zool.* **64**(2).
• Lidicker, W. Z., Jr., 1962, The nature of subspecies boundaries in a desert rodent and its implications for subspecies taxonomy, *Syst. Zool.* **11**(4).
• Lidicker, W. Z., Jr., 1963, The genetics of a naturally occurring coat-color mutation in the California vole, *Evolution* **17**(3).
• Lidicker, W. Z., Jr., 1965, Comparative study of density regulation in confined populations of four species of rodents, *Res. Popul. Ecol.* **7**(2).
• Lidicker, W. Z., Jr., 1966, Ecological observations on a feral house mouse population declining to extinction, *Ecol. Monogr.* **36**(1).
• Lin, T. T., 1954, Hydridisation between *Peromyscus maniculatus oreas* and *P. m. gracilis, J. Mammal.* **35**(3).
• Little, C. C., 1928, Preliminary report on a species cross in rodent *Mus musculus* × *Mus wagneri, Pap. Mich. Acad. Sci.,* No. 8.
Lopyrin, A. I., Loginova, N. V., and Karpov, P. L., 1951, Influence of changed conditions of embryogenesis on the growth and development of lambs, *Sov. Zootekh.,* No. 11.
• Lowther, J. K., 1961, Polymorphism in the white-throated sparrow, *Zonotrichia albicollis* (Gmelin), *Can. J. Zool.* **39**(3).
Lozina-Lozinskii, L. K., 1966, On the reaction of biological systems to extreme influences in connection with tasks of exobiology, *Sb. Rab. Inst. Tsitol. Akad. Nauk SSR,* No. 12.
Lukin, E. I., 1940, *Darwinism and Geographical Rules in the Change of Organisms,* Izd. Akad. Nauk SSSR, Moscow.
Lukin, E. I., 1962, Intraspecific ecological changes of organisms, in: *Problems in Ecology, Material from the Fourth Ecological Conference,* Vol. 4, Kiev.
Lukin, E. I., 1966, Some data and observations on the intraspecific variability of animals, in: *Intraspecific Variability of Terrestrial Vertebrates and Microevolution, Tr. Vses. Soveshchaniya,* Sverdlovsk.
Lukina, A., 1953, On the sex ratio in agricultural animals, in the light of the theory of vitality, *Zh. Obshch. Biol.* **14**(6).
Lysov, A. M., and Pis'mennaya, P. G., 1951, On the experimental breeding of gray Karakul sheep reared under different conditions, *Karakulevod. Zverovod.,* No. 4.
• Mainardi, D., 1959, Immunological distances among some gallinaceus birds, *Nature London* **184.**
• Mainardi, D., 1961, Soluzione del problema del "presuntiibridi" mediante analisis immunogenetica, *Riv. Ital. Ornitol.* **31**(4).
• Mainardi, D., 1962, Immunological data on the phylogenetic relationships and taxonomic position of flamingos (Phoenicopteridae), *Ibis* **104**(3).
• Mainardi, D., 1964, Interazione tra preferenze sessuali delle femmine e predominanza sociale dei maschinel determinismo della sclezione sessuale nel topo (*Mus musculus*), *Atti Accad. Naz. Lincei Cl. Sci. Fis. Mat. Nat. Rend.* **37**(6).
• Mainardi, D., Marsan, M., and Pasquali, A., 1965, Causation of sexual preferences in the

house mouse. The behaviour of mice reared by parents whose odour was artificially altered, *Atti Soc. Ital. Sci. Natur.* **104**(3).

• Mainardi, D., Scudo, F., and Barbieri, D., 1966, Accompiamento preferenziale basato sull'apprediamento infantil in populationi (riassunto), *Atti Assoc. Genet. Ital.* **11.**

• Maldonado, A. A., and Ortiz, E., 1966, Electrophoretic patterns of serum proteins of some West Indean Anolis—(Sauria: Iguanidae), *Copeia,* No. 2.

• Manwell, C., and Kerst, K. V., 1966, Possibilities of biochemical taxonomy of bats using hemoglobin, lactate dehydrogenase, esterases and other proteins, *Comp. Biochem. Physiol.* **17**(3).

Marinina, L. S., 1966, The pre-spring change of the morphophysiological state of populations of the chestnut vole (*Clethrionomys glareolus*), in: *Experimental Study of Intraspecific Variability of Vertebrates, Tr. Inst. Biol. UFAN SSSR,* No. 51.

• Martini, P., Sabels, B., and Shuler, D., 1961, Rampart Cave coprolite and ecology of the Shasta ground sloth, *Amer. J. Sci.,* No. 2.

Mashkovtsev, A. A., 1935, Influence of mountain climate on the constitution of mammals, *Tr. Lab. Evol. Morphol. Akad. Nauk SSSR* **2**(3).

Mashkovtsev, A. A., 1940, Biological and physiological significance of sexual dimorphism in vertebrates, *Dokl. Akad. Nauk SSSR* **28**(1).

Maslennikova, E. M., and Khromach, D. B., 1954, On the different requirements of animals for vitamin B_2, *Theses of the Scientific Session of the Institute of Nutrition,* Medgiz., Moscow.

• Matthey, R., 1954, Analyse caryologique de cinq espèces de *Muridae africains* (Mammalia, Rodentia), *Mammalia* **28**(3).

• Matthey, R., 1955, Nouveaux documents sur les chromosome des Muridae. Problèmes des taxonomie chex les Microtinae, *Rev. Suisse Zool.* **62**(1–5).

• Matthey, R., 1961, La formulechromosomique et la position systematique de *Chamaeleo gallus* Günther (Lacertilia), *Zool. Anz.* **166**(5–6).

• Matthey, R., 1963, La formule chromosomique chez sept espèces et sous-espèces de *Murinae africains, Mammalia* **27**(2).

• Matthey, R., 1964, Evolution chromosomique et speciation chez les mus du sousgenre Leggada Gray 1837, *Experientia* **20**(12).

• Matthey, R., 1966, Cytogenetique et taxonomie des rats appartant au sous-genre Mastomys Thomas (Rodentia–Muridae), *Mammalia* **30**(1).

• Matthey, R., and van Brink, J., 1960, Nouvelle contribution à la cytologie comparée des Chameleontidae (Reptilia—Lacertilia), *Bull. Soc. Vaudoise Sci. Nat.* **67**(6).

• Matveev, B. S., 1963, Role of the work of A. N. Severtsov in the development of evolutionary morphology in the USSR for the last 25 years (1936–1961), *Tr. Inst. Morfol. Zhivotn. A. N. Severtsova,* No. 38.

Matveev, B. S., 1966, Significance of the theoretical heritage of A. N. Severtsov to modern biology, *Zool. Zh.* **45**(9).

Mayr, E., 1942, *Systematics and the Origin of Species* Columbia University Press, New York, 334 pp. [Russian translation: IL, Moscow, 1947].

• Mayr, E., 1954, Change of genetic environment and evolution, in: *Evolution as a Process* (J. Huxley, A. C. Hardy, and E. B. Ford, eds.), George Allen & Unwin, London.

• Mayr, E., 1963, *Animal Species and Evolution,* Harvard University Press, Cambridge, Massachusetts.

• Mayr, E., 1965, Classification and phylogeny, *Amer. Zool.* **5**(1).

• Mayr, E., and Gilliard, J. T., 1952, Altitudinal hybridisation in the New Guinea honeyeaters, *Condor,* No. 54.

Mayr, E., Linsley, E., and Huizinger, R., 1956, *Methods and Principles of Zoological Systematics* [Russian translation], IL, Moscow.

• Mayrat, A., 1965, La loi d'allometrice et les allometries, *C. R. Seances Soc. Biol. Paris* **159**(3).

• McEwen, F. L., and Splittstosser, C. M., 1964, A genetic factor controlling color and its association with DDT sensitivity in the cabbage looper, *J. Econ. Entomol.* **57**(2).

Meladze, D. D., 1954, On the question of the acclimatization of Altai squirrels (*Sciuris vulgaris altaicus*) in the Georgian SSR, *Third Ecological Conference,* Part III, Kiev.

Menard, G. A., 1930, *Pantoeconomy,* Gostorgizdat, Moscow and Leningrad.

• Mertens, R., 1950, Über Reptilienbastarde, I., *Senckenbergiana Biol.* **31.**

• Mertens, R., 1956, Über Reptilienbastarde, II., *Senckenbergiana Biol.* **37.**

• Mertens, R., 1963, Über Reptilienbastarde, III., *Senckenbergiana Biol.* **45.**

• Meyer, J. H., 1955, Die Bluttransfusion als Mittel zur Überwindung letaler Keimkombinationen bei Lepidopteren-Bastarden, *Z. Wien. Entomol. Ges.* **64**(33).

Mezhzherin, V. A., 1964, Dehnel's phenomenon and its possible explanation, *Acta Theriol.* **8.**

• Mihail, N., and Asandei, A., 1961, Stimulation of growth in tadpoles by feeding them with snails, *Naturwissenschaften* **48**(13).

Mikhalev, M. V., 1966, Difference in the mobility of protein fractions in close forms of rodents, *Experimental Study of Intraspecific Variability of Vertebrates, Tr. Inst. Biol. UFAN SSSR,* No. 51.

Mikhalev, M. V., and Sidorkin, V. I., 1966, Electrophoretic examination of serum proteins of two species of the gray vole, *Intraspecific Variability of Terrestrial Vertebrates and Microevolution, Tr. Vses. Soveshchaniya,* Sverdlovsk.

• Milani, R., 1957, Genetic research on the resistance of insects to the action of toxic substances, *Rev. Parasitol.* **18**(1).

• Milkman, R. D., 1965, The genetic basis of natural variation. VI. Selection of a crossveinless strain of *Drosophila* by phenocopying at high temperature, *Genetics* **51**(1).

• Miller, A. H., 1955, A hybrid woodpecker and its significance in speciation in the genus *Dendrocopus, Evolution* **9**(3).

• Miller, A. H., 1956, Ecological factors that accelerate formation of races and species of territorial vertebrates, *Evolution* **10**(3).

• Miller, G. S., and Kellog, R., 1955, *List of North American Recent Mammals,* Smithsonian Inst., Washington, D.C.

• Miller, R. R., 1961, Speciation rates in some fresh-water fishes of Western North America, Symposium on Vertebrate Speciation, University of Texas, Austin.

• Miller, W. J., 1964, First linkage of a species antigen in the genus *Steptopelia, Science* **143**(3611).

Milovanov, V. K., 1950, *Raising the Vitality of the Offspring of Agricultural Animals,* Sel'khozgiz, Moscow.

Mirolyubov, I. I., 1949, Hybridization of the spotted deer with the izyubr, *Karakulevod. Zverovod.,* No. 2.

• Mixner, Y. P., and Turner, C. W., 1957, Strain differences in response of mice to mammary gland stimulating hormones, *Dept. Dairy Husb. Univ. Missouri Columbia Anim. Br. Abstr.* **26**(2).

• Monroe, J. E., 1962, Chromosomes of rattlesnakes, *Herpetologica* **17**(4).

• Montalenti, G., 1938, L'ibridazione interspecifica degli Amfibi anuri, *Ann. Zool.* **4.**

• Moody, R. A., Cochran, V. A., and Drugg, H., 1949, Serological evidence on Lagomorpha relationships, *Evolution,* No. 3.

• Moor, J. A., 1946, Studies in the development of frog hybrids, *J. Exp. Zool.* **101**(2).

• Moor, J. A., 1954, Geographic and genetic isolation in Australian Amphibia, *Amer. Nat.* **88**(893).

• Moor, J. A., 1963, Patterns of evolution in the genus *Rana*, in: *Genetics, Paleontology and Evolution* (G. L. Jepsen, G. G. Simpson, and E. Mayr, eds.), Atheneum, New York.

Morozov, V. F., 1955, Variability of the fur coat of the Ussuriisk raccoon acclimatized to Kalinin Province, *Tr. VNIO*, No. 15.

• Morrison, P., and Elsner, R., 1962, Influence of altitude on heart and breathing rates in some Peruvian rodents, *J. Appl. Physiol.* **17**(3).

• Myers, G. S., 1960, The endemic fish fauna of Lake Lanao and the evolution of higher taxonomic categories, *Evolution* **14**(3).

• Nadler, C. F., 1962, Chromosome studies in certain subgenera of *Spermophilus*, *Proc. Soc. Exp. Biol. Med.* **110**(4).

• Nadler, C. F., 1964, Chromosomes and evolution of the ground squirrel, *Spermophilus richardsonii*, *Chromosoma* **15**(3).

• Nadler, C. F., and Block, M. H., 1962, The chromosomes of some North American chipmunks (Sciuridae) belonging to the genera *Tamias* and *Eutamias*, *Chromosoma* **13**(1).

• Nadler, C. F., and Hughes, C. E., 1966, Serum protein electrophoresis in the taxonomy of some species of the ground squirrel subgenus *Spermophilus*, *Comp. Biochem. Physiol.* **18**(3).

Nasimovich, A. A., 1961, Some general problems and results of the acclimatization of terrestrial vertebrates, *Zool. Zh.* **60**(7).

Nasimovich, A. A., 1966, Ecological consequences of the inclusion of a new species in the mainland biocenosis (the muskrat in Eurasia), *Zool. Zh.* **45**(11).

Nasimovich, A. A., Novikov, G. A., and Semenov-Tyan'Shanskii, O. V., 1948, The Norwegian lemming, *Fauna Ékol. Gryzunov*, No. 3.

Naumov, N. P., 1948, *Sketches of the Comparative Ecology of Mouselike Rodents*, Izd. Akad. Nauk SSSR, Moscow and Leningrad.

Naumov, N. P., 1955, *Ecology of Animals*, Sov. Nauka, Moscow.

Naumov, N. P., 1963, *Ecology of Animals*, Vysshaya Shkola, Moscow.

Naumov, N. P., 1964, Spatial features and mechanisms of the dynamics of numbers in terrestrial vertebrates, in: *Contemporary Problems in the Study of the Dynamics of Numbers in Animal Populations*, Izd. Akad. Nauk SSR, Moscow.

Naumov, N. P., 1965, Some problems of population biology and the hunting economy, in: *Questions of the Hunting Economy of the USSR*, Kolos, Moscow.

Naumov, N. P., and Nikol'skii, G. V., 1962, On some general rules of the population dynamics of animals, *Zool. Zh.* **41**(8).

Naumov, S. P., 1964, On the geographical differentiation of mammalian species, *Theses of the Reports of the Second Scientific Conference of Zoologists*, Ped. Inst. RSFSR, Krasnodar.

Naumov, S. P., 1966, Ecological and geographical differentiation of settlements of game species of Yakut mammals, *Byull. Mosk. O. Ispyt. Prir. Otd. Biol.* **71**(6).

Nazarova, I. V., 1958, Ecological and morphological features of the gray voles from the Volzhskoye–Kamskoye border, *Izv. Kol'skogo Fil. Akad. Nauk SSSR Ser. Biol. Nauk Zool.*, No. 6.

Nenashev, G. A., 1966, Determination of the heritability of different traits in fish, *Genetika*, No. 11.

• Neumann, D., 1959, Experimentelle Untersuchungen des Farbenmusters der Schale von *Theodoxus fluviatilis* L., *Verh. Dtsch. Zool. Ges. Munster*.

Nikol'skii, G. P., 1965, *Theory of the Dynamics of a School of Fish*, Nauka, Moscow.

Nikol'skii, G. V., and Pikuleva, V. A., 1958, On the adaptive significance of the amplitude of specific traits and properties of organisms, *Zool. Zh.* **37**(7).

Nikolyukin, N. I., 1966, Some questions on the cytogenetics, hybridization, and systematics of sturgeon, *Genetika*, No. 5.

• Norris, J. D., 1963, A campaign against the coypus in East Anglia, *New Sci.* **17**(331).

Novikov, B. G., 1949, Regional sensitivity of plumage to hormones, *Tr. Inst. Zool. Akad. Nauk Ukr. SSR* **2.**

Novikov, B. G., 1952, Experimental analysis of the development of seasonal coat color in squirrels, *Nauk Zap.* **2**(12).

Novikov, B. G., and Blagodatskaya, G. I., 1948, Mechanisms in the development of seasonal protective coloration, *Dokl. Akad. Nauk SSSR* **61**(1).

Numerov, K. D., 1964, On changes in the coloring of sables of Yeniseisk Siberia over the years, *Zool. Zh.* **43**(4).

Nutrition Revue, 1954, Pantothenic acid and the pituitary-adrenal axis, **12**(1):20–22.

• O'Donald, P., 1959, Possibility of assortive mating in the Arctic Scua, *Nature London* **183**(4669).

• Odum, E. G., 1963, *Ecology*, Holt, Rinehart and Winston, New York.

• Ogaki, M., and Tsukamoto, M., 1957, Genetic analysis of DDT resistance in some Japanese strains of *Drosophila melanogaster, Japanese Contributions to the Insecticide Resistance Problem*, Nagesawa.

Ognev, S. I., 1928–1950, *Animals of the USSR and Adjoining Countries*, Vols. I–VII, Izd. Akad. Nauk SSSR, Moscow and Leningrad.

Okolovich, A. K., and Korsakov, G. K., 1951, *Muskrats*, Moscow.

Olenev, V. G., 1964, Seasonal changes in some morphological traits of rodents in connection with the dynamics of population age structure, author's Candidate dissertation, Inst. Biologii UFAN SSSR, Sverdlovsk.

Olenov, Yu. M., 1959, On the role of spatial isolation in speciation, *Usp. Sovrem. Biol.* **48**(3).

• Oliver, J., and Schaw, C., 1953, The amphibians and reptiles of the Hawaiian Islands, *Zoologica* **38**(2).

Ol'nyanskaya, R. M., 1949, *On the Physiology of Long Acclimatization of Sheep to Lowered Atmospheric Pressure*, Izd. Akad. Nauk SSSR, Moscow and Leningrad.

• Orians, G., 1962, Natural selection and ecological theory, *Amer. Nat.* **96**(890).

• Osche, G., 1962, Das Praeadaptationsphänomen und seine Bedeutung für die Evolution, *Zool. Anz.* **169**(1–2).

Ovchinnikova, N. A., 1964, Biological features of the nominal and northern subspecies of the *oeconomus* voles and their hybrids, in: *Questions of the Intraspecific Variability of Terrestrial Vertebrates and Microevolution, Theses of Reports of the Conference on Intraspecific Variability and Microevolution*, Sverdlovsk.

Ovchinnikova, N. A., 1966, Biological features of the northern and nominal subspecies of the *oeconomus* vole and their hybrids, in: *Intraspecific Variability of Terrestrial Vertebrates and Microevolution, Tr. Vses. Soveshchaniya*, Sverdlovsk.

Ovchinnikova, N. A., 1967, Studies on two *Microtus oeconomus* subspecies and their hybrids in experimental conditions, *Materialy Otchetnoi Sessii Laboratorii Populatsionnoi Ekologii Pozvonochnych Zhyvotnykh*, Sverdlovsk.

• Packard, G. C., 1966, The influence of ambient temperature and aridity on modes of reproduction and excretion of amniote vertebrates, *Amer. Nat.* **100**(916).

• Pajunen, V. I., 1966, The influence of population on the territorial behaviour of *Leucorrhinia rubicunda* L. (Odon., Libellulidae), *Ann. Zool. Fennica* **3**(1).

• Pasternak, J., 1964, Chromosome polymorphism in the black fly *Simulium vittatum* (Zett.), *Can. J. Zool.* **42**(1).

• Pasteur, G., 1960, A propos des cochevis nord-africains variation clinale et sousespèces (additum à l'article precédent), *Bull. Soc. Sci. Nat. Phys. Maroc.* **40**(2).

• Pasteur, G., and Bons, N., 1962, Note preliminaire sur *Alytes* (*obstetricans*) Maurus. Gomellarite ou polytorisme? Remarques biogeographiques, genetiques et taxonomiques, *Bull. Soc. Zool.* Fr. **87**(1).

Pavlinin, V. N., 1948, Data on the fluctuation of the mole in the Urals, *Zool. Zh.* **27**(6).

Pavlinin, V. N., 1959, Characteristics of the fur coat of Tobol'sk sables in connection with an appraisal of the results of the output of eastern sables in Sverdlovsk Province, in: *Questions of the Acclimatization of Mammals in the Urals, Tr. Inst. Biol. UFAN SSSR*, No. 18.

Pavlinin, V. N., 1966, Experimental study of the genetics of subspecies of the common squirrel, in: *Experimental Study of the Intraspecific Variability of Vertebrates, Tr. Inst. Biol. UFAN SSSR*, No. 51.

Pavlov, B. K., 1965, Variability of the phenotypic structure of squirrel populations of eastern Siberia, *Byull. Mosk. O. Ispyt. Prir. Otd. Biol.* **70**(3).

Pavlov, B. K., and Sorokin, E. P., 1969, Influence of trapping on dynamics in squirrel numbers, *Izv. Vost. Sib. Otd. Geogr. O. SSSR* **666,** Irkutsk.

Pavlov, B. K., and Smyshlyaev, M. I., 1967, Squirrel trapping and change of population structure, in: *Data of the All-Union Scientific–Industrial Conference on Squirrels*, Izd. Tsentrsoyuz, Kirov.

• Peakall, D. B., 1960, Analysis of the proteins of egg-white as an aid to the classification of birds, *J. Bombay Nat. Hist. Soc.* **57**(3).

• Peakall, D. B., 1964, Biochemistry and evolution, *New Sci.* **24**(415).

• Pearson, T., 1938, The Tasmanian brush opossum, its distribution and colour varieties, *Pap. Proc. Soc. Tasmania* **21.**

• Penney, R. L., 1964, The Adelie penguin's faithfulness to territory and mate, *Biol. Antarctique Paris.*

• Perri, I., 1954, Ibridazioni, trapianti e competenza, *Arch. Zool. Ital.* No. 39.

• Petit, C., 1956, L'influence de la temperature sur l'isolement sexual, *C. R. Acad. Sci.* **243**(21).

• Petrusewicz, K., 1958, Differences in male and female quantitative dynamics in confined populations of mice, *Bull. Acad. Pol. Sci. Ser. Biol.* **6**(6).

• Petrusewicz, K., 1959, Teoria evolucji Darwina jest teoria ecologiczna, *Ecol. Pol. Ser. A* **5**(4).

• Petrusewicz, K., 1966, Dynamics, organisation and ecological structure of population, *Ecol. Pol. Ser. A* **14**(25).

• Petrusewicz, K., and Andrzeijewski, R., 1962, Natural history of a free-living population of house mice (*Mus musculus* L.) with particular reference to grouping within the population, *Ecol. Pol. Ser. A* **10**(5).

• Picard, J., Heremans, J., and Vandebroek, G., 1963, Serum proteins found in primates. Comparative analyses of the antigenic structure of several proteins, *Mammalia* **27**(2).

• Pimentel, D., 1965, Population ecology and the genetic feedback mechanism, in: *Genetics Today*, Vol. 2, Pergamon Press, Elmsford, New York.

• Pimentel, R. A., 1959, Mendelian infraspecific divergence levels and their analysis, *Syst. Zool.* **8**(3).

• Pimlott, D. H., 1961, The ecology and management of moose in North America, *Terre Vie* **108**(2–3).

• Pinowski, J., 1965, Overcrowding as one of the causes of dispersal of young tree sparrows, *Bird Study* **12**(1).

Pogosyants, E. E., Prigozhina, E. L., and Egolina, N. A., 1962, Intertwined ascitic tumor in rat's ovaries (strain O. Ya.), *Vopr. Onkol.* **8**(11).

Pokrovskii, A. V., 1962, Individual variability in the rate of sexual maturation of females of the steppe spotted vole (*Lagurus lagurus*), in: *Questions of Intraspecific Variability of Mammals, Tr. Inst. Biol. UFAN SSSR*, No. 29.

Pokrovskii, A. V., 1967*a*, Seasonal changes in the rate of sexual maturation of females of the steppe spotted vole and some other species of voles, *Tr. Mosk. O. Ispyt. Prir. Biol. Ural Otdel.* **15.**

Pokrovskii, A. V., 1967*b*, Growth rate of young voles, depending on the time of birth, *Tr. Mosk. O. Ispyt. Prir. Otd. Biol. Ural. Otdel.* **25.**

Pokrovskii, A. V., Smirnov, V. S., and Shvarts, S. S., 1962, Colorimetric study of variability in rodents' coloration in connection with the problem of hybrid populations, in: *Questions of Intraspecific Variability of Mammals, Tr. Inst. Biol. UFAN SSSR* **29.**

Polyakov, I. Ya., 1956, On the nature of variability of rodent populations, during a change in their numbers, *Zh. Obshch. Biol.* **17**(1).

Polyakov, I. Ya., 1964, *Prognosis of Distribution of Agricultural Pests*, Kolos, Leningrad.

Polyakov, I. Ya., and Pegel'man, S. G., 1950, Some age features of the requirements of social voles (*Microtus socialis*) for thermal conditions, *Zh. Obshch. Biol.* **11**(6).

Polyakov, I. Ya., Kubantsev, B. S., Meier, M. N., and Skholl', E. D., 1958, Some features of the morphological and ecological variability of the little suslik in different parts of its range, in: *Biological Bases of the Control of Rats*, Moscow.

Popov, V. A., 1949, Data on the ecology of the mink (*Mustela vison*) and results of its acclimatization in Tatar ASSR, *Tr. Kaz. Fil. Akad. Nauk SSSR*, No. 2.

Povetskaya, M. A., 1951, Change in the commercial quality of the pelts of squirrels acclimatized to new regions, in: *Questions of the Science of Commodities of Fur-Bearing Stock, Tr. VNIO*, Moscow.

Puzanov, I. I., 1954, Saltation and metamorphosis, *Byull. Mosk. O. Ispyt. Prir. Otd. Biol.* **9**(4).

Pyastolova, O. A., 1964, Specific features of the population age structure of the *oeconomus* vole in the extreme northern limits of its distribution, in: *Contemporary Problems in the Study of the Dynamics of Numbers in Animal Populations, Mater. Soveschaniya Inst. Morf. Zhiv. A. N. Severtsova AN SSSR*, Moscow.

Pyastolova, O. A., 1967, Biological features of subarctic populations of the *oeconomus* vole, in: *Data of the Summary Session of the Laboratory of Population Ecology of Vertebrates for 1966*, Inst. Ékol. Rast. i Zhiv. UFAN, Sverdlovsk.

Pyastolova, O. A., Dobrinskii, L. N., and Ovchinnikova, N. A., 1966, On the question of the specificity of accumulation and expenditure of energy reserves in female and male animals in natural populations and under experimental conditions, *Experimental Study of the Intraspecific Variability of Vertebrates, Tr. Inst. Biol. UFAN SSSR*, No. 51.

• Radovanovic, M. M., 1959, Zum Problem der Speziation bei Inseleidechsen, *Zool. Jahrb. Abt. Syst. Oekol. Geogr. Tiere* **86**(4–5).

• Radovanovic, M. M., 1961, Resultats des recherches faites dans les iles adriatiques sous le jour de l'evolutionnisme, *Bull. Acad. Serbe Sci. Arts* **26**(8).

Raevskii, V. V., 1947, *Life of the Kondo-Sosva Sable*, Moscow.

Rakhmanin, G. E., 1959, The fur trade in the Yamal–Nenetsk region and measures for its rationalization, *Tr. Salikhard. Statsion.* **1**(1).

• Rao, S. R., and Venkatasubba, S., 1964, Somatic chromosomes of the Indian five striped squirrel *Funambulus pennanti* Wr., *Cytogenetics* **3**(5).

• Rasmussen, D. I., 1964, Blood group polymorphism and inbreeding in natural populations of the deer mouse *Peromyscus maniculatus*, *Evolution* **18**(2).

• Rausch, B. R., 1953, On the status of some Arctic mammals, *J. Arctic Inst. North America* **6**(2).

- Rauschert, K., 1963, Sexuelle Affinität zwischen Arten und Unterarten von Rötelmäusen (*Clethrionomys*), *Biol. Zbl.* **82**(6).

Raushenbakh, Yu. O., 1958, On the physiological nature of animals' resistance to the lowered oxygen conditions of high mountains, in: *Experiment on the Study of the Regulation of Physiological Functions,* Izd. Akad. Nauk SSSR, Leningrad.

Raushenbakh, Yu. O., 1959, On different types of adaptations to lowered oxygen conditions of high mountains in animals differing in their ecogenesis, in: *Theses of the Reports of the All-Union Conference on the Physiology and Biochemistry of Domestic Animals,* Moscow.

Raushenbakh, Yu. O., 1966, Genetic and physiological investigations of the resistance of animals to extreme factors of the environment, author's (lecture) doctoral dissertation, Novosibirsk.

- Reinig, W., 1937, *Die Holarktis. Ein Beitrag zur diluvialen und alluvialen Geschichte der Zircumpolaren Faunen und Florengebiete,* Jena.
- Remane, A., 1948, Die Typen der Mutationen, *Verh. Dtsch. Zool. Ges. Kiel.*
- Remane, A., 1949, Die Entstehung der Metamerie der Wirbelosen, *Verh. Dtsch. Zool. Ges. Mainz.*
- Remane, A., 1964, Das Problem Monophylie–Polyphylie mit besonderer Berücksichtigung der Phylogenie der Tetrapoden. Kalin. Discussionsbemerkung, *Zool. Anz.* **173**(1).
- Rensch, B., 1929, *Das Prinzip geographischer Rassenkreise und das Problem der Artbildung,* Berlin.
- Rensch, B., 1959, *Evolution above the Species Level,* London.
- Rensch, B., 1961, *The Laws of Evolution. Evolution after Darwin,* Chicago University Press, Chicago.

Resolution III of the Ecological Conference, 1954, Kiev.

- Reynafarie, B., and Morrison, 1962, Myoglobin levels in some tissues from wild Peruvian rodents native to high altitude, *J. Biol. Chem.* **2**(9).
- Rhoad, A. O., 1938, Some observations on the response of purebred *Bos taurus* and *Bos indicus* and their crossbred types to certain conditions of environment, *Amer. Soc. Animal Produc. Proc.*
- Riddle, Q., 1928, The establishment of races of pigeons characterized by large and small thyroids, *Anat. Rec.* **41**(1).
- Robertson, I. G., 1957, Changes in resistance to DDT in *Macrocentrus ancylivorus* Roh. (Hymenoptera, Braconidae), *Can. J. Zool.* **35.**
- Robertson, F. W., 1960, The ecological genetics of growth in *Drosophila, Genet. Res.,* No. 1.
- Röhrs, M., 1959, Neue Ergebnisse und Probleme der Allometrieforschung, *Z. Wiss. Zool.* **162.**
- Röhrs, M., 1961, Allometrie und Systematik, *Z. Säugetierk.* **26**(3).

Romer, A. M., 1939, *Vertebrate Paleontology,* Gos. Nauch.-Tekhn. Izd., Moscow and Leningrad.

- Romer, A. S., 1965, Possible polyphylety of the vertebrate classes, *Z. Jahrb. Abt. Syst. Oekol. Geogr. Tiere* **92**(1).

Rubtsov, I. A., 1948, *Biological Methods in the Control of Insect Pests,* Sel'khozgiz, Moscow.

- Ruddle, F. H., and Roderick, T., 1965, The genetic control of three kidney esterases in C57 BL/6J and RF/J mice, *Genetics* **51**(3).
- Ruibal, R., 1962, The ecology and genetics of a desert population of *Rana pipiens, Copeia,* No. 1.

- Russel, E. S., 1962, The diversity of animals. An evolutionary study, *Acta Bioteoret.* **13**, *Suppl.*, No. 9.

Ryzhkov, V. L., 1959, Differentiation of cytoplasm during speciation, *Zh. Obshch. Biol.* **20**(1).

- Salthe, S. K., and Kaplan, N. O., 1966, Immunology and rates of enzyme evolution in the amphibia in relation to the origins of certain taxa, *Evolution* **20**(4).
- Sammalisto, L., 1968, Variations in the selective advantage of hybrids in the Finnish population of *Motacilla flava*, *Ann. Zool. Fenn.* **5**(2).
- Sand, S. A., 1965, Position effects and the problem of coding a program for development, *Amer. Nat.* **99**(904).
- Sanders, O., and Cross, J. C., 1964, Relationships between certain North American toads as shown by cytological study, *Herpetologica* **19**(4).
- Sandness, G. C., 1955, Evolution and chromosomes in intergeneric pheasant hybrids, *Evolution*, No. 9.
- Sasaki, K., and Suzuki, S., 1962, Serobiological relationship between jungle fowl, *Proc. 12th World's Poultry Congr.*, Sydney.
- Sawada, S., 1963, Studies on the local races of the Japanese newt, *Triturus pyrrhogaster* Boie. II. Sexual isolation mechanisms, *J. Sci. Hiroshima Univ. Ser. B Div.*
- Schaeffer, B., 1965, The role of experimentation in the origin of higher levels of organization, *Syst. Zool.* **14**(4).
- Schaeffer, B., and Hecht, M. K., 1965, The origin of higher levels of organization. Introduction and historical resume, *Syst. Zool.* **14**(4).
- Schäperclaus, W., 1953, Die Züchtung der Karauschen mit höchster Leistungsfähigkeit, *Z. Fisch.* **2** N/F(1/2).
- Schlager, G., 1963, The ecological genetics of the mutant sooty in populations of *Tribolium castaneum*, *Evolution*, No. 17.

Schmalhausen, I. I., 1946, *Factors of Evolution*, Izd. Akad. Nauk SSSR, Moscow [also in English translation: Blakiston, Philadelphia, 1949].

Schmalhausen, I. I., 1964, *Regulation of Formation in Individual Development*, Izd. Akad. Nauk SSSR.

Schmalhausen, I. I., 1965, Evolution in the light of cybernetics, *Nauka*, No. 13.

Schmalhausen, I. I., 1966*a*, F. Engels on the teachings of C. Darwin, in: *Philosophical Problems in Contemporary Biology*, Nauka, Moscow and Leningrad.

Schmalhausen, I. I., 1966*b*, The problem of adaptation in the work of Darwin and of anti-Darwinists, in: *Philosophical Problems in Contemporary Biology*, Nauka, Moscow and Leningrad.

- Schmidt, S., Spielmann, W., and Weber, M., 1962, Serologische Untersuchungen zur Frage der verwandschaftlichen Beziehungen von *Pan paniscus* Schwarz 1929 zu anderen Hominoiden, *Z. Säugetierk.* **27**(1).
- Schmidtke, C., 1956, Polyploidie und Tierzucht, *Zuchtungskunde* **28**(4).
- Schnetter, M., 1950, Veränderungen der genetischen Konstitution in natürlichen Populationen der polymorphen Bauderschnecken, *Verh. Dtsch. Zool. Marburg.*
- Schock, N. W., 1951, *A Classified Bibliography of Gerontology and Geriatrics*, Stanford University Press, Stanford, California.
- Schweizer, H., 1941, Die Bastardform von *Vipera aspis aspis* × *Vipera ammodytes ammodytes*, *Wochenschr. Aquar. Terrar. Kunde* **38**(3).
- Scossiroli, R. E., 1962, Inincrocio ed omozigosi in popolazioni, *Bull. Zool.* (*1963*) **29**(2).
- Sebek, Z., 1955, O pouziti imunitnich reakci v systematike zoologii vesmir, *Ecol. Polska* **34**(6).

• Sealander, K., and Johnston, R. F., 1963, Geographic variation and evolution in North American house sparrow (*Passer domesticus*), *Proceedings of the XVI International Congress on Zoology*, Vol. 2, Washington, D.C.

• Serafinski, W., 1965, Pochodzenie srodkowoeuropejskich myszy domowych, *Przegl. Zool.* **9**(2).

• Serafinski, W., 1965, The subspecific differentiation of the Central European house mouse (*Mus musculus* L.) in the light of their ecology and morphology, *Ecol. Polska* **13**(7).

• Serafinski, W., 1967, Ecologiczna struktura gatunka u ssakow. I. Wewnatrazgatunkowe kategorie taxonomiczne o jednostki ecologiczne, *Ecol. Polska* **13**(1).

Severtsov, A. N., 1912, Etudes on the theory of evolution, Kiev, in: *Collected Works*, Vol. III, Izd. Akad. Nauk SSSR, Moscow.

Severtsov, A. N., 1939, *Morphological Rules of Evolution*, Izd. Akad. Nauk SSSR, Moscow.

Shaposhnikov, F. D., 1955, On the ecology and morphology of the Altai northern deer, *Zool. Zh.* **34**(1).

Shaposhnikov, G. Kh., 1965, Morphological divergence and convergence in an experiment with aphids (*Homoptera aphidinea*), *Entomol. Obozr.* **44**(1).

Shaposhnikov, G. Kh., 1966, Emergence and loss of reproductive isolation and criteria of species, *Entomol. Obozr.* **45**(1).

Shaposhnikov, L. V., 1958, Acclimatization and formation of groups in mammals, *Zool. Zh.* **37**(9).

Shaposhnikov, L. V., and Shaposhnikov, F. D., 1949, On the joint habitation of the desman, muskrat, and river beaver, *Zool. Zh.* **28**(4).

• Sharma, G. P., Parshad, R., and Kristhan, A., 1961, The chromosome number in pigeons and doves, *Indian J. Vet. Sci.* **31**(4).

Shchepot'ev, N. V., 1952, On the question of the economic significance of the agile lizard (*Lacerta agilis*) in field-protecting forest belts, *Zool. Zh.* **31**(4).

• Sheppard, P. M., 1958, *Natural Selection and Heredity*, London.

• Sheppard, P. M., 1959, *Natural Selection and Heredity*, Hutch. University, London.

• Sheppard, P. M., 1965, Mimicry and its ecological aspects, *Genetics Today*, Vol. 3, Pergamon Press, Elmsford, New York.

Shevareva, T. P., 1965, Populational features of the migration of birds, *Ornitologiya*, No. 7 (MGU).

Shkorbatov, G. L., 1964, On the theory of the acclimatization of aquatic animals, *Zool. Zh.* **43**(7).

Shtegman, B. K., 1948, On the functional significance of subspecific traits in oats, *Zool. Zh.* **27**(3).

Shvarts, S. S., 1954, On the question of the specificity of vertebrate species, *Zool. Zh.* **33**(3).

Shvarts, S. S., 1959, Some questions on the problem of species in terrestrial vertebrates, *Tr. Inst. Biol. UFAN SSSR*, No. 11.

Shvarts, S. S., 1960*a*, The role of the endocrine glands in mammalian adaptations to a change in the conditions of existence, *Tr. Mosk. O. Ispyt. Prir. Otd. Biol. Ural. Otdel.*, No. 2.

Shvarts, S. S., 1960*b*, Principles and methods of modern animal ecology, *Tr. Inst. Biol. UFAN SSSR*, No. 21.

Shvarts, S. S., 1961*a*, On the means of adaptation of terrestrial vertebrates to subarctic conditions, *Probl. Severa*, No. 4.

Shvarts, S. S., 1961*b*, Study of the correlation of rodents' morphological characteristics with their growth rate in connection with some questions of intraspecific systematics, in:

Questions of Intraspecific Variability of Mammals, Tr. Inst. Biol. UFAN SSSR, No. 29.

Shvarts, S. S., 1961*c*, Intraspecific variability of mammals and the methods of its study, in: *Theses of the Reports on the First All-Union Conference on Mammals*, Izd. MGU.

Shvarts, S. S., 1963*a*, Ecological and geographical bases of the process of acclimatization, in: *Acclimatization of Animals in the USSR*, Izd. Akad. Nauk Kaz. SSR.

Shvarts, S. S., 1963*b*, Intraspecific variability of mammals and methods of its study, *Zool. Zh.* **12**(3).

Shvarts, S. S., 1963*c*, *Means of Adaptations of Terrestrial Vertebrates to Conditions of Existence in the Subarctic*, Vol. 1, *Mammals, Tr. Inst. Biol. UFAN SSSR*, No. 33.

Shvarts, S. S., 1964, Experimental methods of investigation in theoretical taxonomy, in: *Problems of Intraspecific Changes in Terrestrial Vertebrates and Microevolution*, Moscow.

Shvarts, S. S., 1965, Age structure of animal populations and problems of microevolution (theoretical analysis of problems), *Zool. Zh.* **44**(10).

Shvarts, S. S., and Pokrovskii, V. A., 1966, Experiment on the convergence of the specific subspecies coloration of two sharply differentiated subspecies by means of selection in laboratory populations, *Zool. Zh.* **45**(1).

Shvarts, S. S., Pavlinin, V. N., and Syuzyumova, L. M., 1955, Theoretical bases for constructing forecasts of the numbers of mouselike rodents, in: *Rodents of the Urals, Tr. Inst. Biol. UFAN*, No. 8.

Shvarts, S. S., Kopein, K. I., and Pokrovskii, A. V., 1960, Comparative study of some biological features of *M. g. gregalis, M. g. major*, and their hybrids, *Zool. Zh.* **39**(6).

Shvarts, S. S., Bol'shakov, V. N., and Pyastolova, O. A., 1964*a*, New data on different ways of animals' adapting to change in habitat, *Zool. Zh.* **13**(4).

Shvarts, S. S., Pokrovskii, A. V., Ishchenko, V. G., Olenev, V. S., Ovchinnokova, N. A., and Pyastolova, O. A., 1964*b*, Biological pecularities of seasonal generations of rodents with special reference to the problem of senescence in mammals [in English], *Acta Theriol.* **8.**

Shvarts, S. S., Dobrinskaya, L. A., and Dobrinskii, L. N., 1966*a*, On the principal differences in the character of evolutionary reorganizations in fish and higher vertebrates, in: *Intraspecific Variability of Terrestrial Vertebrates and Microevolution, Tr. Vses. Soveshchaniya* UFAN SSSR, Sverdlovsk.

Shvarts, S. S., Dobrinskii, L. N., Bol'shakov, V. N., and Birlov, R. I., 1966*b*, Experiment on the elaboration of methods to determine the direction of natural selection in natural animal populations, *Tr. Inst. Biol. UFAN SSSR*, No. 51.

Shvarts, S. S., Pavlinin, V. N., Dobrinskii, L. N., Gashev, N. S., Boikov, V. N., and Boikova, F. I., 1966*c*, Morphophysiological features of tundra populations of hares in connection with the cyclicity of their life activity, in: *Fifty-Year Anniversary Collection in Honor of A. G. Bannikov.*

Shvarts, S. S., Pokrovskii, V. A., and Ovchinnokova, N. A., 1966*d*, Experimental investigation of the founder principle, in: *Experimental Study of Intraspecific Variability in Vertebrates, Tr. Inst. Biol. UFAN SSSR*, No. 51.

• Sibley, C. G., 1954, Hybridization in the red-eyed towhees of Mexico, *Evolution* **8**(3).

• Sibley, C. G., 1960, The electrophoretic patterns of avian egg-white proteins as taxonomic characters, *Ibis* **102**(2).

• Sidorowicz, S., 1960, Problems of the morphology and zoogeography of representatives of the genus *Lemmus* Link 1795 from the Palaearctic, *Acta Theriol.* **4**(5).

• Simmonds, F. I., 1947, Improvement of the sex-ratio of a parasite by selection, *Can. Entomol.* **95.**

• Simpson, G. G., 1956, Symposium on evolution held in Spain, *Evolution* **10**(3).

• Simpson, G. G., 1961, Historical zoogeography of Australian mammals, *Evolution* **15**(4).

• Simpson, G. G., 1963, Rates of evolution in animals, in: *Genetics, Paleontology and Evolution* (G. L. Jepsen, G. G. Simpson, and E. Mayr, eds.), Atheneum, New York.

Skalon, N. N., 1960, Distribution and life form of wild ungulates in the basins of the Olekma River, *Biol. Sb. Vostochno-Sibirsk. Otd. Georgrafichesk. Obshch. SSSR, Protivochymn. Inst. Sibiri Dal'nego Vostoka,* Irkutsk.

• Skaren, U., and Halkka, O., 1966, The karyotype of *Sorex caecutiens* Laxmann, *Hereditas* **54**(3).

Skvortsova, T. A., 1956, Size and structural features of the brain of some bird species in correlation with other organs and in connection with the life form and motor activity, Candidate's dissertation, Leningrad.

Slonim, A. D., 1950, Physiological bases for the study of ecological features of mammals, in: *Works of the Second Ecological Conference,* Vol. 2, Kiev.

Slonim, A. D., 1952, *Animal Heat and Its Regulation in the Mammalian Organism,* Izd. Akad. Nauk SSSR, Moscow and Leningrad.

Slonim, A. D., 1961, *Bases of the General Ecological Physiology of Mammals,* Izd. Akad. Nauk SSSR, Moscow and Leningrad.

Slonim, A. D., 1962, *The Particular Ecological Physiology of Mammals,* Izd. Akad. Nauk SSSR, Moscow and Leningrad.

Sludskii, A. A., 1948, *The Muskrat,* Alma-Ata.

Sludskii, A. A., Strautman, E. I., and Afanas'ev., Yu. G., 1962, Fur resources of Kazakhstan, *Collected Data on the Fauna and Ecology of Terrestrial Vertebrates of Kazakhstan* (*Mammals*), *Tr. Inst. Zool. Akad. Nauk Kaz. SSR* **17,** Izd. Akad. Nauk Kaz. SSR, Alma-Ata.

Snegirevskaya, E. M., 1961, Observations on the *oeconomus* voles on the islands of the Central Volga, in: *Morphology and Ecology of Vertebrates, Tr. Zool. Inst. Akad. Nauk SSSR* **29.**

Sokolov, I. I., 1965, On some principles and methods of systematics, *Tr. Zool. Inst. Akad. Nauk SSSR* **35.**

Sokolovskaya, N. I., 1936, Precipitation reaction in hybridization. Investigation of the precipitation reaction in some Lamellirostres (Lamellirostris), *Izv. Akad. Nauk SSSR Otd. Mat. Estestv. Nauk,* No. 2–3.

Sorokin, M. G., 1953, Acclimatization of the raccoon dog in Kalinin Province, *Priroda,* No. 6.

Sorokin, M. G., 1956, Biological and morphological changes in the raccoon dog acclimatized to Kalinin Province, *Uch. Zap. Kalinin. Gos. Pedagog. Inst.,* No. 20.

Sosin, V. F., 1967, On the dynamics of the numbers of muskrats in isolated settlements within a region, in: *Materials of the Summary Session of the Laboratory of the Population Ecology of Vertebrates for 1966,* Inst. Ékol. Rast. i Zhiv. UFAN SSSR, Sverdlovsk.

• Soule, M., 1966, Trends in the insular radiation of a lizard, *Amer. Nat.* **100**(910).

• Stallcup, W. B., 1954, Myology and serology of the avian family Fringillidae. A taxonomic study, *Univ. Kans. Publ. Nat. Hist. 8/29; B. A.* **30**(3).

Starkov, I. D., 1947, *Biology and Breeding of Sables and Martens,* V/o Mezhdunarodnaya Kniga.

Starkov, I. D., 1952, The influence of age and multiple births of sables on the fecundity of their progeny, *Zh. Obshch. Biol.* **12**(6).

- Stebbins, G. L., 1965, The experimental approach to problems of evolution, *Folia Biol. USSR* **11**(1).
- Stehr, G., 1964, The determination of sex and polymorphism in microevolution, *Can. Entomologist* **96**(1–2).
- Stein, G. H. W., 1956, Sippenbildung bei der Feldmaus, *Microtus arvalis* L., *Z. Säugetierk.* **21.**
- Stein, G. H. W., 1961, Beziehungen zwischen Bestandsdichte und Vermehrung bei der Waldspitzmaus, *Sorex araneus,* und weiteren Rotzahnspitzmäusen, *Z. Säugetierk.* **26**(1).

Stepanov, D. L., 1959, Political conceptions of species in paleontology, *Paleont. Zh.* **1**(3).

- Stern, C., 1963, Gene and character, in: *Genetics, Paleontology and Evolution* (G. L. Jepsen, G. G. Simpson, and E. Mayr, eds.), Atheneum, New York.
- Stormont, C., 1965, Mammalian immunogenetics, in: *Genetics Today,* Vol. 3, Pergamon Press, Elmsford, New York.

Strautman, F. I., Sukhomlinov, B. F., Kushniruk, I. F., Kushniruk, V. A., and Chugunov, N. D., 1963, Electrophoretic analysis of the blood serum proteins in the crow family, in: *Theses of the Reports of the Fifth Baltic Ornithological Conference,* Tartu.

Strautman, F. I., Sukhomlinov, B. F., and Kushniruk, V. A., 1964, The use of biochemical methods for elucidation of the formative role of the environment, in: *Questions of Intraspecific Variability of Terrestrial Vertebrates and Microevolution, Theses of the Reports of the Conference on Intraspecific Variability and Microevolution,* Sverdlovsk.

Stroganov, S. U., and Yudin, V. S., 1956, On the systematics of some rodent species of Western Siberia, *Tr. Tomsk. Gos. Univ.,* No. 12.

- Suchetet, A., 1890, Les oiseaux hybrides recontés à l'état sauvage, Paris.

Sukhomlinov, B. F., Viktorenko, V. G., Kushniruk, V. A., Chugunov, N. D., and Kushniruk, I. F., 1962, Electrophoretic characteristics of the hemoglobin of different bird species, in: *Materialy III Vses. Ornitolog. Konf.,* Vol. 2, L'vov.

Sukhomlinov, B. F., Kushniruk, V. A., and Chugunov, M. D., 1966, Electrophoretic analysis of the hemoglobin of different bird species, *Dokl. Akad. Nauk SSSR,* No. 12.

- Sumner, F. B., 1915, Genetic studies of several geographic races of California deer-mice, *Amer. Nat.,* No. 49.
- Sumner, F. B., 1932, Genetic, distributional and evolutionary studies of the subspecies of deer-mice (*Peromyscus*), *Biblogr. Genet.* **9.**

Svetozarov, E., and Shtraukh, G., 1936, Factors determining the sexual seasonal dimorphism of plumage in ducks, *Izv. Akad. Nauk SSSR Ser Biol.,* No. 3.

- Svihla, A., 1931, Change in a captive red squirrel, *Amer. Nat.* **65**(696).
- Symposium, Institute of Biology (University College, London), Biology and ageing, *Nature London* **178**(4543).

Syuzyumova, L. M., 1966, On the study of the antigenic characteristics of serum proteins of two subspecies of the *oeconomus* vole by the method of precipitation reaction in agar, in: *Experimental Study of Intraspecific Variability in Vertebrates, Tr. Biol. UFAN,* No. 51.

Syuzyumova, L. M., 1967, On the application of the methods of homotransplantation of skin and the hemoagglutination reaction for determination of intraspecific differentiation in mammals, *Tr. Mosk. O. Ispyt. Prir. Otd. Biol. Ural. Otdel.* **25.**

- Taber, R., and Dasmann, R. F., 1954, A sex difference in mortality in young Columbian black-tailed deer, *J. Wildl. Manag.* **18**(3).

Takhtadzhyzn, A. L., 1955, Some questions on the theory of species in the systematics of modern and fossil plants, *Bot. Zh.* **40**(6).

Taliev, D. N., 1935, Toward apprehending the hemagglutination reaction in fish, *Tr. Baik. Limnol. Stn.* **6,** Izd. Akad Nauk SSSR, Moscow.

• Tamsitt, J. R., 1961, Morphological comparison of P and F_1 generations of three species of the *Peromyscus truei* species group of mice, *Tex. J. Sci.* **13**(2).

• Tappen, N. C., 1960, Promising developments in primatology, *S. Afr. J. Sci.* **56**(3).

• Taylor, H. L., and Medica, P. A., 1966, Natural hybridisation of the bisexual tejid lizard *Chemidophorus inornatus* and the unisexual *Cnemidophorus perplexus* in Southern New Mexico, *Univ. Colo. Stud. Ser. Biol.,* No. 22.

• Tenczar, P., and Bader, R. S., 1966, Maternal effect in dental traits of the house mouse, *Science* **152**(3727).

Terent'ev, P. V., 1946, Experiment on the application of mathematical statistics to zoogeography, *Vestn. LGU,* No. 2.

Terent'ev, P. V., 1951, The influence of ambient temperature on the dimensions of snakes and tailless amphibians, *Byull. Mosk. O. Ispyt. Prir. Otd. Biol.* **56**(2).

Terent'ev, P. V., 1957, On the applicability of the concept "subspecies" in the study of intraspecific variability, *Vestn. LGU,* No. 221.

Terent'ev, P. V., 1961, Microclines as a form of adaptation, *Tr. Leningr. Ova. Estestvoispyt.* **72**(1).

Terent'ev P. V., 1962, The nature of geographical variability in the green frog, *Tr. Petergof. Biol. Inst.,* No. 19.

Terent'ev, P. V., and Chernov, S. A., 1949, *Definer of Reptiles and Amphibians,* Svetskaya Nauka, Moscow.

• Test, F. H., 1954, Seasonal differences in populations of the red-backed salamander in Southern Michigan, *Acad. Sci. Arts Lett.* **40**(2).

• Thenius, E., and Hofer, H., 1960, *Stammesgeschichte der Säugetiere,* Springer-Verlag, Berlin.

• Thornton, W. A., 1955, Interspecific hybridization in *Bufo woodhousi* and *Bufo valliceps, Evolution* **9**(4).

Timofeev-Resovskii, N. V. [Timofeeff-Ressovsky, N. W.], 1940, Zur Analyse des Polymorphismus bei *Adalia bipunctata* L., *Biol. Zentralbl.* **60**(3–4).

Timofeev-Resovskii, N. V., 1964, On polymorphism, in: *Questions of Intraspecific Variability of Terrestrial Vertebrates and Microevolution, Theses of the Reports of the Conference on Intraspecific Variability and Microevolution,* Sverdlovsk.

Timofeev-Resovskii, N. V., and Svirezhev, Yu. M., 1966, On adaptational polymorphism in populations of *Adalia bipunctata* L., *Probl. Kibern.,* No. 16, Nauka, Moscow.

• Tinbergen, N., 1965, Behaviour and natural selection, *Ideas Mod. Biol. N.Y.*

• Tomich, P. Q., and Kami, H. T., 1966, Coat colour inheritance of the roof rat in Hawaii, *J. Mammal.* **47**(3).

Tomilin, A. G., 1938, On the biology of cetaceans, *Priroda,* No. 7–8.

• Tomlinson, J. 1966, The advantage of hermaphroditism and parthenogenesis, *J. Theor. Biol.* **11**(1).

Toporkova, L. Ya., and Shvarts, S. S., 1960, Amphibians beyond the polar circle, *Priroda,* No. 10.

Toporkova, L. Ya., and Zubareva, É. L., 1965, Data on the ecology of the grass frog in the Polar Urals, in: *Ecology of Vertebrates of the Extreme North, Tr. Inst. Biol. UFAN SSSR,* No. 38.

• Tower, W. L., 1906, An investigation of evolution in Chrysomelid beetles of the genus *Leptinostera, Carnegie Inst. Publ.,* No. 48.

Tsalkin, V. I., 1945, On the vertical distribution of wild sheep, *Byull. Mosk. O. Ispyt. Prir.* **50**(1–3).

Tsalkin, V. I., 1950, The Siberian mountain billy goat, *Tr. Mosk. O. Ispyt. Prir.*, Moscow.

Tserevitinov, B. F., 1951, Variability of muskrat fur in connection with its acclimatization in the USSR, *Vopr. Tovarovedeniya Pushno-Mekhovogo Syr'ya*, Zagotizdat, Moscow.

Tserevitinov, B. F., 1953, Variability of the fur coat of the Ussuriisk raccoon during acclimatization, *Sb. Nauchn. Rab. Mosk. Inst. Nar. Khoz.*, No. 3.

Tsetsevinskii, L. M., 1940, Data on the ecology of the polar fox of northern Yamal, *Zool. Zh.* **19**(1).

• Twitty, V. C., 1964, Fertility of *Taricha* species hybrids and viability of their offspring, *Proc. Nat. Acad. Sci. U.S.A.* **5**(2).

• Uda, H., 1957, Sex ratio and the "sexual age." Idengaku dsasu, *J. Genet. Jpn.* **32**(2).

• Udagawa, T., 1955, Karyogram studies in birds. VI. The chromosomes of five species of the Turdidaé. Nihon dabuzugazu, *Annot. Zool. Jpn.* **28**(4).

• Uhlenhuth, E., 1919, Relation between metamorphosis and other developmental phenomena in amphibia, *J. Gen. Physiol.* **1**, 525.

• Ullrich, F. H., 1965, Unterschiede in DNS-Gehalt der Genome von *Bufo bufo* und *Bufo viridis*, *Z. Naturforsch.* **20b**(7).

Ushakov, B. P., 1955, Heat resistance of the somatic musculature of amphibians in connection with the conditions of existence for the species, *Zool. Zh.* **34**(3).

Ushakov, B. P., 1958, On the conservatism of protoplasmic proteins in species of poikilothermic animals, *Zool. Zh.* **37**(5).

Ushakov, B. P., 1964, *Report on the Competition for the Scientific Degree of Doctor of Biological Sciences*, Zool. Inst. Akad. Nauk SSSR, Leningrad.

Vakhrushev, I. I., and Volkov, M. G., 1945, *Eskimo Hunting Dogs*, Zagotizdat, Moscow.

• Valentine, D., and Löve, A., 1958, Taxonomy and biosystematic categories, *Brittonia* **10**(153).

• Van Harriș, T., 1954, Experimental evidence of reproductive isolation between two subspecies of *Peromyscus maniculatus*, *Contrib. Lab. Vertebr. Biol. Univ. Mich.*, No. 70.

• Van Valen, L., 1965*a*, The study of morphological integration, *Evolution* **19**(3).

• Van Valen, L., 1965*b*, Selection in natural populations. III. Measurement and estimation, *Evolution* **19**(4).

• Van Valen, L., 1965*c*, Morphological variation and width of ecological niche, *Amer. Nat.* **99**(908).

• Van Valen, L., 1965–1966, Selection in natural populations. IV. British house mice (*Mus musculus*), *Genetica* **36**(2).

Vasil'ev, V. N., 1963, Populations and their role in the life of the species, *Bot. Zh.* **48**(3).

Vasil'ev, V. N., 1964, Rates of evolution, in: *Second Moscow Conference on the Phylogeny of Plants*, Moscow.

Vasil'ev, V. N., 1965, On the duration of the existence of species, in: *Problems of Modern Botany*, Vol. 1, Nauka, Moscow and Leningrad.

Vinberg, G. G., 1950, The intensity of metabolism and the dimensions of crustaceans, *Zh. Obshch. Biol.* **11**(5).

Vinberg, G. G., 1966, Growth rate and intensity of metabolism in animals, *Usp. Sovrem. Biol.* **61**(2).

Vinogradov, B. S., and Argiropulo, A. I., 1941, *Definer of Rodents*, Izd. Akad. Nauk SSSR, Moscow.

• Vojtiskova, M., 1959–1960, Zur Frage des Mechanismus der Befruchtungsinkompatibilitaät bei der entfernten Kreuzung des Gefluügels, Workshop on Questions of Evolution, Jena.

• Von Wahlert, G., 1965, The role of ecological factors in the origin of higher levels of organization, *Syst. Zool.* **14**(4).

Vorontsov, N. N., 1958, Significance of karyotype studies for the systematics of mammals, *Byull. Mosk. O. Ispyt. Prir. Otd. Biol.* **63**(2).

Vorontsov, N. N., 1960, Species of Palearctic hamsters (Cricetinae: Rodentia) *in statu nascendi, Dokl. Akad. Nauk SSSR* **132**(6).

Vorontsov, N. N., 1967, Evolution of lower hamster forms, author's doctoral dissertation, Moscow.

Vorontsov, N. N., Radzhabli, S. I., and Lyapunova, K. L., 1967, Karyological differentiation of allopatric forms of the hamsters of the superspecies *Phodopus sungarus* and heteromorphism of the sex chromosomes in the males, *Dokl. Akad. Nauk SSSR* **172** (1–3).

Vyazov, O. E., 1962, *Immunology of Embryogenesis*, Medgiz, Moscow.

Vyazov, O. E., Konyukhov, B. V., Averkina, R. F., and Titova, I. I., 1965, Application of immunological methods to the study of questions of tissue evolution, *Izv. Akad. Nauk SSSR Ser. Biol.*, No. 1.

• Waddington, C. H., 1956, Genetic assimilation of the bithorax phenotype, *Evolution* **10**(1).

• Waddington, C. H., 1958, Inheritance of acquired characters, *Proc. Linn. Soc. London* **196**(1–2).

• Waddington, C. H., 1959, Evolutionary systems, animal and human, *Proc. R. Inst.* **37**(5).

• Waddington, C. H., 1960, Experiments on canalizing selection, *Genet. Res.* **1**(1).

Waddington, K., 1964, *Morphogenesis and Genetics*, Mir, Moscow.

• Wallace, B., 1955, Inter-population hybrids in *D. melanogaster, Evolution* **9**(2).

• Warburton, T., 1955, Feedback in development and its evolutionary significance, *Amer. Nat.* **89**(846).

• Watson, D. M. S., 1963, The evidence afforded by fossil vertebrates on the nature of evolution, in: *Genetics, Paleontology and Evolution* (G. L. Jepsen, G. G. Simpson, and E. Mayr, eds.), Atheneum, New York.

• Wecker, S., 1964, Habitat selection, *Sci. Amer.* **211**(4).

• Weinbach, E. C., and Garbus, J., 1956, Age and oxidative phosphorylation in rat liver and brain, *Nature London* **178**(4544).

• Weiss, I., 1960, Umwandlung von Dysnothoi der Bastardkombination *Bufo calamita* × *Bufo viridis* in Eunothoi durch experimentelle Verdoppelung des mütterliche Genoms, *Roux. Arch. Ent. Organism.* **152**(4).

• Westphal, F., 1962, Der Übergang vom Reptil zum Säugetier, *Nature London* **70**(5–6).

• Whirter, M., 1956, Control of sex ratio in mammals, *Nature London* **178**(4538).

• White, M. I., 1958, Specification in animals, *Aust. J. Sci.* 22.

• White, M. J. D., Carson, H. L., and Cheney, I., 1964, Chromosomal race in the Australian grasshoper *Maraba viatica* in a zone of geographic overlap, *Evolution* **18**(3).

Wiener, N., 1964, *I Am a Mathematician* [Russian translation], Nauka, Moscow [original English: MIT Press, Cambridge, Massachusetts, 1956].

• Withman, C., 1919, Inheritance, fertility and the dominance of sex and colour in hybrids of wild species of pigeons, Washington.

• Wilkes, A., 1947, The effects of selective breeding on the laboratory propagation of insect parasites, *Proc. R. Soc. London* **134**.

• Williamson, M. H., 1957, An elementary theory of interspecific competition, *Nature London* **80**(4583).

• Willmer, E. N., 1956, Factors which influence the acquisition of flagella by the amoeba *Naegleria gruberi, J. Exp. Biol.* **33**(4).

• Wilson, O. E., and Brown, W. L., 1953, The subspecies concept, *Syst. Zool.* **2**.

• Wince, M. A., and Warren, R. P., 1963, Individual differences in taste discrimination in the great titmouse (*Parus major*), *Anim. Behav.* **11**(4).

• Winterbert, P., 1962, *Le Vivant Createur de Son Evolution,* Masson et Cie., Paris.

• Wood, J. E., 1959, Age structure and productivity of a gray fox population, *J. Mammal.* **39**(1).

• Wright, P., 1953, Intergradation between *Martes americana* and *Martes caurina* in western Montana, *J. Mammal.* **34**(1).

• Wright, S., 1945, Tempo and mode in evolution: A critical review, *Ecology,* No. 26.

• Wright, S., 1948, On the roles of directed and random changes in gene frequency in the genetics of populations, *Evolution* **2**(2).

• Wright, S., 1955, Classification of the factors of evolution, *Cold Spring Harbor Symp. Quant. Biol.* **20**.

• Wright, S., 1959, Physiological genetics, ecology of populations and natural selection, *Perspect. Biol. Med.* **3**(1).

• Wright, S., 1965, Genetics and twentieth century Darwinism. A review and discussion, *Amer. J. Hum. Genet.* **12**(2).

• Wright, S., and Dobzhansky, T., 1946, Genetics of natural populations, XII, *Genetics* **31**.

• Wynne-Edwards, V. C., 1962, Animal dispersion in relation to social behaviour, Oliver and Boyd, Edinburgh and London.

Yablokov-Khnzaryan, S. M., 1963, On the rhythm of evolution, *Zool. Zh.* **42**(10).

Yazan, Yu. P., 1964, On some morphological and ecological displacements in beavers in connection with their reacclimatization in the Pechoro-Ilych Preserves, *Tr. Pechoro-Ilychskogo Gos. Zapov.,* No. 11.

• Yosida, H. T., 1955, Origin of V-shaped chromosomes occurring in tumor cells of some ascites sarcomas in the rat, *Proc. Jpn. Acad.* **31**(237).

Zablotskii, M. A., and Flerov, K. K., 1963, The past of the bisons, *Priroda,* No. 7.

• Zarron, M. X., and Denison, M. E., 1956, Sexual difference in the survival time of rats exposed to a low ambient temperature, *Amer. Physiol. Am. Br. Abstr.* **25**(2).

• Zdansky, R., 1962, Die Evolution im Lichte experimenteller Mutationsforschung und biologischer Statistik, *Strahlentherapie* **117**(1).

• Zeller, C., 1956, Über den Ribonukleinsäurestoffwechsel des Bastardmerogons *Triton palmatus* × *Triturus cristatus, Roux. Arch.* **148** (3).

• Zeuner, F. E., 1955, Time rates of organic evolution, *Bull. Nat. Inst. Sci. India,* No. 7.

Zhukov, V. V., 1966, Study of the antigenic properties of erythrocytes in two subspecies of the *oeconomus* voles, in: *Experimental Study of the Intraspecific Variability of Vertebrates, Tr. Inst. Biol. UFAN SSSR,* No. 51.

Zhukov, V. V., 1967*a*, On the study of intergeneric immunological interrelationships in the subfamily Microtinae, in: *Materials of the Summary Session of the Laboratory of the Population Ecology in Vertebrates for 1966,* Inst. Ékol. Rast. i Zhiv. UFAN SSSR, Sverdlovsk.

Zhukov, V. V., 1967*b*, Immunological interrelationships of some forms of the voles of the genus *Microtus,* in: *Materials of the Summary Session of the Laboratory of the Population Ecology of Vertebrates for 1966,* Inst. Ékol. Rast. i Zhiv. UFAN SSSR, Sverdlovsk.

Zil'ber, L. A., 1962, On the defense of an organism against alien genetic information, *Vestn. Akad. Med. Nauk SSSR,* No. 4.

• Zimmermann, K., 1965, Art-Hybriden bei Rotelmäusen, *Z. Säugetierk.* **30**(5).

• Zimmermann, W., 1962, Kritische Beiträge zu einigen biologischen Problemen. IV. Die Ursachen der Evolution, *Acta Biotheor.* **14**(3).

• Zimmermann, W., 1963, Die Chinchillazucht muss eine Pelztierzucht werden. Erfolg in der Chinchillazucht, *Chinchilla-Post* **8**(6).

Zverev, M. D., 1930, Experimental study of the biology of Siberian predatory birds, in: *Material on the Ornithology of the Siberian Territory, Tr. Ova. Izucheniya Sibiri i Ee Proizvoditel'nykh Sil.*, Novosibirsk.

Appendix I

The Species Problem and New Methods of Systematics

S. S. Shvarts

Published as the Preface to *Experimental Investigations of the Species Problem*, Works of the Institute of the Ecology of Plants and Animals, No. 86, Akad. Nauk SSSR, Ural'skii Nauchnyi Tsentr, Sverdlovsk (1973), pp. 3–18.

The species is a basic biological concept, a real unit of the structure of nature, a real unit of taxonomic systematics with objectivity adequate to modern natural science and reflecting the phylogenetic interrelationships of animals and plants. The theory of species could have arisen only on the basis of a materialistic view of nature (Charles Darwin). It was developed primarily on the basis of morphology, embryology, and zoogeography—the ruling biological sciences of the second half of the nineteenth and the first half of the twentieth century. The outcome of these investigations is the well-known biological criterion of species, based on the notion of its genetic isolation (in contrast to all intraspecific forms). It is symptomatic that this conclusion was formulated long before systematics had armed itself with the genetic experiment.

Classic systematics was able to understand correctly the true causes of the zoogeographical and morphological differences between species and intraspecific forms. Many unsolved questions remained, however. Disputes concerning the species problem did not die out. They continue even now. It is natural, therefore, that the possibility of utilizing new methods, new approaches to the theory and practice of taxonomy and, in particular, to the species problem, was met with enthusiasm by systematists. As a matter of fact, the "new" approaches to the solution of taxonomic problems are not new. The significance for systematics of investigating the biochemical specificity of organisms was noted by Blagoveshchensky as early as the twenties (Blagoveschensky, 1929), and Komarov (1944) wrote: "Biochemical differences lie at the basis of all traits, even purely morphological ones, on the basis of which we classify plants and establish species." In order to realize these profound positions, however, the introduction into investigative practice of new methodologies accessible to systematics and their mastery was necessary. The most important of these methods proved to be chromatography, the electrophoresis of tissue proteins, and a group of immunological tests. At approximately the same time, karyological investigations became widespread, as well as methods that allowed observed differences between compared forms to be described with a great deal of objectivity (colorimetry of coloration, allometric relationships, and so forth) or the scale of observed differences to be evaluated with a high degree of precision, based on modern mathematical techniques (discriminant analysis, numerical taxonomy).

It is difficult, of course, to overestimate the significance of the "new methods," since they allow us to use those animal traits that were inaccessible in the past for the construction of taxonomic systems. This led to the conviction on the part of some biologists, unfortunately quite a few, that the "new methods" could work on the principle of a "magic wand," infallibly deciding the central biological question: "Is it or is it not a species?"

In order to evaluate the true significance of the new methods for solving principal questions of evolutionary studies and systematics, it seemed advisable to us to include them in complex investigations, utilizing an experimental approach to the species problem. Over a period of many years, we have conducted a comparative experimental and ecological study of a series of closely related forms: *Microtus oeconomus oeconomus* Pall. and *M. o. chahlovi* Scalon; *M. gregalis gregalis* Pall. and *M. g. major* Ogn; *M. juldaschi juldaschi* Sev. and *M. j. carruthersi* Thomas; *M. middendorffi middendorffi* Poljakov and *M. m. hyperboreus* Vinogr.; *M. arvalis transuralensis* Serebre. and *M. transcaspicus* Satunin.

Obviously, intraspecific forms of different ranks and close forms were subjected to pairwise comparisons. *Lagurus lagurus, Lemmus obensis, Clethrionomys frater, C. rufocanus, Alticola roylei, A. strelzovi,* and mixed populations of *M. arvalis* and *M. oeconomus* were also investigated. The results of these investigations appeared in a large number of publications (Kopein, 1958; Shvarts *et al.,* 1960; Ovchinnikova, 1966, 1968; Mikhalev, 1970; Syuzyumova, 1969; Zhukov, 1967; Bol'shakov *et al.,* 1969; Pokrovskii *et al.,* 1970; Ishchenko, 1966; and others) and in a number of articles in contemporary collections.

The overall results of this comparative study of the forms enumerated are given in Table 1, which clearly shows that the resolution of the question "Is it a species or not?" is possible only on the basis of a complex investigation of the animals. Let us turn our attention to the following facts. On the basis of karyological differences and some distributional features alone, *juldaschi* and *carruthersi* could be assigned to independent species. But the study of other characteristics of these forms (including hybridization experiments) indicates that this assignment is absolutely mistaken. The notion of the species independence of *middendorffi* and *hyperboreus* also proved to be mistaken.

Morphological and morphophysiological differences between *M. g. gregalis* and *M. g. major* are fully commensurate with distinctions between any close (and even not very close) species of *Microtus*, but a complex of other indices attests to their membership in the same species. During a comparison of *M. o. oeconomus* and *M. o. chahlovi*, on one hand, and of *M. g. gregalis* and *M. g. major*, on the other, it was found that the degree of their morphophysiological differentiation did not correspond to the overall index of electrophoretic distance between the subspecies compared. A comparison of the forms studied on the basis of immunological criteria leads to analogous conclusions.

Among the new methods, immunological tests have special significance. An analysis of the enormous amount of factual material collected by investigators of different schools leads to the concept of an immunological unit of the species. L. M. Syuzyumova's research, which is presented in this collection, also indicates this concept. Against a background of clear interpopulational differences, species-specificity in heterotransplants is confirmed by all experiments. Every subspecies, every form, every population of a species distinguishes "its" species from "another's." The unit of the species at the tissue level, the species-specificity of tissues, unites all levels of intraspecific differentiation. This is confirmed both by the nature of the manifestation of tissue incompatibility in heterotransplants (without regard to dependence on intraspecific differences between donors) and by the species-specificity of humoral changes.

The observations of L. M. Syuzyumova and her student, V. V. Zhukov, however, show very clearly that at both the intraspecific and the supraspecific level, immunological distance, reflecting the biochemical specificity of organisms, does not correspond to the degree of morphophysiological differences of the forms compared.

That it does not means that the biochemical characteristics of animals are determined not only by the adaptations of different forms to different environmental conditions, but also by genetic differences that are not directly connected with adaptive characteristics of populations, subspecies, or species. It would be easiest to attribute these differences to the action of neutral mutations, which even under conditions of not very strict isolation should inevitably accumulate in different quantities and with a different quality in the compared forms.* Diverse investigations show, however, that there are no neutral mutations, in the strict sense of the word.

Even in those cases in which the mutation leads to a change in a chemically inactive part of the protein molecule and does not have a signifi-

* This viewpoint has adherents. It is symptomatic, however, that it leads the majority of investigators to anti-Darwinist and essentially antievolutionary conclusions, reflected in the term *non-Darwinian evolution*. A critique of these ideas was given recently by Richmond (1970).

TABLE 1. Comparative Characteristics of Closely

Forms Compared	Morphological differences	Ecological differences	Morphophysiological characteristics
Mixed populations of *Microtus arvalis* and *M. oeconomus*	None	None	None
Microtus oeconomus oeconomus and *M. o. chahlovi*	Insignificant differences in body dimensions and proportions. Striking differences in tail length are determined by different morphogenetic reactions to temperature change. Differences in color are slight, but statistically significant (measured by colorimetry). They are polygenic.	*M. o. oeconomus* is a forest steppe form. *M. o. chahlovi* is subarctic. Even so, there are no differences in seasonal cyclicity of life activity. Basic circadian rhythms coincide, but periods of generative rest are greater in *M. o. chahlovi,* and periods of increased fertility occur earlier. In nature, *M. o. chahlovi* is bigger, is distinguished by greater fertility and by a longer period of maximal growth rate for the young.	The general direction of morphophysiological differences indicates a more economical type of metabolism in *M. o. chahlovi.* These features are significantly less expressed than in typical subarctic rodents (lemmings, Middendorf voles). In experimental conditions, *M. o. oeconomus* is distinguished by larger testes and smaller adrenals.
Microtus gregalis gregalis and *M. g. major*	Morphophysiological differences are commensurate with differences between species (differences in dimensions and proportions of the body and skull and in coloring). There is a hiatus in a number of traits. Sharp differences in the nature of the allometric growth of parts of the skull indicate a very substantial morphological divergence.	*M. g. gregalis* is a forest steppe form. *M. g. major* is subarctic. There are differences in the biology of reproduction, growth rate and development of young, seasonal dynamics of growth rate, reaction to temperature change, and others.	Morphophysiological characteristics of *M. g. major* indicate a more economical type of metabolism.
Microtus middendorffi middendorffi and *M. m. hyperboreus.* (Considered independent species by all authors.)	Morphological differences are insignificant. Among *M. m. hyperboreus,* a polymorphism in color is noted; melanists are often encountered.	Subarctic voles. *M. m. hyperboreus* gravitates to mountainous regions. *M. m. middendorffi* is not found in the mountains.	There are no essential differences.
Microtus juldaschi juldaschi and *M. j. carruthersi.* (Many authors regard them as independent species.)	Differences in color are slight, but statistically significant (measured by colorimetry). Often-noted differences in the	Essential ecological differences are not noted in nature, nor are they observed in confined animals.	There are no essential differences.

Related Forms of Rodents According to a Complex of Indices

Overall electrophoretic characteristic of serum blood proteins	Immunological distance	Karyological differences	Results of hybridization experiments
Differences are insignificant.	Significant differences in mean indices. Immunological differences among individuals commensurate with differences between subspecies.	None	Complete hybrid fertility.
Differences slightly exceed differences between mixed populations of the same species.	Does not exceed the usual interdeme differences.	None	Complete hybrid fertility.
Overall indices of electrophoretic distance are less than in close species, but greater than in other intraspecific forms investigated.	Not studied	None	Complete fertility of initial forms and the hybrids.
Electrophoretic differences at the level of usual subspecies.	Not studied	Karyotypes are identical (NF of males is 57–59, of females 58–60). Polymorphism of the largest autosomes is characteristic for both forms.	Complete fertility of the initial forms and the hybrids.
Differences at the level of usual subspecies.	Not studied	Intrademe chromosomal polymorphism is characteristic for both forms. Karytopic differences in a number of	Complete fertility of the initial forms and the hybrids.

TABLE 1. (Continued)

Forms Compared	Morphological differences	Ecological differences	Morphophysiological characteristics
M. j. j. and *M. j. c.* (cont.)	proportions of the body and skull are the consequence of differences in body size. Allometric curves of parts of the body and skull and of internal organs differ at the level of "average" subspecies; changes of allometric indices do not occur during hybridization. Essential differences in the distribution curves of a number of traits are observed.		
Microtus arvalis and *M. transcaspicus*	Essential differences in the dimensions and proportions of the body and in color.	*M. arvalis* is characterized by a wide distribution. It achieves maximal numbers in forest steppe regions. *M. transcaspicus* is met primarily in riparian valleys of desert and semidesert zones. The fertility of *caspicus* is lower than that of *M. arvalis*. The growth rate of the young is significantly higher, but growth is completed at an earlier age.	There are numerous morphophysiological differences. One of the most essential is the exceedingly large dimensions of *M. transcapicus'* adrenals.

cant influence on its functioning, it cannot be acknowledged as neutral, since the amino acid that is being substituted should be present in the cell in the same quantity as the replaced amino acid, and its synthesis and transmission should demand identical expenditures of energy (Richmond, 1970). In the opposite case, a disturbance of the mutual conformation of proteins and a disturbance of the organism's energy balance are inevitable.* The notion of "neutral mutations" as an independent factor of the evolutionary

* There are also more specific indications of the extreme doubtfulness of the existence of neutral mutations. One of the most interesting is the following: Chance inversions could transfer a neutral allele to a different genic environment (gene arrangements). Precise observations showed that, in reality, this does not occur in the course of a million years (Prakash and Lewontin, 1968). Hence, the conclusion follows that natural selection does not remain indifferent to "neutral" mutations, but maintains the gene arrangement at an optimal level.

Overall electrophoretic characteristic of serum blood proteins	Immunological distance	Karyological differences	Results of hybridization experiments
		indices (morphology of the X chromosome and three pairs of autosomes) are so essential that they could serve as a basis for acknowledging species independence of the compared forms.	
Differences at the species level.	Not studied	$2n = 54$ for both forms.	Sexual attraction of the initial forms is disturbed. Hybrids have lower viability. Hybrids are sterile.

reorganizations of organisms, therefore, does not withstand criticism. This viewpoint seems well founded to us. On the basis of a complex morphophysiological study of a large number of forms in different environmental conditions, we arrived at the conclusion that the degree of adaptability of any organism is determined not only by its ecological and physiological perfection, but also by the energy cost of those physiological, biochemical, and morphogenetic reactions that ensure its viability* (Shvarts, 1968). This system of views helps explain the origin of genetic differences among morphophysiologically indistinguishable populations on the basis of I. I. Schmalhausen's theory of stabilizing selection and the principle of optimal phenotype developed from this theory (Shvarts, 1968).

* The fundamental similarity of conclusions from investigations conducted at the morphophysiological and molecular levels deserves attention.

The prolonged existence of a species in a certain environment leads to such a complete balancing of its biological features with the environmental conditions, that any deviation from the norm is cut off by selection. If two populations are under identical environmental conditions (more precisely, so close that stabilizing selection turns out to be more advantageous than directional), selection will work in the direction of maintaining and stabilizing the optimal phenotype. But since the genetic structure of populations is different from the start (the result of genetic automatic processes*) and the differences are gradually strengthened due to chance mutations, selection for the stabilization of an optimal phenotype should lead inevitably to the rearrangement of genotypes so that the action of new mutations is neutralized. This leads to the origin of genetic differences among populations that possess a large number of morphophysiological similarities: the degree of genetic differences does not coincide with the degree of morphophysiological differences.

It seems to us that the data obtained and their analysis shed light on the nature of so-called "twin species." Roughly speaking, twin species are species that cannot be differentiated with the aid of the usual morphological methods. Mayr (1968) writes (p. 42): twin species can be defined as morphologically similar or identical, but reproductively isolated, natural populations. In those cases in which essential ecological and physiological or internal anatomical differences are hidden behind identical exteriors, the twin species are in no fundamental way distinguishable from ordinary species, since natural selection works just as successfully at all levels of organization and the genetic defense of the species' physiological specificity is just as necessary as the defense of its morphological specificity. In a number of cases, however, the morphophysiological differences between twin species do not exceed the average differences between populations whose membership in the same species is not subject to doubt. In these cases, their origin becomes a problem of primary importance.

At present, there is no theory on the origin of twin species, although this question has been touched on by many researchers. Mayr (1968) writes: "Speciation among twin species is in no way distinguished from speciation in other species." Timofeev-Resovskii *et al.* (1969) link the origin of twin species with sympatric speciation on the basis of the chance restructuring of karyotypes of an initially small number of individuals of the population. The authors cite hypothetical examples to show that the isolation of a twin species within the ancestral population by this means has "low probability, but is not excluded" (p. 294). It cannot be categorically denied that other paths of speciation, besides the magistral path of evolution—the origin of

* Due to sampling phenomena such as the founder's principle and genetic drift—Ed.

new species in the process of the progressive mastery of new ecological niches, of progressive adaptability under the influence of natural selection—may exist, including those based on the chance origin of the genetic isolation of some of the individuals in a single population. It appears to us, however, that the origin of twin species is explained naturally by well-founded theoretical concepts, in particular, I. I. Schmalhausen's theory of stabilizing selection.

All populations, even partially isolated ones, are genetically specific. To explain this specificity, as we attempted to show, there is no necessity to resort to the dubious hypotheses of neutral mutations. It is superbly explained on the basis of the theory of stabilizing selection and the principle of optimal phenotype (Shvarts, 1968) developed from this theory, and is corroborated by the results of investigations that show that the degree of immunological and the degree of morphophysiological divergence do not coincide. It is theoretically possible that the origin of differences among populations under the influence of stabilizing selection will lead to the formation of a genotype so distinct from the genotypes of other populations of the species that genetic isolation will arise. Undoubtedly, this process is slow and is impeded by inevitable interdeme hybridization. True twin species, the differences between which do not exceed the usual interdeme differences, are therefore probably not encountered very often, but in those cases in which the separated populations acquire ecological and physiological characteristics slightly greater than the average in magnitude, the process of interdeme hybridization should be weakened, and the genetic isolation that arises will sharply accelerate the process of speciation.

Thus, the origin of twin species is naturally explained on the basis of the theory of stabilizing selection. It seems to us that Mayr arrived at a similar idea. In another section of the book already cited (Mayr, 1968), he ties the origin of twin species with "the selective advantage of preserving the invariability of the phenotype" (p. 60). Unfortunately, this thought, which is more profound in comparison with the categorical statement of his we have already cited, is not developed further. It remains unclear, therefore, how Mayr reconciles this idea with the assertion that "speciation among twin species is in no way distinguished from speciation in other species." As we tried to show, these differences are essential: "usual" speciation occurs on the basis of the directional form of natural selection; "doubles" speciation, on the basis of its stabilizing form. In both cases, however, the motive force of speciation is the main factor of the evolutionary process, i.e., natural selection. The investigations cited showed that the "average" immunological differences among populations do not exceed the differences among extreme variants of the same population. Considering that immunological distance reflects the degree of biochemical differences among compared organisms, which in turn reflects the degree of genetic differences (hereditary information is realized in the process of the biosynthesis of proteins), it is correct to assume that the immunological unity of the species excludes the possibility of a hiatus arising between any intraspecific forms. The formal criterion of species (hiatus) noted long ago by systematists thus reflects its biological essence, which is the genetic unity displayed in the character of intrademe variability of biochemical traits.

Immunological investigations also provide interesting material for an analysis of the question of the equivalence of species in different taxa.

Are species of mammals, fish, flowering plants, and bacteria fully comparable? From a general biological point of view, they are unconditionally comparable, since within each of these groups the species represents a relatively isolated genetic system, characterized by morphophysiological distinctiveness. Any zoologist knows, however, that morphological differences between species of mammals and of birds are immeasurably less than differences among many intraspecific forms of fish. These differences are determined by a different range of intraspecific variability. Only through the benefit of century-old experience and by intuition does the systematist avoid mistakes issuing from the principal differences in the character of intraspecific variability within different taxa. The material cited in this collection allows us to hope that immunological research will provide an objective underpinning for the systematists' intuition. Let us illustrate this position by means of a comparison of research on mammals (L. M. Syuzyumova's) and fish (L. A. Dobrinskii's). Numerous experiments (Gotronei and Perri, 1946; Vojtiskova, 1960) and also theoretical considerations indicate the close and unconditionally causal relationship between tissue compatibility and the ability of organisms to produce fertile offspring when crossed. The deepest biochemical distinctions among intraspecific forms can therefore be detected in those groups in which interspecific hybridization is more widely prevalent. The difference between mammals and fish is significant, in this regard. Interspecific hybridization is encountered more often among fish than among mammals. Separate populations of fish that remain within the limits of a species can therefore "permit themselves the luxury" of diverging at the biochemical level: the immunological distance of some exceeds the species norm.

The work of L. A. Dobrinskii (the article in this collection) shows that the tissue compatibility of fish of different species may be higher than that among individuals of different populations of the same species. The result is admittedly paradoxical, but it finds its explanation in the following system of views, which we (in a very general form) have already had the opportunity to argue (Shvarts, 1969).

The species is morphophysiologically unique. It is characterized by defined interrelationships with its environment and maintains its biological specificity in all its manifestations. It therefore occupies a unique position in the ecological systems of every rank: from individual ecosystems to the biosphere as a whole. Many observations to which we attach the significance of *experimentum crucis,* in particular, testify to this very clearly. The substitution of one form of a species for another does not lead to essential biocenotic consequences. If a new form succeeds in passing through the first period of acclimatization, the ecosystem does not note the substitution. Numerous commercial experiments on the reacclimatization of game species attest to this. The substitution of one species for another, however, inevitably elicits a change in the character of the biocenotic interrelationships. It follows that precisely those traits of the organism that determine its biological uniqueness should be considered as specific traits in the strict sense of the word. As a rule, the differentiation of individual populations at the tissue level changes the character of the organism's relationship with the environment and disturbs the species integrity. Biochemical differences, which are more clearly shown by immunological reactions, are therefore more substantial among species than among intraspecific forms. That they are explains better the correlation between genetic compatibility of the forms compared and their immunological distance. In addition, the investigations show that there is no direct dependence between immunological distance, which reflects the genetic difference among organisms, and the degree of their morphophysiological divergence.

This finds a natural explanation in the diverse genetic value of different traits (Bremermann, 1967). The quantity of genetic information in various anatomical structures and physiological processes is different. It is therefore theoretically conceivable that in individual cases, intraspecific differentiation, which does not disturb the morphological specificity of the species, demands a greater change of the genetic information than differentiation at the species level. In these cases, the immunological distance among species may be less than among intraspecific forms.* For the reasons on which we

* The same thought may be expressed in another way. Immunological differences establish only the degree of genetic differences, not their quality. They fix the degree of biochemical differences (the primary gene products), which are first of all differences in protein composition. Changes in a set of biologically active proteins, however, can in the process of epigenesis lead to change in a trait that radically changes the relationship of the organism to its environment, with all the ensuing consequences, while in a different case, analogous to the immunological viewpoint, the changes will have only insignificant morphophysiological consequences. The immunological distance between two substantially differentiated forms may therefore turn out to be equal to the immunological distance between indistinguishable populations, while in some groups the degree of morphophysiological differentiation to which the century-old experience of zoologists attaches "species rank" may be realized on the basis of less substantial genetic differences than those that characterize the average immunological distance between populations of the same species.

have dwelled, such a situation should be rare and observed more often among fish than in the higher vertebrates.

Nevertheless, it may be hoped that the application of still finer methods of investigation will permit us to establish essentially similar phenomena among the mammals as well. It was tempting to conduct a direct comparative study of the structure of the genome of two ecologically sharply differentiated subspecies of the same species and of an independent species of the same genus. Our choice fell on the Arctic (Yamal) and forest steppe (southern Transurals) subspecies of the *oeconomus* voles, which were compared with the common vole.

We analyzed the DNA of *M. o. oeconomus, M. o. chahlovi,* and *M. arvalis.*

The research material (liver specimens fixed in ethanol) was transferred to the Interdepartmental Laboratory of Bioorganic Chemistry of Moscow State University (MGU). G. P. Miroshnichenko, K. M. Val'skho-Roman, and N. B. Petrovym, colleagues from the laboratory, determined the nucleotide composition of the DNA, the frequency of pyrimidine sequences, and the "gene spectra." The methods used were described previously (Antonov and Belozerskii, 1972).

It was found that the nucleotide composition of the DNA of all three forms of voles is practically identical (42–43 mole % GC). This is not surprising, since the DNA composition of mammals as a whole varies within a very narrow range (Antonov, 1972). Analysis of the degree of linkage of pyrimidine nucleotides in DNA (β) showed that according to this DNA characteristic, *M. o. oeconomus* and *M. arvalis* are indistinguishable ($\Delta\beta = 0.03$), but both these forms differ appreciably from *M. o. chahlovi* ($\Delta\beta = 0.23$ and 0.26, respectively). Basic differences are noted in the fraction of monopyrimidine fragments. Similar results are obtained in a comparison of the "gene spectra" of the DNA of the forms studied. Statistically significant differences (using Fisher's criteria) were found both in a comparison of the "gene spectra" of *M. o. oeconomus* and *M. arvalis* (7.5) and in a comparison of both these forms with *M. o. chahlovi* (19.7 and 27.0). These results are preliminary, since only one DNA preparation of each of the forms of voles studied was used in the experiments.

Nevertheless, the data obtained permit us to assume that the genic material of *M. o. oeconomus* is more similar to the DNA of *M. arvalis* than to the DNA of *M. o. chahlovi.* The greatest differences are detected in a comparison of the primary structures of the DNA of *M. arvalis* and *M. o. chahlovi.* The degree of genetic differences between the two subspecies turned out to be greater than the differences between species: the morphological characteristics of the Arctic *oeconomus* voles coded into the

DNA demanded greater changes at the genome level than the differences between the initial (forest steppe) forms of *M. oeconomus* and *M. arvalis* (the genetic value of the subspecific differences was found to be greater than the specific differences). Immunological investigations and their analysis showed that this result could be predicted on theoretical grounds, and served as the basis for the construction of appropriate experiments. Species independence is determined by the quality, not the quantity, of the animals' genetically controlled characteristics. The question "Is it a species or not?" is resolved on an ecological level, not on a physiological or genetic level.

This indicates that the process of speciation, in its biological essence, consists in the progressive morphophysiological differentiation of closely related forms, which at a certain stage leads to the ecological and genetic isolation of the parental species. The morphophysiological definition and genetic isolation of the species are two steps of a single process. Such a statement of the question does not place in doubt the exceptional significance of the criterion of "fertile crossing." Furthermore, we suggest that if the earth's biosphere did not consist of genetically isolated units, evolution would be impossible (Shvarts, 1969). We think, however, that "genetic isolation" and "morphophysiological definition" should be regarded as a dialectical unit,* the major element of which is the morphophysiological definition, which stipulates the biological (ecological) uniqueness of the species. It is symptomatic that in those cases in which obvious biological uniqueness is not accompanied by genetic isolation, the systematist without hesitation gives preference to the morphological criterion. We will list just two examples for illustration.

Hybridization of the spotted deer with the izyubr (*Cervus elaphus xantopygus*) has long been known (Menard, 1930; Mirolyubov, 1949). In captivity, the hybrids of these species are easily obtained and are fully fertile. Full fertility is also shown by natural hybrids, which are known by the Chinese under the name *chin-da-guiza*. The izyubr takes over the "harem" of the spotted deer, which leads to the appearance of a large number of hybrid individuals. It is interesting that both R. K. Maak and N.

* The existence of difficult cases (genetically isolated species that are not distinguished by substantial morphophysiological characteristics or, on the other hand, sharply differentiated but genetically compatible species) does not in the least contradict this assertion. Let us recall the words of Lenin: ". . . the quasirealistic, but in fact eclectic, pursuit of a full enumeration of all the individual traits and individual 'factors.' As a result, of course, this senseless attempt to include in a general concept all the partial characteristics of solitary phenomena or, on the other hand, 'to avoid a clash between extremely diverse phenomena'—an attempt that attests simply to an elementary misunderstanding of what science is—leads the theorist to the point where he cannot see the forest for the trees" (Lenin, V. I., *Complete Collected Works*, Vol. 5, p. 142).

M. Przheval'skii considered them an independent species. Not one modern systematist, however, unites the izyubr and the spotted deer in one species.

We have borrowed another example from the most recent literature. After the invasion of Iceland by the Herring gull, numerous cases of its hybridization with the glaucous gull (the burgomaster) were observed. The hybrid individuals constituted up to 50% in some populations, and it was even suggested that they should be named *Larus argentatus hyperboreus* (Ingolfsson, 1970). This nomenclatural monster is eloquent enough, but it does not undermine confidence in the species independence of *L. argentatus* and *L. hyperboreus*.

Analysis shows that utilization of the new methods in systematics allows one to approach a deeper understanding of the complex questions of the species problem. They are also useful during the resolution of certain questions in population ecology.

L. M. Syuzyumova began an investigation with the study of related animals—siblings. This organization of the work at first seemed to us somewhat formal (a series of intraspecific comparisons was desired, beginning with minimal genetic differentiation of animals and leading up to the species), but it led to unexpected and interesting results. It was found that although the mean immunological distance among populations is greater than among individuals of the same population, even among siblings, cases were observed of immunological dissimilarity just as marked as among different populations. The dissimilarity of the parents (from the same or different populations), therefore, does not always increase the genetic heterogeneity of their offspring. According to the mean indices of interpopulation crossing, however, as a rule the immunologically different quality of the offspring is increased. During this, the effect of hybridization is distinctly detected, independent of the degree of separateness of the populations and of their morphophysiological characteristics. These conclusions have fundamental ecological significance, since they uncover yet another aspect of the phenomenon of population homeostasis.

A sharp drop in numbers leads to the situation that on large territories occupied by a certain species, only separated foci of its survivors remain. Under these conditions, the growth in numbers inevitably begins with the crossing of closely related individuals. The high genetic capacity of the parents is maintained by the heterogeneity of their offspring. Following a period of reduction in numbers, migration (the consequence of a local increase in the density of a settlement) inevitably leads to the crossing of totally unrelated animals and an increase in the heterogeneity of their offspring. This is the consequence of the maintenance of a sufficiently high species viability, despite constant change in its numbers and consequent change in its system of pair formation.

References

As in the reference list starting on page 223, those references that were originally listed separately from the Russian-language references are here designated by bullets (•).

Antonov, A. S., 1972, in: *The Structure of DNA and the Position of Organisms in the System,* Izd. MGU, Moscow.

Antonov, A. S., and Belozerskii, A. N., 1972, in: *The Structure of DNA and the Position of Organisms in the System,* Izd. MGU, Moscow.

• Blagoveschensky, A. V., 1929, On the relation between the biochemical properties and the degree of evolutionary development of organisms, *Biol. Gen. 5.*

Bol'shakov, V. N., Rossolimo, O. L., and Pokrovskii, A. V., 1969, Systematic status of the Pamir-Alaiskii Mountain voles of the *Microtus juldaschi* group (Mammalia, Cricetidae), *Zool. Zh.* **48**(7).

• Bremermann, H., 1967, Quantitative aspects of goal-seeking, self-organizing systems, *Prog. Theor. Biol.* **1,** Academic Press, New York.

Gileva, É. A., and Pokrovskii, A. V., 1970, Karyotype features and chromosome polymorphism in the Pamir-Alaiskii Mountain voles of the *Microtus juldaschi* group (Mammalia, Cricetidae), *Zool. Zh.* **49**(8).

• Gotronei, G., and Perri, T., 1946, I trapianti studiati in rapporto con le ibridazioni interspecifiche, *Bull. Zool.,* No. 12.

• Ingolfsson, A., 1970, Hybridization in glaucous gulls, *Larus hyperboreus,* and Herring gulls, *L. argentatus,* in Iceland, *Ibis* **112**(3).

Ishchenko, V. G., 1966, The application of allometric levels for the study of morphophysiological differentiation, *Tr. Inst. Biol. UFAN SSSR,* No. 5.

Komarov, V. L., 1944, *Studies of Plant Species,* Izd. AN SSSR, Moscow.

Kopein, K. I., 1958, Data on the biology of the obensis lemming and the large narrow-skulled vole, *Byull. Ural. Otd. Mosk. O. Ispyt. Prir. UFAN SSSR,* No. 1, Sverdlovsk.

Mayr, E., 1968, *Zoological Species and Evolution* [Russian translation], Mir, Moscow.

Menard, G. A., 1930, *The Deer Antler Industry,* Gostorgizdat, Moscow and Leningrad.

Mikhalev, M. V., 1970, Analysis of the taxonomic relationships within the Microtinae on the basis of electrophoretic investigations, author's Candidate dissertation, Sverdlovsk.

Mirolyubov, I. I., 1949, Hybridization of the spotted deer with the izyubr, *Karakulevod. Zverovod.,* No. 2.

Ovchinnikova, N. A., 1966, Biological features of the northern and nominal subspecies of the *oeconomus* voles and their hybrids, in: *Intraspecific Variability of Terrestrial Vertebrates and Microevolution,* UFAN SSSR, Sverdlovsk.

Ovchinnikova, N. A., 1968, Comparative study of the Transcaspian and common vole under laboratory conditions, in: *Reports of the Summary Session of the Laboratory of the Population Ecology of Animals of the Institute of the Ecology of Plants and Animals, UFAN SSSR,* No. 2, Sverdlovsk.

Pokrovskii, A. V., 1969, Fertility of the Pamir and Archevaya voles and their hybrids under laboratory conditions, in: *Reports of the Summary Session of the Laboratory of the Population Ecology of Animals of the Institute of the Ecology of Plants and Animals, UFAN SSSR,* No. 3, Sverdlovsk.

Pokrovskii, A. V., 1971, Reproduction and hybridization of the Middendorf voles and the northern Siberian voles, in: *Reports of the Summary Session of the Laboratory of the Population Ecology of Animals of the Institute of the Ecology of Plants and Animals, UFAN SSSR,* No. 4, Sverdlovsk.

Pokrovskii, A. V., Krivosheev, V. G., and Gileva, É. A., 1970, Experimental study of the ecology and degree of reproductive isolation of two close forms of the northern voles (*Microtus middendorffi* Poljakov, 1881, and *M. hyperboreus* Vinogradov, 1933), *Ékologiya,* No. 1.

• Prakash, S., and Lewontin, R., 1968, *Proc. Nat. Acad. Sci. U.S.A.* **59**(398).

• Richmond, R. C., 1970, Non-Darwinian evolution: A critique, *Nature London* **225**(5237).

Shvarts, S. S., 1968, The principle of optimal phenotype (toward a theory of stabilizing selection), *Zh. Obshch. Biol.* **29**(1).

Shvarts, S. S., 1969, *The Evolutionary Ecology of Animals, Tr. Inst. Ekol. Rast. Zhivotn. Ural. Fil. Akad. Nauk SSSR,* No. 65.

Shvarts, S. S., Kopein, K. I., and Pokrovskii, A. V., 1960, Comparative study of some biological features of the voles, *Microtus gregalis gregalis* Pall, *M. g. major* Ogn., and their hybrids, *Zool. Zh.* **39**(6).

Syuzyumova., L. M., 1969, *Species Specificity in the Manifestation of Antigenic Properties of the Tissues During Reaction to Heterotransplantation,* in: *Tr. Inst. Ekol. Rast. Zhivotn. Ural. Fil. Akad. Nauk SSSR,* No. 68.

Timofeev-Resovskii, N. V., Vorontsov, N. N., and Yablokov, A. V., 1969, *A Brief Essay on the Theory of Evolution,* Nauka, Moscow.

• Vojtiskova, M., 1960, Zur Frage des Mechanismus der Befruchtungskompatibilität bei der entfernten Kreuzung des Geflügels, in: *Arbeitstagung: Fragen der Evolution,* Jena.

Zhukov, V. V., 1967, Immunological relationships of some forms of the voles of the genus *Microtus,* in: *Reports of the Summary Session of the Laboratory of the Population Ecology of Animals of the Institute of the Ecology of Plants and Animals, UFAN SSSR,* No. 1, Sverdlovsk.

Appendix II

Recent Work on the Evolutionary Ecology of Animals

A. E. Gill

The purpose of this chapter is to indicate the directions in which the area of the evolutionary ecology of animals has moved since the publication of the original Russian edition of this book in 1969. A sampling of recent literature pertinent to the subjects discussed by Shvarts is presented. The preface by S. S. Shvarts to the 1973 work, *Experimental Investigations of the Species Problem,* which is included in this volume as Appendix I, elaborates the more recent work conducted by him and his colleagues at the Institute of the Ecology of Plants and Animals in Sverdlovsk on the species problem and systematics. Here, some recent work of other authors (especially American) indicative of current activity in this field is described.

A major purpose of this book is to explain the key significance of ecological mechanisms in the evolutionary process. At the beginning, Shvarts states three questions that must be investigated in order to synthesize research on morphological and physiological traits with population ecological studies. These questions are:

1. The evolutionary significance of the dynamics of population structure in space.
2. The correlation of micro- and macroevolution.
3. The reality of taxa above the species level.

The citation of recent literature will follow the order in which the topics are discussed in the book.

In the Preface to the 1969 publication, Shvarts points out some of the successes and weaknesses of the synthetic theory of the neo-Darwinists. He states that although neo-Darwinism explains many phenomena in wholly or partially isolated populations, it does not deal satisfactorily with the question of speciation. Similar views are expressed by Lewontin (1974) in *The Genetic Basis of Evolutionary Change* regarding population genetics (p. 12):

"While population genetics has a great deal to say about changes or stability of the frequencies of genes in populations and about the rate of divergence of gene frequencies in populations partly or wholly isolated from each other, it has contributed little to our understanding of speciation and nothing to our understanding of extinction." He notes that population genetics is an essential part of evolutionary studies, but not the whole story. An interdisciplinary approach to population studies has developed (see, for example, Lewontin, 1968; Levins, 1968; MacArthur, 1972), in which the dynamic nature of both the environment and the genetic structure is recognized.

Not everyone uses the term *population* in the same way, and the question of how a population is defined is pertinent to this discussion. Shvarts considers the population to be a unit that has an independent existence and development over a long period of time. In his view, the ecological structure of a population is extremely important. It consists of a certain ratio of age classes, a certain sex ratio, a combination of resident and migrant animals, and the presence of special groupings such as family, herd, and so forth. The definition given by Shvarts and a similar definition presented by Petrusewicz (1974) at the First International Theriological Congress stress the importance of the ecological structure of the population to its long-term existence as an entity. Their definitions differ from that usually given by population geneticists in the United States. The following quote from one of the best recent American textbooks on general biology is representative of the common definition used in the United States: "The basic unit of evolutionary studies is the population—a group of individuals of the same species who exchange genes by interbreeding" (Wilson *et al.*, 1973, p. 767). This is purely a genetic definition that does not directly address the effects of ecological structure on the exchange of genes. Anderson (1970) discusses some of the Russian work on the ecological structuring of local populations and notes especially the distinction made between stable and unstable habitats. He, however, considers the *deme*—a much smaller unit than Shvarts's population—to be the self-perpetuating unit. The deme is closer to Wilson and co-workers' definition of a population, including only the members of the interbreeding group and not members of adjacent groups with which it may occasionally exchange genes. If the deme is regarded as the population, attention is focused on factors more effective at small population sizes—genetic drift and interdeme selection—as seen in Anderson's emphasis on the evolutionary significance of these factors.

Chapter I describes the contribution of population genetics to an understanding of the ways in which the genetic organization of populations can change. The important book by Lewontin (1974) on the present state of affairs in population genetics and assessment of the potentials and theoretical problems in its development amplifies and brings this subject up to date. His contrast of the classic and balance schools of population genetics is related to Shvarts's discussion of selection in a number of instances in this chapter. The early reaction of Darwinists to the emergence of Mendelian genetics and the role to which it seemed to relegate selection is described by Shvarts (p. 21): "The role of sieve is not a very honorable role for the leading mechanism of the evolutionary process." This is essentially the classic view of selection as a purifying process to rid the population of deleterious genes. On the other hand, the creative role played by selection in producing balanced genotypes and a balanced gene pool, as described by Shvarts, corresponds to the balance school, as discussed by Lewontin.

It is pointed out in Chapter I that the value of a mutation is determined by the properties of the whole genome and, further, of the population gene pool as a whole. Shvarts stresses the importance of heterozygosity. This position is buttressed by the earlier work of Lerner (1954), in which he describes the value of heterozygosity both for the buffering afforded the heterozygous individual and for its role in population genetic homeostasis. Berry (1974) gives an example of homeostatic fluctuations in nonmetrical skeletal characters of Skokholm Island mice over a period of 10 years. This adaptation to minor environmental fluctuations is made possible by a high level of genetic variability.

As the book points out, the gap between population genetic models and reality remains great (see also Lewontin, 1974). For example, selection has been found to be more effective than the models had indicated (Dobzhansky, 1970; Levins, 1968; Dowdeswell and others in Ehrlich *et al.*, 1969, Section 1); i.e., the selection coefficients found in nature are sometimes much higher than those postulated in the models. Rates of evolution present theoretical difficulties in population genetics and are currently a subject of great debate. Selectionists generally hypothesize variable rates of evolution due to differential selection (e.g., Gillespie and Kojima, 1968), while others cite evidence for the apparent constancy of evolutionary rates (J. L. King and Jukes, 1969; Kimura and Ohta, 1972). Most of this evidence deals with rates at the molecular level. Wilson and his co-workers have presented evidence for a statistically constant evolutionary rate of serum albumin and some other proteins (Sarich and Wilson, 1967; Prager *et al.*, 1972; and others). It has been suggested that evolutionary rates appear to be constant as the result of averaging rates of genetic change over a long period of time (Nevo *et al.*, 1974; Ayala *et al.*, 1974). The latter authors, however, apply the measure of genetic distance devised by Nei (1971, 1972), which is based on assumed constant rates, to the estimation of divergence times between groups. Despite their reservations (see their discussions of assumptions and possible inaccuracies of the measure), they found that the divergence times calculated according to Nei's genetic distance measure corresponded well with known evolutionary relationships based on, respectively, the fossil record for pocket gophers and on studies of reproductive isolation, chromosomal polymorphism, morphological biometry, sexual behavior, and geographic distribution for the *Drosophila*. It has become clear, however, that rates of evolution may differ, depending on which types of traits (biochemical, morphological, karyotypic) are being considered, as Shvarts points out in Chapter VI. Recent research along this line will be cited in connection with references for Chapter VI.

In Chapter III, Shvarts discusses the contrast between microevolution (intraspecific reorganizations) and macroevolution (processes leading to the origin of higher taxa) in the modern synthetic theory of evolution. He believes that followers of the synthetic theory feel that different laws govern micro- and macroevolution and cites the evidence for this belief. Wilson and co-workers, however, in answer to the question, "Can macroevolution be explained as an outcome of microevolutionary shifts?," respond (Wilson *et al.*, 1973, p. 792): "The answer is that it is entirely plausible, and no one has come up with a better explanation consistent with the known biological facts."

Shvarts does not feel that all changes in the genetic structure of a population should be equated with microevolution. Some reorganizations of genetic structure are homeostatic in nature and function to maintain the dynamic equilibrium of the population's genetic structure in a variable environment. Only those changes that are irreversible and lead to speciation should be considered microevolutionary. The distinction between homeostatic reorganizations of a population and microevolution (used in the sense of the beginning of the evolutionary process) is blurred by current practices in subspecific nomenclature. Shvarts gives examples of intraspecific groups that have been termed subspecies on the basis of purely formal criteria, without evidence that they are evolving to the species level. Anderson (1970) gives additional examples of this practice and warns against the application of subspecific nomenclature on inadequate grounds. An example of a group that was greatly subdivided is that of the African gerbils, genus *Taterillus*. Robbins (1974) suggests a revision of this genus based on univariate and multivariate computer analysis of morphological measurements of more than 3000 museum specimens. Rather than the 18 species and 23 subspecies originally indicated, his data suggest only 5–8 distinct species. The criteria for application of subspecific and specific names is at issue here. Shvarts stresses the importance of the evolutionary and phylogenetic significance of the nomenclature used, especially to prevent the proliferation of subspecific names.

Chapter IV deals with ecological mechanisms that may be involved in the maintenance of a population's genetic heterogeneity. Heightened viability and reproductive success of heterozygotes is one of the mechanisms maintaining heterogeneity. Gill (1977) found evidence of heterozygote superiority in several factors relating to reproductive success in a laboratory study of California voles from a polymorphic population. Another major mechanism maintaining heterogeneity is differential selection of age and sex groups in the population. Johnston *et al.* (1972) reexamined the data of Bumpus (1899, cited in Johnston *et al.*, 1972) on the selective action of a severe winter storm on house sparrows, *Passer domesticus*. The size relationships that affected the sexual dimorphism of the survivors as compared with the nonsurvivors of the storm had not been analyzed previously. Using multivariate analysis of nine characters measured by Bumpus, these authors found that large males had a survival advantage over small males, whereas intermediate-sized females had a selective advantage over larger and smaller females. The authors concluded that on the basis of these and other differences found in the response of the two sexes, they would expect abiotic stresses to consistently operate differently on the sexes. They thought that "each sex represents a distinct, though related, adaptive system." In a study of Arctic fox (*Alopex lagopus*) specimens collected over a 16-year period in northern European USSR, Shilyaeva (1974) found differential selection of sexes and generations, in accordance with cyclical environmental changes. Sufficiently long-term studies with large samples (in this case, 937 skulls and 567 carcasses) are made possible by the use of data from breeding and hunting industries in the Soviet Union.

Many of the examples of differential selection of age groups or sex groups given in Chapter IV might well be fitted by models of selection in different niches developed by Levene (1953) and further explored by Levins and MacArthur (1966). These models or variations of them may provide a mathematical basis for determining the conditions for polymorphism when different segments of the population undergo differential selection at certain stages of their life history, but remain part of one breeding group.

The importance of sexual reproduction in the maintenance of population genetic heterogeneity is stressed by Shvarts. A pertinent discussion of sexual reproduction and mate selection, with selected references, is given in the book by Pianka (1974) entitled *Evolutionary Ecology* (see Chapter 4).

In Chapter V, Shvarts details how (in addition to individual selection) changes in age structure, mortality due to apparently nonselective factors, and the spatial structure of populations affect the genetic structure of populations. Bellamy *et al.* (1973) give evidence for the dependence of the genotypic structure on the age structure of the population of house mice living on Skokholm Island. In this well-studied population, it was found that the allele frequencies at certain loci changed progressively over the summer, with a change in age structure and selection pressures.

The intensity of reproduction may vary in natural populations, dependent on their density. This variation affects both the age structure and the genetic structure of the population. Additionally, there may be differences in selective forces in different phases of population density and reproductive effort. Carson (1968; see also the earlier work of Ford, 1964) suggests that a lessening of environmental pressures may lead to an increase in population size and release of genetic variability. Chitty (1970) provides a model for population cycles in voles in which different genotypes are favored at different phases of the cycle. Such a model receives some support from data on genetic changes found in fluctuating vole populations (Tamarin and Krebs, 1969; Gaines and Krebs, 1971; Krebs *et al.,* 1973). Despite the association of genetic changes with density fluctuations, however, it is not clear at present what is causing the cycles to occur.

Another explanation for fluctuations in gene frequencies in populations with overlapping generations is given by Charlesworth and Giesel (1972). They postulate that although fluctuations in age structure that accompany density cycles may be nonspecific with respect to genotypes, they can result in gene frequency changes. Specifically, if generations overlap, genotypes reproducing early are favored in an increasing population (weighted toward young individuals), but are at a disadvantage in a decreasing population (weighted toward old individuals). The results of a computer model showing genotypic changes resulting in populations undergoing genetically *nonselective* changes in population age structure are presented. They conclude their discussion with the following remarks (p. 399): "Study of natural selection in populations with overlapping generations emphasizes the importance of taking ecological and demographic variables into account in population genetics. Phenomena described here are impossible with discrete generations and depend on explicit incorporation of ecological variables into genetic models." Giesel (1972) highlights an important implication of their model, namely, that changes in population age structure (through a change in r, the population's rate of increase) will maintain genetic variability in natural populations with essentially no cost in terms of "genetic load." The change in gene frequencies depends on the change in age structure associated with density cycles, not on environmental selective pressures that are genotype-specific. Therefore, no "load" exists.

In considering the significance of spatial structuring in populations, it is important to recall Shvarts's definition of a population. As he indicates, the groups termed populations by some researchers would be considered subpopulations by his definition. The spatial structure of populations will affect the exchange of genes within and among populations. Here, we are concerned with the literature on two phenomena—migration (and, correspondingly, partial and complete isolation of groups) and group selection (as contrasted with individual selection). Recent mathematical studies of the effect of migration on variability are given by Maruyama (1971), employing a stepping-stone model, and by Latter (1973), using an island model to study genetic differentiation in a subdivided population. Endler (1973) tests a theoretical stepping-stone model of gene flow with experimental *Drosophila* populations and arrives at the conclusion that gene flow may be unimportant in the differentiation of populations along environmental gradients. With relatively uniform selection gradients, considerable gene flow did not prevent local differentiation in his model. Slatkin (1973) points out that Endler's conclusion depends heavily on the linearity of the selection pressures. Slatkin develops a model of the effect of gene flow and natural selection in a continuously distributed, infinite population. He arrives at conclusions about the population's response to selection that are independent of the pattern of variation in selection and the exact form of dispersal of the population.

Varying views on group selection are presented by G. C. Williams (1971), Lewontin (1970), Gadgil (1975), and A. C. Wilson (1975). Gilpin (1975) has written on group selection in predator–prey communities. Much attention has been given recently to kin selection (Maynard Smith and Ridpath, 1972; Hamilton, 1964, 1972; E. D. Wilson, 1971). Additional references on group and kin selection are given by Emlen (1973). As far as individual selection is concerned, most models of population genetics have dealt with single loci or their combinations, ignoring interaction, whereas the genome as a whole or at least groups of loci may well be the unit of selection (Franklin and Lewontin, 1970; Lewontin, 1970).

In Chapter V, it is pointed out that smaller animals are more likely to adapt to microclimates than are larger ones. Larger animals adapt to major differences in climate. An example of adaptation to microclimates is afforded by the work of Jones (1974) on natural populations of the snail, *Cepaea vindobonensis,* in Yugoslavia. In two topographically similar areas, he found that snails with less band pigment are common only outside large basins that act as frost hollows. He showed experimentally that the darker snails absorb more solar energy and can become active in the sun more quickly than the light ones. Hence, the dark snails can occupy the frosty basins without being at a disadvantage.

Just as adaptations to microclimates may affect the genetic makeup and spatial structure of populations, predator–prey relationships may also affect these features, as well as the numbers of populations. Khlebnikov *et al.* (1974) conducted a 10-year study that showed predation by sable (*Martes zibellina*) to be an important factor in the numbers and distribution of the Pika (*Ochotona alpina*). Pimentel (1961, 1968) has proposed a genetic feedback mechanism to describe the controls that have evolved over the numbers of the populations involved in parasite–host and predator–prey systems. Pimentel and Soans (cited in Pimentel, 1973) designed a laboratory experiment to study the genetic feedback mechanism in the coevolution of plant and herbivore. A balance was reached between a resistant laboratory-designed "plant" and its herbivore, in which only the surplus energy of the plant was removed and the numbers of the herbivore population were controlled. Levin (1972) has also analyzed a mathematical model of the functioning of the genetic feedback mechanism.

The species problem is discussed by Shvarts in Chapter VI. Controversy continues over the definition of a species, with researchers generally recognizing the problems inherent in any given definition. Wilson *et al.* (1973) present criteria similar to those given by Shvarts for defining a species, including his fourth criterion, which they state as ". . . it (species) adapted to the environment in ways peculiar to itself" (p. 828). At the International Congress on Systematic and Evolutionary Biology in 1973, a symposium was devoted to contemporary systematic philosophies. Papers by Griffiths (1974), Nelson (1974), Sokal (1974), and Bock (1974) dealt with various types of biological classification and problems inherent in the classification systems. Sokal, in particular, addressed himself to the species problem—both its definition and its genesis. He described four common definitions of species—classic phenetic, numerical taxonomic, biological, and evolutionary—and noted the problems in each. He concluded his discussion with this statement (Sokal, 1974, p. 372): "I simply wished to deemphasize the single-minded focus on reproductive isolation to the exclusion of other factors, which has for so long dominated the speciational literature." Dobzhansky (1972), in considering speciation in *Drosophila,* concludes that there are several kinds of species and types of speciation in *Drosophila,* and even more in the biotic world at large. He also notes that reproductive isolation may sometimes follow and sometimes precede the adaptive divergence of population gene pools. Scudder (1974) concurs with Dobzhansky in emphasizing that there are many kinds of species with different inherent characteristics and strategies of evolution. He stresses that allopatric speciation should no longer be considered the only or the main mode of speciation. Scudder gives many examples from the literature on research regarding the speciation process.

Use of new methods (cited by Shvarts in Chapter VI) in population biology has increased enormously the data base for consideration of variation in natural populations and the beginning of the speciation process, but it has not led to a resolution of the species problem (see, for example, Lewontin, 1974). Biochemical techniques and multivariate analysis are among the most important of the new methods discussed in Chapter VI. A symposium entitled *Molecular Evolution* (Ayala, 1976) was held in June 1975 in Davis, California, at which recent advances and overviews of this area were presented. Pertinent here are the work of Selander and associates on electrophoretic variation in natural populations (see Selander *et al.*, 1971, for a discussion of electrophoretic techniques), Ayala and co-workers on variation in *Drosophila* (Ayala *et al.*, 1970, 1974, and others), and Johnson (1973, 1976) on the meaning of electrophoretic variants and the control of variable conditions in electrophoretic work. Avise (1975) describes the value of electrophoretic data as a tool in systematics. Good bibliographies for literature on electrophoretic variation are given in the papers by Yamazaki and Maruyama (1974) and Johnson (1974). Enzyme variations found in mammalian populations were reviewed by Lush (1970), and genetic variation in vertebrate species by Selander and Johnson (1973).

The techniques of protein sequencing and microcomplement fixation have been used to study protein changes that may have evolutionary significance. It is well known that various proteins evolve at different rates, depending on their function and importance. Cyctochrome C, for example, is very conservative (Fitch and Margoliash, 1967), whereas the fibrinopeptides evolve rapidly (Dayhoff, 1972). Chirpich (1975) reviews the available evidence and suggests that the rates of protein evolution are a function of amino acid composition. He argues that the conservation of the secondary and tertiary structure of proteins restricts the total number of amino acid replacements compatible with preservation of structure. The rates of replacement, therefore, depend on whether the constituent amino acids are more or less subject to evolutionary substitution. The work of Wilson and co-workers indicating constant evolutionary rates for certain proteins, using microcomplement fixation techniques, has already been mentioned. M.-C. King and Wilson (1975) analyze differences between chimpanzees and humans, which are the only two species that have been compared by all the various biochemical methods (electrophoretic, immunological, protein sequencing, annealing of nucleic acids), as well as at the organismal level (in regard to anatomy, physiology, behavior, and ecology). While chimpanzees and humans are quite similar at the biochemical level (genetic distances between them are on the order of sibling species), they differ markedly at the organismal level. The authors suggest that changes in a small number of regulatory genes may be responsible for the organismal differences. A model for gene regulation in eukaryotes proposed by Britten and Davidson (1969) shows the importance of pattern and chromosomal rearrangement in the regulation of gene transcription. The significance of regulatory genes is apparent in their model. Wilson (1975, 1976) reviews the evidence on which he bases his hypothesis that the evolution of structural genes is of secondary importance to that of regulatory genes. The rate of evolution of albumin, for example, is the same in frogs and mammals (Wallace *et al.,* 1971; Maxson and Wilson, 1974), so that one finds much greater differences among the albumins of frogs that look very similar but have been separated a long time than between the albumins of chimpanzees and humans. Wilson (1976) points out that the ability to hybridize is quickly lost in mammals, but retained much longer in frogs, and suggests that the ability to produce viable hybrids is a measure of the compatibility of the parental regulatory systems of gene activation.

Shvarts has stressed the importance of immunological methods in taxonomy. Besides the microcomplement fixation studies of the Wilson group, other useful immunological methods are hemagglutination tests, time of rejection of transplanted skin grafts, and reaction of lymph nodes (Syuzyumova, 1973, 1974). Micková and Iványi (1974) have used histocompatibility tests to define demes in wild house mice.

Another new method in the study of phylogenetic relationships is the use of chromosome banding techniques, such as G (giemsa), Q (quinacrine fluorescent), and C constitutive heterochromatin) banding. The method has been applied to the study of the Cetacea by Arnason (1974*a,b*) and Duffield (1976), and of primates by workers at the University of Barcelona (Miro and Goday, 1974; Caballin *et al.,* 1974; and others). Fredga (1974) used banding patterns to identify races of the common shrew.

Morphometric analysis remains an important method in evolutionary studies, and its value is greatly enhanced by the use of advanced numerical techniques and computer technology. Choate and Genoways (1974), for example, applied cluster, principal component, and discriminant function analyses to morphological measurements of two groups of shrews and found them to be distinct and separable on these bases. Multivariate analysis of 32 morphological variables in a heteromyid rodent and its correlation with geographical and climatic variables was conducted by D. F. Williams (1974). Multiple regression analysis has been used in a number of studies to assess the relationship between environmental heterogeneity and genic variability. Nevo *et al.* (1974) used this method to find the best predictors of heterozygosity in populations of pocket gophers. Bryant (1974), using data from the literature, applied stepwise multiple regression to test Levins's theory that genetic heterozygosity should be related to the heterogeneity of the environment. Taylor and Mitton (1974) used a sophisticated multivariate analysis to clarify complex relationships between the physical environment and geographic variation in isozyme frequencies of ants.

The various methods used in the study of taxonomic relationships often do not lead to similar conclusions, especially as regards rates of divergence (Lewontin, 1974; Sokal, 1974). Ecologically marginal populations, for example, are more likely to be monomorphic as regards morphological and chromosomal traits than populations in the center of the distribution, but there is no correlation of biochemical variation at the genic level with different environments (Lewontin, 1974). In an evolutionary study of the *Anolis* lizards of the eastern Caribbean, it was found that their morphological evolution was not at all coupled with their biochemical evolution, as measured electrophoretically (Yang *et al.*, 1974). There are many such examples in the literature. It has been suggested by Shvarts and others (e.g., Berry, 1974) that diverse methods must be used and a range of traits studied if we are to understand the meaning of the variation found in natural populations. Nadler *et al.* (1974) have used a multiple-method approach to the study of the evolution of Holarctic ground squirrels, analyzing their data by means of mutivariate methods. The work of Shvarts and his colleagues presented in this book is, of course, a prime example of such a comprehensive approach.

An important aspect of Shvarts's work is his emphasis and illustration of the differences between species' adaptations to specialized conditions and the adaptations of highly specialized subspecies. He points out that a specialized species will be able to adapt with a lower energy cost than a specialized subspecies. A number of papers were presented at the First International Theriological Congress on adaptations to high altitudes—work that will help to evaluate the energy cost of such adaptations. Belkin (1974) showed that the reconstruction that takes place in the endocrine glands of mammals depends on their duration of stay at different altitudes. Under experimental conditions, Blessing (1974) found coordinated anatomical and physiological changes in female Wistar rats to accommodate a lack of oxygen. It was found by Korzhuyev (1974) that the bone outgrowths at the base of the well-developed horns of some mountain species of sheep and goats have a function similar to bone marrow as foci of hemoglobin synthesis. In this way, the larger horns of these high-altitude mammals are adaptations to reduced oxygen pressure.

Chapter VII stresses the importance of ecological factors in the process of macroevolution. Shvarts argues that macroevolution is predictable. The initial physiological and morphological characteristics of animals and their adaptations will strongly influence the course of macroevolution. Shvarts points out the importance of preadaptations for a change to an ecologically different form of life. Further, he states that the same mechanisms govern micro- and macroevolution, but that macroevolution is the result of adaptive radiation in a new environment. Very similar views were expressed by Wright (in Jepsen *et al.*, 1949, p. 387): "Nevertheless the critical event in the appearance of a higher category seems to be a major ecological opportunity rather than any sort of mutation." According to Wright, the ways in which this opportunity may arise are:

1. A character or character complex may open up a relatively unexploited way of life. This is often due to the presence of a preadaptation.
2. A form may reach a relatively unoccupied territory and gain a major ecological opportunity.

In the last section of his book, Shvarts presents the conclusions derived from the ecological and evolutionary studies he has analyzed. It is hoped that the literature references given here will prove useful to the reader in supplementing the excellent and voluminous bibliography cited by the author in the original.

References

Anderson, P. K., 1970, Ecological structure and gene flow in small mammals, *Symp. Zool. Soc. London* **26**:299–325.

Arnason, U., 1974*a*, Karyotypic evolution and speciation in Cetacea and Pinnipedia, in: *Transactions of the First International Theriological Congress,* Vol. I, Nauka, Moscow (abstract).

Arnason, U., 1974*b*, Comparative chromosome studies in Pinnipedia, *Hereditas* **76**:179.

Avise, J. C., 1975, Systematic value of electrophoretic data, *Syst. Zool.* **23**:465–481.

Ayala, F. J. (ed.), 1976, *Molecular Evolution,* Sinauer, Sunderland, Massachusetts.

Ayala, F. J., Mourão, C. A., Pérez-Salas, S., Richmond, R., and Dobzhansky, T., 1970, Enzyme variability in the *Drosophila willistoni* group. I. Genetic differentiation among sibling species, *Proc. Nat. Acad. Sci. U.S.A.* **67**:225–232.

Ayala, F. J., Tracey, M. L., Hedgecock, D., and Richmond, R. C., 1974, Genetic differentiation during the speciation process in *Drosophila, Evolution* **28**:576–592.

Belkin, V. Sh., 1974, Morphological aspects of the endocrinic system adaptations in some mammals to high altitudes, in: *Transactions of the First International Theriological Congress,* Vol. I, Nauka, Moscow (abstract).

Bellamy, D., Berry, R. J., Jakobson, M. E., Lidicker, W. Z., Jr., Morgan, J., and Murphy, H. M., 1973, Ageing in an island population of the house mouse, *Age Ageing* **2**:235–250.

Berry, R. J., 1974, Variability in mammals—concepts and complications, in: *Transactions of the First International Theriological Congress,* Vol. I, Nauka, Moscow (abstract).

Blessing, M. H., 1974, Experimental adaptation to simulated high altitudes, in: *Transactions of the First International Theriological Congress,* Vol. I, Nauka, Moscow (abstract).

Bock, W. J., 1974, Philosophical foundations of classical evolutionary classification, *Syst. Zool.* **22**:375–392.

Britten, R. J., and Davidson, E. H., 1969, Gene regulation for higher cells: A theory, *Science* **165**:349–357.

Bryant, E. H., 1974, On the adaptive significance of enzyme polymorphisms in relation to environmental variability, *Amer. Nat.* **108**:388–401.

Bumpus, H. C., 1899, The elimination of the unfit as illustrated by the introduced sparrow, *Passer domesticus, Biol. Lect., Mar. Biol. Lab. Woods Hole,* pp. 209–226.

Caballin, M. R., Rubio, A., and Egozcue, J., 1974, Banding patterns of the chromosomes of the Papinae, in: *Transactions of the First International Theriological Congress*, Vol I, Nauka, Moscow (abstract).

Carson, H., 1968, The population flush and its genetic consequences, in: Lewontin, R. C. (ed.), *Population Biology and Evolution*, Syracuse.

Charlesworth, B., and Giesel, J. T., 1972, Selection in populations with overlapping generations. II. Relations between gene frequency and demographic variables, *Amer. Nat.* **106**:388–401.

Chirpich, T. P., 1975, Rates of protein evolution: A function of amino acid composition, *Science* **188**:1022–1023.

Chitty, D., 1970, Variation and population density, *Symp. Zool. Soc. London* **26**: 327–333.

Choate, J. R., and Genoways, H. H., 1974, Biosystematic investigations of shrews (Insectivora) of the genus *Blarina*, in: *Transactions of the First International Theriological Congress*, Vol. I, Nauka, Moscow (abstract).

Dayhoff, M. O. (ed.), 1972, *Atlas of Protein Sequence and Structure*, National Biomedical Research Foundation, Washington, D.C.

Dobzhansky, T., 1970, *Genetics in the Evolutionary Process*, Columbia University Press, New York.

Dobzhansky, T., 1972, Species of *Drosophila*, *Science* **177**:664–669.

Duffield, D. A., 1976, Phylokaryotypic relationships within the Cetacea: Evolution of certain of the modern species, Ph.D. dissertation, University of California at Los Angeles.

Ehrlich, P. R., Holm, R. W., and Raven, P. R. (eds.), 1969, *Papers on Evolution*, Little, Brown & Co., Boston.

Emlen, J. M., 1973, *Ecology: An Evolutionary Approach*, Addison-Wesley, Reading, Massachusetts.

Endler, J. A., 1973, Gene flow and population differentiation, *Science* **179**:243–250.

Fitch, W. M., and Margoliash, E., 1967, A method for estimating the number of invariant amino acid coding positions in a gene using cytochrome c as a model case, *Biochem. Genet.* **1**:65–71.

Ford, E. B., 1964, *Ecological Genetics* Methuen & Co., London.

Franklin, I., and Lewontin, R. C., 1970, Is the gene the unit of selection, *Genetics* **65**:707–734.

Fredga, K., 1974, Chromosome identification in races of the common shrew (*Sorex araneus*) based on banding patterns, in: *Transactions of the First International Theriological Congress*, Vol. I, Nauka, Moscow (abstract).

Gadgil, M., 1975, Evolution of social behavior through interpopulation selection, *Proc. Nat. Acad. Sci. U.S.A.* **72**:1199–1201.

Gaines, M., and Krebs, C. J., 1971, Genetic changes in fluctuating vole populations, *Evolution* **25**: 702–723.

Giesel, J. T., 1972, Maintenance of genetic variability in natural populations—an alternative implication of the Charlesworth-Giesel hypothesis, *Amer. Nat.* **106**:412–414.

Gill, A. E., 1977, Maintenance of polymorphism in an island population of the California vole, *Microtus californicus*, *Evolution* (in press).

Gillespie, J., and Kojima, K., 1968, The degree of polymorphism in enzymes involved in energy production compared to that in nonspecific enzymes in two *D. ananassae* populations, *Proc. Nat. Acad. Sci. U.S.A.* **61**:582–585.

Gilpin, M. E., 1975, *Group Selection in Predator–Prey Communities*, Monograph in Population Biology, Princeton University Press, Princeton, New Jersey.

Griffiths, G. C. D., 1974, Some fundamental problems in biological classification, *Syst. Zool.* **22**:338–343.

Hamilton, W. D., 1964, The genetical evolution of social behavior (two parts), *J. Theor. Biol.* **7**:1–52.

Hamilton, W. D., 1972, Altruism and related phenomena, mainly in insects, *Ann. Rev. Ecol. Syst.* **3**:193–232.

Jepsen, G. L., Simpson, G. G., and Mayr, E. (eds.), 1949, *Genetics, Paleontology, and Evolution,* Princeton University Press, Princeton, New Jersey.

Johnson, G. B., 1973, Enzyme polymorphism and biosystematics: The hypothesis of selective neutrality, *Annu. Rev. Ecol. Syst.* **4**:93–116.

Johnson, G. B., 1974, Enzyme polymorphism and metabolism, *Science* **184**:28–37.

Johnson, G. B., 1976, Genetic polymorphism and enzyme function, in: Ayala, F. J. (ed.), *Molecular Evolution,* Sinauer, Sunderland, Massachusetts.

Johnston, R. F., Niles, D. M., and Rohwer, S. A., 1972, Herman Bumpus and natural selection in the house sparrow *Passer domesticus, Evolution* **26**:20–31.

Jones, J. S., 1974, Experimental selection in the snail *Cepaea vindobonensis* in the Lika area of Yugoslavia, *Heredity* **32**:165–170.

Khlebnikov, A. I., Khlebnikov, I. P., and Antipov, E. I., 1974, Eliminating action of sable, *Martes zibellina* L., on population of *Ochotona alpina,* in: *Transactions of the First International Theriological Congress,* Vol. I, Nauka, Moscow (abstract).

Kimura, M., and Ohta, T., 1972, Population genetics, molecular biometry, and evolution, in: *Proceedings of the Sixth Berkeley Symposium on Mathematics, Statististics, and Probability,* Vol. 5, pp. 43–68.

King, J. L., and Jukes, T. H., 1969, Non-Darwinian evolution: Random fixation of selectively neutral mutations, *Science* **164**:788–798.

King, M.-C., and Wilson, A. C., 1975, Evolution at two levels in humans and chimpanzees, *Science* **188**:107–116.

Korzhuyev, P. A., 1974, Morphological adaptation of high altitude mammals to the reduced partial pressure of oxygen, in: *Transactions of the First International Theriological Congress,* Vol. I, Nauka, Moscow (abstract).

Krebs, C. J., Gaines, M. S., Keller, B. L., Meyers, J. H., and Tamarin, R. H., 1973, Population cycles in small rodents," *Science* **179**:35–44.

Latter, B. D. H., 1973, The island model of population differentiation: A general solution, *Genetics* **73**:147–157.

Lerner, I. M., 1954, *Genetic Homeostasis,* John Wiley & Sons, New York.

Levene, H., 1953, Genetic equilibrium when more than one ecological niche is available, *Amer. Nat.* **87**:131–133.

Levin, S., 1972, A mathematical analysis of the genetic feedback mechanism, *Amer. Nat.* **106**:145–164.

Levins, R., 1968, *Evolution in Changing Environments,* Monograph in Population Biology, Princeton University Press, Princeton, New Jersey.

Levins, R., and MacArthur, R., 1966, The maintenance of genetic polymorphism in a spatially heterogeneous environment: Variations on a theme by Howard Levene, *Amer. Nat.* **100**:585–589.

Lewontin, R. C., 1970, The units of selection, *Annu. Rev. Ecol. Syst.* **1**:1–18.

Lewontin, R. C., (ed.), 1968, *Population Biology and Evolution,* Syracuse University Press, Syracuse.

Lewontin, R. C., 1974, *The Genetic Basis of Evolutionary Change,* Columbia University Press, New York and London.

Lush, I. E., 1970, The extent of biochemical variation in mammalian populations, *Symp. Zool. Soc. London* **26**:43–71.

MacArthur, R. H., 1972, *Geographical Ecology,* Harper & Row, New York.

Maruyama, T., 1971, Speed of gene substitution in a geographically structured population, *Amer. Nat.* **105**:253–265.

Maxson, L. R., and Wilson, A. C., 1974, Convergent morphological evolution detected by studying proteins of tree frogs in the *Hyla eximia* species group, *Science* **185**:66–68.

Maynard Smith, J., and Ridpath, M. G., 1972, Wife sharing in the Tasmanian native hen, *Tribonyx mortierii*: A case of kin selection?, *Amer. Nat.* **106**:447–452.

Micková, M., and Iványi, P., 1974, Polymorphism of histocompatibility systems of wild mice (*Mus musculus* L.), in: *Transactions of the First International Theriological Congress*, Vol. I, Nauka, Moscow (abstract).

Miro, R., and Goday, E., 1974, Quinacrine fluorescence studies of primate chromosomes, in: *Transactions of the First International Theriological Congress*, Vol. I, Nauka, Moscow (abstract).

Nadler, C. F., Hoffmann, R. S., Sukernick, R. I., and Vorontsov, N. N., 1974, A comparison of biochemical and morphological evolution in Holarctic ground squirrels (*Spermophilus*), in: *Transactions of the First International Theriological Congress*, Vol. II, Nauka, Moscow (abstract).

Nelson, G., 1974, Classification as an expression of phylogenetic relationships, *Syst. Zool.* **22**:344–359.

Nei, M., 1971, Interspecific gene differences and evolutionary time estimated from electrophoretic data on protein identity, *Amer. Nat.* **105**:385–398.

Nei, M., 1972, Genetic distance between populations, *Amer. Nat.* **106**:283–292.

Nevo, E., Kim, Y. J., Shaw, C. R., and Thaeler, C. S., Jr., 1974, Genetic variation, selection and speciation in *Thomomys talpoides*, pocket gophers, *Evolution* **28**:1–23.

Petrusewicz, K., 1974, Ecological organization of a population, in: *Transactions of the First International Theriological Congress*, Vol. II, Nauka, Moscow (abstract).

Pianka, E. R., 1974, *Evolutionary Ecology*, Harper & Row, New York.

Pimentel, D., 1961, Animal population regulation by the genetic feedback mechanism, *Amer. Nat.* **95**:65–79.

Pimentel, D., 1968, Population regulation and genetic feedback, *Science* **159**:1432–1437.

Pimentel, D., 1973, *Genetics and Ecology of Population Control*, Addison-Wesley Module in Biology No. 10.

Prager, E. M., Arnheim, N., Mross, G. A., and Wilson, A. C., 1972, Amino acid sequence studies on bobwhite quail egg white lysozyme, *J. Biol. Chem.* **247**:2905–2916.

Robbins, C. B., 1974, The systematics, ecology, and zoogeography of the African gerbils, *Taterillus* (Rodentia: Cricetidae), *Transactions of the First International Theriological Congress*, Vol. II, Nauka, Moscow (abstract).

Sarich, V. M., and Wilson, A. C., 1967, Immunological time scale for hominid evolution, *Science* **158**:1200–1203.

Scudder, G. G. E., 1974, Species concepts and speciation, *Can. J. Zool.* **52**:1121–1134.

Selander, R. K., and Johnson, W. E., 1973, Genetic variation among vertebrate species, *Annu. Rev. Ecol. Syst.* **4**:75–91.

Selander, R. K., Smith, M. H., Yang, S. Y., Johnson, W. E., and Gentry, J. B., 1971, Biochemical polymorphism and systematics in the genus *Peromyscus*. I. Variation in the old-field mouse (*Peromyscus polionotus*), *Stud. Genet.* **6**:49–90.

Shilyaeva, L. M., 1974, The variability of some morphological characters of different generations of *Alopex lagopus* L. (Carnivora), in: *Transactions of the First International Theriological Congress*, Vol. II, Nauka, Moscow (abstract).

Slatkin, M., 1973, Gene flow and selection in a cline, *Genetics* **75**: 733–756.

Sokal, R. R., 1974, The species problem reconsidered, *Syst. Zool.* **22**:360–374.

Syuzyumova, L. M., 1973, Intraspecific features of tissue incompatibility in voles, in: *Experi-*

mental Investigations of the Species Problem, Tr. Inst. Ekol. Rast. Zhivotn., No. 86, Akad. Nauk SSSR, Sverdlovsk.

Syuzyumova, L. M., 1974, Use of immunological methods for the settlement of disputable taxonomic problems, in: *Transactions of the First International Theriological Congress*, Vol. II, Nauka, Moscow (abstract).

Tamarin, R. H., and Krebs, C. J., 1969, *Microtus* population biology. II. Genetic changes at the transferrin locus in fluctuating populations of two vole species, *Evolution* **23:**183–211.

Taylor, C. E., and Mitton, J. B., 1974, Multivariate analysis of genetic variation, *Genetics* **76:**575–585.

Wallace, D. G., Maxson, L. R., and Wilson, A. C., 1971, Albumin evolution in frogs: A test of the evolutionary clock hypothesis, *Proc. Nat. Acad. Sci. U.S.A.* **68:**3127–3129.

Williams, D. F., 1974, A multivariate analysis of the geographic variation of *Perognathus flavescens* (Rodentia: Heteromyidae) and its correlation to climatic and geographic variables, in: *Transactions of the First International Theriological Congress*, Vol. II, Nauka, Moscow (abstract).

Williams, G. C., 1971, *Group Selection*, Aldine-Atherton, Chicago.

Wilson, A. C., 1975, Relative rates of evolution of organisms and genes, *Stadler Genet. Symp.* **7:**117–134.

Wilson, A. C., 1976, Gene regulation in evolution, in: Ayala, F. J. (ed.), *Molecular Evolution*, Sinauer, Sunderland, Massachusetts.

Wilson, E. O., 1971, *The Insect Societies*, Belknap Press, Cambridge, Massachusetts.

Wilson, E. O., Eisner, T., Briggs, W. R., Dickerson, R. E., Metzenberg, R. L., O'Brien, R. D., Susman, M., and Boggs, W. E., 1973, *Life on Earth*, Sinauer Publ., Stamford, Connecticut.

Yamazaki, T., and Maruyama, T., 1974, Evidence that enzyme polymorphisms are selectively neutral, but blood group polymorphisms are not, *Science* **183:**1091–1092.

Yang, S. Y., Soulé, M., and Gorman, G. C., 1974, *Anolis* lizards of the Eastern Caribbean: A case study in evolution. I. Genetic relationships, phylogeny, and colonization sequence of the *roquet* group, *Syst. Zool.* **23:**387–399.